Aquaculture Development in Archipelago

Estu Nugroho • Imron
Bambang Gunadi • Rosmiati
Khairul Syahputra
Editors

Aquaculture Development in Archipelago

Domestication and Cultivation of Freshwater and Marine Fishes in Indonesia

Springer

Editors
Estu Nugroho
Research Center for Fishery
Research Organization for Earth Sciences and Maritime
National Research and Innovation Agency
Cibinong, Indonesia

Imron
Research Center for Fishery
Research Organization for Earth Sciences and Maritime
National Research and Innovation Agency
Cibinong, Indonesia

Bambang Gunadi
Research Center for Fishery
Research Organization for Earth Sciences and Maritime
National Research and Innovation Agency
Cibinong, Indonesia

Rosmiati
Research Center for Fishery
National Research and Innovation Agency
Cibinong, Indonesia

Khairul Syahputra
Research Center for Fishery
Research Organization for Earth Sciences and Maritime
National Research and Innovation Agency
Cibinong, Indonesia

ISBN 978-981-95-1083-2 ISBN 978-981-95-1084-9 (eBook)
https://doi.org/10.1007/978-981-95-1084-9

Jointly published with BRIN Publishing

This Springer imprint is published by the registered company Springer Nature Singapore Pte Ltd.
The registered company address is: 152 Beach Road, #21-01/04 Gateway East, Singapore 189721, Singapore

Contents

Introduction: Overview of the Fish Genetic Resources Utilization in Archipelago

1

Estu Nugroho

The contribution of the fisheries sector to gross domestic product (GDP) in 2023 is targeted to be between 4% and 6% or the equivalent of IDR 259.78 trillion. The role of aquaculture in achieving the national fishery production target is increasing every year. In 2018, aquaculture production reached 17.2 million tons, accounting for 60% of the total national production and increased to 21.8 million tons, or 66%, in 2020. The total value of aquaculture production in 2020 reached IDR 400 billion (Pusdatin, 2022). Aquaculture is also expected to be a source of healthy protein for the community, contributing to the success of the stunting reduction program. The national stunting rate is currently 24.4%, or approximately 5.33 million children under five, which is expected to decrease to around 14% in 2024 (www.setneg.go.id).

This achievement in aquaculture is closely tied to the rapid development of aquaculture technology in the archipelago. Three key factors influence success in aquaculture: the availability of sufficient and high-quality water, feed that meets the nutritional needs of fish, and the use of quality and superior seeds for certain characters (Nugroho, 2010; Nugroho & Haryadi, 2017). In general, the development of freshwater aquaculture commodities is the fastest, followed by brackish and sea water aquaculture commodities. In the last few periods, aquaculture has also expanded to include local and native commodities in the archipelago (Nugroho et al., 2012).

The comprehensive information presented in the sections of this book details the development of aquaculture in the archipelago, from domestication to the latest technology, including the breeding of native Indonesian fish commodities and introduced fish in freshwater, brackish, and marine ecosystems. This knowledge is crucial for all stakeholders in the field of fisheries, as it can be used in designing and

---

E. Nugroho (✉)
Research Center for Fishery, Research Organization for Earth Sciences and Maritime, National Research and Innovation Agency, Cibinong, Indonesia
e-mail: estu003@brin.go.id

E. Nugroho et al. (eds.), *Aquaculture Development in Archipelago*,
https://doi.org/10.1007/978-981-95-1084-9_1

directing further developments and future developments in aquaculture for the welfare and health of the community.

1.1 Utilization of Germplasm

Indonesia has a diversity of fish species, comprising approximately 8500 of the estimated 20,000 fish species worldwide. Aquaculture activities utilizing fish germplasm resources have been carried out for hundreds of years, beginning with the cultivation of milkfish (*Chanos chanos*), which were reared in ponds during the Majapahit Kingdom era. The cultivation of carp (*Cyprinus carpio*) is believed to have started in the nineteenth century through the domestication of fish from China and Hungary (Nugroho et al., 2012). To date, most freshwater fish production of cultivation (75%) in Indonesia is contributed by introduced fish commodities that have been domesticated over time, such as catfish, tilapia, carp, and jambal catfish (Pusdatin, 2022). The production of native Indonesian fish species has not been recorded in detail and is still grouped under other fish groups, accounting for approximately 20% of national aquaculture production in 2020.

This condition arises from the rapid expansion of introduced fish cultivation, which has been more widely adopted and previously domesticated. The ease of cultivation technology has also influenced people's preferences, leading to a greater interest in introduced fish cultivation over local fish, especially freshwater fish species. There are about 21 types of local fish with potential for germplasm use in aquaculture, including local catfish, Asian redtail catfish, gourami, and stripped snakehead fish (Nugroho et al., 2012).

Additionally, the use of germplasm in brackish and seawater fish has become widespread among cultivators in Indonesia. The most commonly cultivated brackish water commodities include shrimp, both vanamei (an introduced species) and tiger prawns, as well as milkfish, which is native to Indonesia. Meanwhile, well-developed seawater commodities are grouper fish and seaweed.

1.2 Breeding

Breeding programs play a very important role in providing superior seeds, both in terms of growth and adaptation to environmental fluctuations as well as disease resistance. These superior seeds can be produced through a series of stages, starting with collection activities, followed by characterization and evaluation activities, and continuing with the selection of either conventional or unconventional breeding methods. The use of superior seeds in cultivation can increase production or reduce production costs, thereby increasing margins (Nugroho et al., 2013).

Collection is the initial step in forming a population for breeding purposes. Collections can be made from natural populations or cultivated varieties that meet specific parameters. Collection activities for fish breeding were initiated in 1985, focusing on collecting carp varieties from various isolated areas in Sumatra, Java, and Bali. Four selected carp varieties were obtained, namely Sutisna, Rajadanu, Wildan, and Domas (Nugroho & Wahyudi, 1991).

Characterization and evaluation are carried out to obtain genetic background information, including diversity or variation, fitness conditions, and kinship among fish population stocks intended for breeding. To accelerate the production of superior seeds, several genetic markers can be used, such as isozymes (Nugroho & Kusmini, 2007), RAPD (Nugroho, 2013; Nugroho et al., 2006, 2016), Mt DNA (Nugroho et al., 2008, 2019, 2020a), and microsatellites (Nugroho et al., 1998, 1999, 2000, 2001).

The results of the characterization and evaluation show that freshwater fish exhibit relatively lower variation value than marine fish. One reason for this is inbreeding and geographic barrier, which are more likely to occur in freshwater fish (Taniguchi et al., 2003; De Bruyn et al., 2005). Additionally, marine fish tend to have extensive migratory characteristics and large breeding units, thereby increasing the potential for gene flow between distantly related populations (Nugroho, 2001). A decrease in genetic variation is observed in cultivated fish compared to their wild counterparts (Nugroho, 2018). This reduction is due to the limited numbers of broodstock in aquaculture, resulting in the loss of certain genes.

Conventional breeding methods, such as selection and hybridization, are applied based on genetic variation data. The higher the genetic variation of a fish population stock used as breeding material, the more successful the selection program will be. Conversely, if the variation value is low, hybridization or crossbreeding between individuals from distantly related populations is more appropriate for producing superior broodstock (Nugroho, 2018). In addition to conventional breeding, another way to produce superior fish varieties is through unconventional methods, including chromosome set engineering (Nugroho, 1994; Nugroho et al., 1994), maleization using sex-switching hormones (Nugroho, 2004), and production of growth hormone for gene transfer (Nugroho et al., 2008, 2009).

Recent developments in producing high-yielding seeds involve genome engineering, specifically genome editing methods like clustered regularly interspaced short palindromic repeats (CRISPR/Cas9) (Zhu & Ge, 2018). This technology can be used precisely and efficiently to target genome editing across different species. However, further studies are required to categorize whether these products produced fall under the category of genetic modified organisms (GMO) or other classifications.

1.3 Agribusiness

Initially, food fish agribusiness, particularly freshwater fish, was largely focused on species already well-established within the community. However, a persistent issue emerged: the continuous rise in feed prices, while the selling price of fish could not be increased. This led to natural selection, where only commodities that were easy to cultivate and had lower operational costs—by utilizing various alternative feed sources became the focus of development (Nugroho & Kristanto, 2008). Meanwhile, agribusiness for brackish water and marine commodities is in relatively better condition, as their selling prices are higher than their production costs.

The government has supported the development of fish farming-based agribusiness through the smart aquaculture program. One key initiative of this program is the formation of one village for one superior commodity. It is projected that by 2024, aquaculture production will reach 22.65 million tons, comprising 10.32 million tons of fish and 12.33 million tons of seaweed (MMAF, 2019).

Broadly, fish agribusiness in Indonesia can be divided into two segments, namely hatchery and rearing businesses. Hatchery business includes broodstock rearing activities, spawning, and continued larval rearing to reach the seed size, which will be used in the rearing segment. Rearing continues growing fish from seed size to market size. For certain commodities, both freshwater and seawater, the procurement of seeds still relies on wild catches due to undeveloped hatchery technology. Examples include freshwater eel and marine Napoleon grouper. In contrast, for brackish water commodities, there have been many developments in hatchery technology, except for mangrove crabs.

1.4 Industry

Several freshwater fish commodities have been developed under an industrial concept. One notable example is tilapia aquaculture, which can serve as a model for real-world business operations. The most recent advancements in tilapia farming involve the use of the biofloc system. Optimization of biofloc technology, along with the use of high-yield seeds, is continuously pursued to control production factors and ensure sustainability (Nugroho & Aisyah, 2021; Nugroho et al., 2020a, b, 2022, 2023).

The application of appropriate technology, combined with superior seeds, can increase catfish farming productivity by up to 800% compared to the use of local seeds and existing cultivation technology in cultivators (Nugroho, 2021). Nugroho et al. (2022) reported that cultivating tilapia with superior seeds in the biofloc system achieved a productivity level of 40 kg/m^3. The use of superior tilapia seeds in the floating net cages system yielded a productivity level of 0.345 kg/m^3, while the use of local seeds in the same system exhibited a productivity of 0.708 kg/m^3 (Nugroho et al., 2013).

References

de Bruyn, M., Nugroho, E., Hossain, M. M., Wilson, J. C., & Mather, P. B. (2005). Phylogeographic evidence for the existence of an ancient biogeographic barrier: The Isthmus of Kra Seaway. *Heredity, 94*, 370–378. https://doi.org/10.1038/sj.hdy.6800613

Ministry of Marine Affairs and Fisheries [MMAF]. (2019). *Pengembangan komoditas unggulan strategis perikanan budidaya, dan tata kelola perizinan untuk memacu investasi* [*Development of strategic superior aquaculture commodities and licensing management to boost investment*].

Workshop pembangunan perikanan budidaya berkelanjutan [Sustainable aquaculture development workshop]. BAPPENAS.

Nugroho, E. (2004). *Jantanisasi ikan dengan sistem resirkulasi tertutup* [*Masculinization of fish using a closed recirculation system*]. TROBOS, 56/Tahun ke-V/Mei.

Nugroho, E. (2010). *Nila unggul #1* [*Superior Tilapia #1*]. Penebar Swadaya.

Nugroho, E. (2011). Kajian lapang budidaya keramba jaring apung ikan nila "Mandiri" di Waduk Cirata dan Jatiluhur [Field study on "Mandiri" Tilapia floating net cage aquaculture in Cirata and Jatiluhur reservoirs]. *Media Akuakultur, 6*(1), 54–58.

Nugroho, E. (2013). Genetic variability of giant gouramy strain revealed by random amplified polymorphism DNA. *Indonesian Aquaculture Journal, 8*(1), 107–111.

Nugroho, E. (2018). *Aplikasi ilmu genetika dalam pemuliaan perikanan air tawar* [*Application of genetic science in freshwater fish breeding*]. Amafrad Press.

Nugroho, E., & Kusmini, I. I. (2007). Evaluasi variasi genetik tiga ras ikan gurame (*Osphronemus gouramy*) dengan menggunakan isozyme [Evaluation of genetic variation in three strains of gourami (*Osphronemus gouramy*) using isozymes]. *Jurnal Riset Akuakultur, 2*(1), 51–57.

Nugroho, E., & Wahyudi, N. A. (1991). Evaluasi beberapa strain ikan mas koleksi dengan menggunakan "skor-z" [Evaluation of several strains of collected carp using "z-scores"]. *Bulletin Penelitian Perikanan Darat, 10*(2), 49–54.

Nugroho, E., Takagi, M., Sugama, K., & Taniguchi, N. (1998). Detection of GT repeats microsatellite loci and their polymorphism for grouper of the genus Epinephelus. *Fisheries Science, 65*(3), 836–837.

Nugroho, E., Taniguchi, N., Kato, K., & Miyashita, S. (2000). Genetic difference among seed population of greater amberjack used in aquaculture farm of Japan. *Suisanzoshoku, 48*(4), 665–674.

Nugroho, E., Ferrel, D. J., Smith, P., & Taniguchi, T. (2001). Genetic divergence of kingfish from Japan, Australia and New Zealand inferred by microsatellite DNA and mitochondrial DNA control region markers. *Fisheries Science, 67*(5), 843–850.

Nugroho, E., Subagja, J., Asih, S., & Kurniasih, T. (2006). Evaluasi keragaman genetik ikan kancra dengan menggunakan marker MtDNA D-loop dan Random Amplified Polymorphism DNA (RAPD) [Evaluation of genetic diversity in kancra fish using MtDNA D-loop markers and Random Amplified Polymorphic DNA (RAPD)]. *Jurnal Riset Akuakultur, 1*(2), 211–217.

Nugroho, E., Alimuddin, Kristanto, A. H., Carman, O., & Megawati, N. (2008). Kloning cDNA hormon pertumbuhan dari ikan gurame (*Osphronemus gouramy*) [Cloning of growth hormone cDNA from gourami fish (*Osphronemus gouramy*)]. *Jurnal Riset Akuakultur, 3*(2), 183–190.

Nugroho, E., Alimuddin, Kristanto, A. H., & Charman, O. (2009). Kloning Promoter beta aktin dari ikan gurame (*Osphronemus gouramy*) [Cloning of beta-actin promoter from gourami fish (*Osphronemus gouramy*)]. *Jurnal Riset Akuakultur, 4*(1), 23–29.

Nugroho, E., Saepudin, & Bajar, M. (2013). Kajian lapang penggunaan benih nila (*O. niloticus*) hasil pemuliaan di keramba jaring apung Jatiluhur [Field study on the use of selectively bred tilapia (*O. niloticus*) seeds in floating net cages at Jatiluhur]. *Jurnal Riset Akuakultur, 8*(1), 43–49.

Nugroho, E., Azrita, Syandri, H., & Refilza. (2016). Evaluasi keragaman genetik ikan kalui (*Osphronemus goramy*) dari Kabupaten Lima Puluh Kota Sumatera Barat berdasarkan marka Random Amplified Polymorphism DNA (RAPD) [Evaluation of genetic diversity in kalui fish (*Osphronemus goramy*) from Lima Puluh Kota Regency, West Sumatra, based on Random Amplified Polymorphic DNA (RAPD) markers]. *Jurnal Riset Akuakultur, 11*(4), 313–319.

Nugroho, E., Azrita, Syandri, H., & Dewi, R. R. S. P. (2019). DNA barcoding of giant gourami (*Osphronemus goramy*) from West Sumatra, Indonesia. *AACL Bioflux, 12*(4), 1074–1079.

Nugroho, E., Astuti, A., Yulaeni, F., Praninda, P. Y., Putra, R. K., Suprayitno, S., Hariyanti, D. A., Sutandi, L., & Irvani, F. (2020a). Sekuense Mt-DNA CO-1, karakter reproduksi, dan toleransi terhadap lingkungan dari udang galah Bengawan Solo [Mt-DNA CO-1 sequencing, reproductive characteristics, and environmental tolerance of giant freshwater prawns from Bengawan Solo]. *Jurnal Riset Akuakultur, 15*(1), 1–8.

Nugroho, E., Khakim, A., & Dewi, R. S. S. P. (2020b). The performance of tilapia culture in biofloc technology (BFT) and floating net cage (KJA) system as a candidate for the next tilapia culture in Indonesia. *IOP Conference Series: Earth and Environmental Science, 521*, 012020-1–8. https://doi.org/10.1088/1755-1315/521/1/012020

Nugroho, E., Zulkarnaen, R., Kristanto, A. H., Dewi, R. S. S. P., Gunadi, B., Robisalmi, A., Pamungkas, W., Setijaningsih, L., & Rifaldi, M. (2022). Assembling the superior biofloc tilapia strains for industry. *The Israeli Journal of Aquaculture – Bamidgeh, 74*, 1787805.

Nugroho, E., Kristanto, A. H., Pamungkas, W., Dewi, R. S. S. P., & Rifaldi, M. (2023). Optimizing tilapia biofloc technology systems and its economic profitability on an industrial scale in Indonesia. *IOP Conference Series: Earth and Environmental Science, 1137*(1), 012061.

Nugroho, E. & Haryadi, J. (2017). Budidaya lele dengan sistem total akuakultur [Catfish farming with a total aquaculture system]. Penebar Swadaya.

Nugroho, E., Sudarto, & Suseno, D. (1994). Aplikasi kejutan suhu pada produksi benih ikan mas triploid [Application of Temperature Shock in the Production of Triploid Common Carp Seed]. Balai Penelitian Perikanan Air Tawar. 178–183.

Nugroho. E., Taniguchi, N., & Takagi, M. (1999). Isolation of greater amberjack microsatellite DNA and their application as genetic marker species of genus Seriola from Japan. Fisheries Science, 65(3), 353–357.

Nugroho, E., Sukadi, M.F., & Huwoyon, G.H. (2012). Beberapa Jenis ikan Lokal yang Potensial untuk Budidaya:Domestikasi, Teknologi Pembenihan, dan Pengelolaan Kesehatan Lingkungan Budidaya [Potential Local Fish Species for Aquaculture: Domestication, Hatchery Technology, and Environmental Health Management]. Media Akuakultur, 7(1), 52–57.

Nugroho, E. (1994).Teknologi triploid dapat memacu laju pertumbuhan ikan mas [Triploid technology can accelerate the growth rate of common carp]. Warta Litbangtan, XVI(6),1.

Nugroho, E. (2001). Practical manual on detection of DNA polymorphism in fish population study. Bull. Marine Sciences and Fisheries, 7 (1), 1–21.

Nugroho, E. (2021). Benih unggul hasil pemuliaan untuk peningkatan produktifitas air tawar [Improved Strains from Breeding for Enhancing Freshwater Productivity]. Amafrad Press.

Nugroho, E., & Kristanto, A.H. (2008). Panduan lengkap ikan konsumsi air tawar [Comprehensive Guide to Freshwater Food Fish Culture]. Penebar Swadaya.

Nugroho, E. & Aisyah. (2021). Budidaya ikan nila bio: Kiat panen berkesinambungan budidaya ikan nila dengan aplikasi teknologi bioflok [Tilapia Farming with Biofloc: Tips for Sustainable Tilapia Harvesting through Biofloc Technology Application]. Amafrad Press.

Pusdatin. (2022). *Marine and fisheries in figures 2022* (Vol. 1). The Center for Data, Statistical and Information.

Taniguchi, N., Perez-Enriquez, R., & Nugroho, E. (2003). DNA Marker as a tool for genetic management of brood stock for aquaculture. In: Shimizu N, Aoki T, Takashima F, editor. Aquatic genomics-step toward a great future. Springer-Verlag.

Zhu, B., & Ge, W. (2018). Genome editing in fishes and their applications. *General and Comparative Endocrinology, 257*, 3–12.

Development of Asian Redtail Catfish (*Hemibagrus nemurus*) Farming in Indonesia

2

Erma Primanita Hayuningtyas, Sawung Cindelaras, and Eni Kusrini

2.1 Introduction

Asian redtail catfish (*Hemibagrus nemurus*) is one of the native freshwater fish of Indonesia. In addition to natural sources like rivers, reservoirs, and swamps, the Asian redtail catfish can be domesticated and cultivated in ponds, floating net cages, pen cultures, and polyculture systems (Fig. 2.1) (Aryani, 2017; Suhenda et al., 2010). Consumer demand for Asian redtail catfish is relatively high in some of its producing regions, with selling prices ranging from 50,000 to 70,000 rupiah per kilogram of fresh fish. In some areas with high production rates, this fish is processed into smoked fish, which sells for 150,000 to 200,000 rupiah per kilogram (Heltonika & Karsih, 2017).

Several breeding technologies have been applied to the Asian redtail catfish to increase its production in Indonesia. Cultivation methods for this fish include growth hormone-enriched feed (Subagja & Prakoso, 2018), hatchery operations with photoperiod treatments (Heltonika & Karsih, 2017), rearing in floating nets with high-quality feed (Suhenda et al., 2010), and egg-hatching technology with varied media (Kristanto et al., 2016; Prakoso et al., 2018; Subagja & Prakoso, 2018). These technologies significantly support increased production of the Asian redtail catfish. However, much of the supply still depends on naturally harvested fish from regions such as South Sumatra, Riau, Jambi, and South Kalimantan, which impacts the consistency and size uniformity compared to cultured fish (Subagja in Tirtana, 2022).

E. P. Hayuningtyas (✉) · S. Cindelaras · E. Kusrini
Research Center for Fishery, Research Organization for Earth Sciences and Maritime, National Research and Innovation Agency (BRIN), Jakarta, Indonesia
e-mail: erma009@brin.go.id

E. Nugroho et al. (eds.), *Aquaculture Development in Archipelago*,
https://doi.org/10.1007/978-981-95-1084-9_2

Fig. 2.1 Asian redtail catfish (*Hemibagrus nemurus*). (Source: Photo by J. Subagja, personal collection)

2.2 Asian Redtail Catfish Distribution and Habitat

As a native freshwater fish of Indonesia, the Asian redtail catfish (*Hemibagrus nemurus*) is distributed across regions such as Riau, where it can be found along the Kampar River extending to the Koto Panjang reservoir (Aryani, 2017; Heltonika & Karsih, 2017), as well as Kalimantan along the Barito River (Rosadi et al., 2020), and Java and northern Sumatra (Prianto et al., 2015). Beyond Indonesia, the Asian redtail catfish also inhabits other tropical countries, including regions in Africa, Southeast Asia, and East Asia (Kottelat et al., 1993; Rainboth, 1996).

According to the latest revision by Ng and Kottelat (2013), the distribution of *H. nemurus* is specifically limited to Java Island. Other "nemurus" species found in the Sunda region (including the Malay Peninsula, Sumatra, and Kalimantan) represent distinct species. The habitat range of *H. nemurus* begins from the Cisadane River and extends eastward to the Bengawan Solo River on Java Island (Ng & Kottelat, 2013). The distribution map of the Asian redtail catfish is shown in Fig. 2.2.

The Asian redtail catfish is found in various habitat types, but it most commonly inhabits large, muddy rivers with slow currents and soft bottoms. It also enters flooded forests, where it feeds on aquatic insects, insect larvae, shrimp, crustaceans, and fish. The fish moves to flooded forests to spawn, and young fish are typically first observed in August (Rainboth, 1996). *Hemibagrus nemurus* primarily inhabits large rivers and has been described as an opportunistic predator that feeds on both aquatic and terrestrial invertebrates as well as fish (Rachmatika et al., 2002). According to Mustafa-Kamal et al. (2012), the stomach contents of the Asian redtail catfish consist of approximately 77.0% aquatic insects, 7.5% crustaceans, 2.9% plant material, 7.5% detritus, and 1.7% unidentified matter.

2.3 Taxonomy and Morphology

The classification of the Asian redtail catfish is as follows:

Fig. 2.2 Distribution map of Asian redtail catfish (*Hemibagrus nemurus*) in Indonesia. (Source: Google maps with modification, based on IUCN data)

Phylum: Chordata
Class: Actinopterygii
Order: Siluriformes
Family: Bagridae
Genus: *Hemibagrus*
Species: *Hemibagrus nemurus* (Valenciennes, 1840) according to Fishbase

2.4 Morphometric Characters of the Asian Redtail Catfish

Morphometric character measurements of the Asian redtail catfish (*Hemibagrus nemurus*) have been conducted in populations from the Koto Panjang Reservoir and Rantau Baru Village along the Kampar River, revealing 24 distinct characters (80%). In contrast, measurements in populations from Langgam Village and Rantau Baru Village along the Kampar River show 18 distinct characters (60%) out of a total of 30 characters measured (Aryani et al., 2013). The genetic diversity of the Asian redtail catfish in Riau is relatively high, with a diversity value ranging from 60% to 80%. This high diversity is likely because the cultivation of *Hemibagrus nemurus* is still not widely practiced, resulting in lower inbreeding depression compared to other, more extensively developed fish commodities (Aryani et al., 2013).

2.5 Domestication of the Asian Redtail Catfish

The Asian redtail catfish have been successfully domesticated and cultivated in Indonesia. The cultivation of local fish species is part of broader domestication efforts aimed at maintaining biodiversity and supporting food security. Asian redtail catfish can be cultivated in midland areas (200–400 m above sea level) and lowland areas (<200 m above sea level). These different altitude locations influence water quality parameters, which, in turn, affect fish growth. Fry reared in lowland environments exhibit greater weight gain and biomass compared to those in midland environments. The water temperature in the lowlands was significantly higher than in the midlands ($P < 0.05$) (Subagja et al., 2018).

Two primary temperature-related factors influence growth, namely (1) adaptation to temperature, leading to differences in growth, and (2) adaptation to the length of the growing season, which enables one population to achieve a higher growth rate within a temperature range that supports growth (Conover, 1992; Subagja et al., 2018).

2.6 Reproduction

2.6.1 Reproductive Biology

The reproductive aspects of the Asian redtail catfish (*Hemibagrus nemurus*) in natural settings have been studied, particularly in the Ogan Ilir area of South Sumatra. Asian redtail catfish in the Lubuk Lampam region of Ogan Ilir District, South Sumatra Province, displayed reproductive characteristics that include an annual sex ratio of 0.44:1. Monthly observations, however, showed a varying sex ratio range of 0.13–2.50, indicating an imbalance.

As shown in Fig. 2.3, the sex ratio of the Asian redtail catfish exceeds 1 when water levels begin to rise (water rise, WR) and continue to the maximum flood stage (maximum flood, MF). This suggests that during the spawning season, female catfish are fertilized by more than one male (Prianto et al., 2015).

Other reproductive aspects include gonad maturity stages (GMS), which reach stages GMS III and IV, particularly during the rainy season from November to January. The gonad maturity index ranges from 1.8% to 14.3% in female fish and 1.3% to 3.9% in male fish. The length at first gonad maturity is 232 mm in males and 332 mm in females. The average fecundity of the Asian redtail catfish is approximately 47,882 ± 13,624 eggs, with egg production ranging from 15,386 to 104,758 eggs, depending on the fish's size and age. The Asian redtail catfish follows a reproductive pattern typical of total spawners, spawning once a year. This is evident from the single-mode egg diameter distribution and synchronized gonad maturity (Prianto et al., 2015).

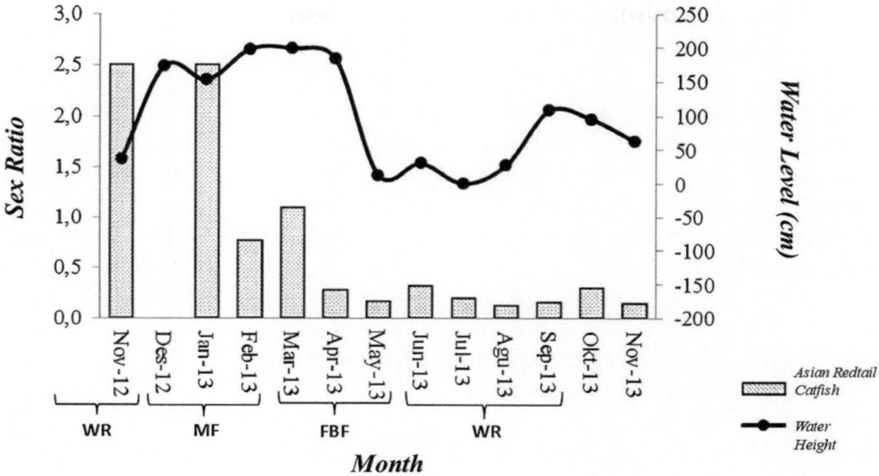

Fig. 2.3 Sex ratio graph of Asian redtail catfish in Lubuk Lampam, Ogan Ilir Regency. Remarks: WR = water rise; MF = maximum flood; FBF = floods begin to fall; NK = sex ratio. (Source: Prianto et al. (2015)https://drive.google.com/file/d/1JQ3ihjEtHLKuH0aTMyujslzc-LJC1bDg/view?usp=sharing)

2.6.2 Reproductive Engineering with the Addition of GnRH Hormone

The artificial spawning of Asian redtail catfish was conducted using GnRH hormone (Ovaprim® Syndel Laboratories, Canada) at three different doses for female broodstock (0.3, 0.5, and 0.7 mL/kg body weight). Meanwhile, a single dose of 0.4 mL/kg body weight was used for male broodstock injections (Slembrouck et al., 2005). Male brood stock was injected once, while female broodstock received two injections, with the first injection at 35% and the second at the remaining 65% of the total hormone dose, administered at an 8 h interval (Legendre et al., 2000; Subagja et al., 2018). Results from spawning trials using GnRH hormone manipulation identified that the optimal dose for hatching rates was 0.5 mL/kg of body weight, yielding a higher hatching rate than doses of 0.3 and 0.7 mL/kg. However, the highest survival and fertilization rates were achieved at a dose of 0.7 mL/kg body weight for the Asian redtail catfish (Subagja & Prakoso, 2018). For comparison, artificial spawning with a dose of 0.6 mL/kg of body weight resulted in a somatic ovo index (SOI) and fecundity of 14.2 ± 2.87% and 46,333 ± 9685 eggs, with a fertilization rate of 85.5 ± 8.67% and a hatching rate above 70.2 ± 19.0% (Kristanto et al., 2016).

2.7 Development of Asian Redtail Catfish Cultivation Technology

Asian redtail catfish cultivation technology has seen significant advancement in Indonesia, encompassing hatchery technology, artificial spawning (Subagja & Prakoso, 2018), crossbreeding (Subagja et al., 2015), hatcheries using photoperiod applications (Heltonika & Karsih, 2017), egg-hatching technology (Prakoso et al., 2018), growth optimization in floating net cages (KJA) (Suhenda et al., 2010), and hormone applications.

2.8 Hatchery Technology with Photoperiod (Lighting) Application

In hatchery technology for Asian redtail catfish, photoperiod application has been employed to address one of the key challenges: reducing cannibalism during early life stages. Cannibalism, which begins at the larval stage and continues into the fry stage, significantly increases fry mortality. The application of photoperiods (controlled lighting conditions) aims to mitigate cannibalism by exploiting the nocturnal behavior of Asian redtail catfish, which are more active at night. On photoperiod applications has examined various lighting treatments, including continuous 24 h light, a 12 h light/12 h dark cycle, and continuous darkness. Findings indicate that catfish reared in constant darkness exhibit the best growth, both in length and weight, and experience a lower mortality rate compared to those in lighted environments. This is attributed to the fish's nocturnal nature, with less light exposure reducing stress and promoting natural behavior (Heltonika & Karsih, 2017).

2.9 Egg-Hatching Technology with Different Hatching Media

Egg hatching is a critical stage in Asian redtail catfish cultivation, requiring tailored technology to optimize hatch rates. Different media are used for hatching to align with the physical characteristics of fish eggs, which can be floating, sinking, adhesive, or nonadhesive. To accommodate these egg types, hatching technology can employ stagnant water, flowing water, substrates, or funnel systems. The selection of hatching media impacts the hatching success rate, hatching duration, and larval survival rate, which are essential for effective hatchery operations (Kristanto et al., 2016; Prakoso et al., 2018; Subagja & Prakoso, 2018).

2.10 Asian Redtail Catfish Farming Technology in Floating Net Cages and Protein-Based Feeding

Asian redtail catfish farming can also be conducted in floating net cages (FNC), primarily aims at fish growing out. Proper nutrition with adequate protein levels is essential for optimal growth in FNC. For this purpose, two different protein levels were tested, namely, 27% (providing 2840 kcal digestible energy (DE) per kg feed) and 31% (providing 2890 kcal DE per kg). These feeds were administered at 4–8% of the fish's total body weight, with a feeding frequency of four times per day. The results from rearing Asian redtail catfish with varied protein levels revealed that a diet with 31% protein produced the best outcomes, including a specific growth rate of 2.67 ± 0.04, a weight gain of 80.48 ± 2.09, a feed conversion ratio of 2.28 ± 0.07, and fat retention of 29.48 ± 0.73. These values significantly surpassed those achieved with the 27% protein feed (Suhenda et al., 2010).

References

Aryani, N. (2017). *Teknologi tepat guna budidaya ikan baung* [*Appropriate technology for the Asian redtail catfish cultivation*]. Bung Hatta University Press. https://www.pustakadunia.com/budidaya-ikan-baung/

Aryani, N., Nuraini, & Suherman, I. (2013). Morphological characterization of Asian redtail catfish (*Hemibagrus nemurus*) aquatic habitat on the different method-based truss morfometrics. *Journal of Fisheries and Aquaculture, 4*(3), 139–142. http://www.bioinfopublication.org/jouarchive.php?opt=&jouid=BPJ0000265

Conover, D. O. (1992). Seasonality and the scheduling of life history at different latitudes. *Journal of Fish Biology, 41*(SB), 161–178. https://doi.org/10.1111/j.1095-8649.1992.tb03876.x

Heltonika, B., & Karsih, O. R. (2017). Pemeliharaan benih ikan baung (*Hemibagrus nemurus*) dengan teknologi photoperiod [Maintenance of Asian redtail catfish (Hemibagrus nemurus) fry with photoperiod technology]. *Jurnal Berkala Perikanan Terubuk, 45*(1), 125–137.

Kottelat, M., Whitten, A. J., Kartikasari, S. N., & Wirjoatmodjo, S. (1993). *Freshwater fishes of Western Indonesia and Sulawesi* (p. 344). Periplus Ed.

Kristanto, A. H., Subagja, J., Ath-thar, M. H. F., Arifin, O. Z., Prakoso, V. A., & Cahyanti, W. (2016). Pengaruh suhu inkubasi induk dan pemberian naungan pada larva terhadap produksi benih ikan baung [The effect of incubation temperature and shading on larvae on Asian redtail catfish fry production]. *Prosiding Forum Inovasi Teknologi Akuakultur*, 163–167.

Legendre, M., Slembrouck, J., Subagja, J., & Kristanto, A. H. (2000). Ovulation rate, latency period and ova viability after GnRH-or hCG-induced breeding in the Asian catfish *Pangasius hypophthalmus* (Siluriformes, Pangasiidae). *Aquatic Living Resources, 13*, 145–151.

Mustafa-Kamal, A. S., Kamarudin, I. S., Christianus, A., Dad, S. K., & Yu-Abit, L. (2012). Feeding habits of fishes in the Pengkalan Gawi–Pulau Dula Section of Kenyir Lake Terengganu, Malaysia. *Asian Fisheries Science, 25*(2), 144–157. https://doi.org/10.33997/j.afs.2012.25.2.004

Ng, H. H., & Kottelat, M. (2013). Revision of the Asian catfish genus *Hemibagrus* Bleeker, 1862 (Teleostei: Siluriformes: Bagridae). *The Raffles Bulletin of Zoology, 61*(1), 205–291.

Prakoso, V., Subagja, J., Radona, D., Kristanto, A., & Gustiano, R. (2018). Derajat penetasan dan sintasan larva ikan baung (*Hemibagrus nemurus*) dalam dua sistem penetasan berbeda [Hatching and survival rates of Asian redtail catfish larvae (*Hemibagrus nemurus*) in two different hatching systems]. *LIMNOTEK Perairan Darat Tropis Di Indonesia, 25*(2), 58–64.

Prianto, E., Kamal, M. M., Muchsin, I., & Kartamihardja, S. (2015). Aspek reproduksi ikan baung (*Hemibagrus nemurus*) di paparan banjiran Lubuk Lampam Kabupaten Ogan Komering Ilir

[Reproductive aspect of green catfish (*Hemibagrus nemurus*) in Lubuk Lampam floodplain, Ogan Komering Ilir Regency]. *BAWAL, 7*(3), 137–146.

Rachmatika, I., Sjafei, D. S., & Nurcahyadi, W. (2002). Fauna ikan Taman Nasional Gunung Halimun: Informasi tambahan tentang pemanfaatannya [Fish fauna of Gunung Halimun National Park Region: Additional information on the utilization]. *Berita Biologi, 6*(April), 13–24. https://doi.org/10.14203/beritabiologi.v6i1.1166

Rainboth, W. (1996). *Fishes of the Cambodian Mekong*. Food Agriculture Organization species identification field guide for fisheriespurposes. FAO, Rome, 265 p.

Rosadi, E., Makmur, S., Subagdja, S., & Fatah, K. (2020). Population dynamics and fishing status of baung (*Hemibagrus nemurus* C.V) in Barito upstream, Central Kalimantan, Indonesia. *Fish Scientiae, 10*(1), 23–33. https://doi.org/10.20527/fishscientiae.v10i1.152

Slembrouck, J. J., Subagja, D. D., & Legendre, M. (2005). *Technical instructions for Indonesian catfish hatchery, Pangasius djambal*. Lembaga Penelitian Perancis Untuk Pembangunan.

Subagja, J., Cahyanti, W., Nafiqoh, N., & Arifin, O. Z. (2015). Bioreproduction and growth performance from threedifferent population of Hemibargus fish (Hemibagrus nemurus, Val. 1840). *Jurnal Riset Akuakultur*, 10(1), 25.https://doi.org/10.15578/jra.10.1.2015.25-32

Subagja, J., & Prakoso, V. A. (2018). Reproduksi ikan baung (*Hemibagrus nemurus*) dengan perlakuan dosis hormon GnRH-a [Reproduction of Asian redtail catfish (*Hemibagrus nemurus*) using different doses of GnRH-a hormone]. *Jurnal Riset Akuakultur, 13*(3), 213. https://doi.org/10.15578/jra.13.3.2018.213-218

Subagja, J., Prakoso, V. A., Arifin, O. Z., Suparyanto, Y., & Suhud, E. H. (2018). Pertumbuhan benih ikan baung (*Hemibagrus nemurus*) hasil domestikasi pada lokasi dengan ketinggian berbeda [Growth performance of domesticated Asian redtail catfish fingerlings (*Hemibagrus nemurus*) reared at different altitude locations]. *Media Akuakultur, 13*(2), 59. https://doi.org/10.15578/ma.13.2.2018.59-65

Suhenda, N., Samsudin, R., & Nugroho, E. (2010). Pertumbuhan benih ikan baung (*Hemibagrus nemurus*) dalam keramba jaring apung yang diberi pakan buatan dengan kadar protein berbeda [Growth of green catfish (*Hemibagrus nemurus*) fry in floating net cage feed by artificial food with different protein content]. *Jurnal Iktiologi Indonesia, 10*(1), 65–71. https://doi.org/10.32491/jii.v10i1.179

Tirtana, A. (2022). Teknologi budidaya ikan lokal BRIN dukung mandiri protein nasional [Local fish farming technology by BRIN supports national protein self-sufficiency]. *ValidNews*, 5 September 2022, 13:41 WIB. https://validnews.id/kultura/teknologi-budidaya-ikan-lokal-brin-dukung-mandiri-protein-nasional

Barred Loach Nemacheilus fasciatus (Valenciennes, 1846) in Indonesia: Domestication and Aquaculture Development

3

Jojo Subagja, Vitas Atmadi Prakoso, Brata Pantjara, and Anang Hari Kristanto

3.1 Introduction

The barred loach *Nemacheilus fasciatus* (Valenciennes, 1846), is an Indonesian native freshwater fish, with economically valuable as potential aquaculture commodity (Hadie et al., 2011). This fish belongs to the Nemacheilidae family and is distributed across rivers in Eurasia and Ethiopia, from lowlands up to altitude of 5200 m above sea level (Kottelat & Freyhof, 2007; Kottelat, 2013). In Indonesia, the population of barred loach is found in Java and Sumatra (Hadiaty & Yamahira, 2014; Prakoso et al., 2016; Ath-thar et al., 2018). Typically, the Nemacheilidae group resides in clean water environments with calm to slightly swift currents, prefers gravel substrates, and tends to lives in groups. The species thrives in water with good quality, oxygen levels above 5 mg/L, pH 6.5–7, and temperature between 24 and 27 °C. Their main diet includes phytoplankton, zooplankton, and detritus, classifying them as mixed eaters or omnivores (Husnah, 2015; Wibisono, 2014). Observation on the stomach contents of barred loach revealed a variety of organisms, including *Cocconeis* sp., *Nebella* sp., *Bulbochaete* sp., *Cymbella* sp., *Navicula* sp., *Surirella* sp., *Anabaena* sp., *Merismopedia* sp., *Cenosphaera* sp., *Lecane* sp., *Trichoptera* sp., and other phytoplankton, with a dietary composition of 58%

J. Subagja (✉) · A. H. Kristanto
Research Center for Applied Zoology, Research Organization for Life Sciences and Environment, National Research and Innovation Agency, Cibinong, Indonesia
e-mail: jojo004@brin.go.id

V. A. Prakoso
Research Center for Conservation of Marine and Inland Water Resources, Research Organization for Earth Sciences and Maritime, National Research and Innovation Agency, Cibinong, Indonesia

B. Pantjara
Research Center for Fishery, Research Organization for Earth Sciences and Maritime, National Research and Innovation Agency, Cibinong, Indonesia

E. Nugroho et al. (eds.), *Aquaculture Development in Archipelago*,
https://doi.org/10.1007/978-981-95-1084-9_3

phytoplankton and 42% zooplankton. Furthermore, the observed sex ratio of barred loach population in the Lekso River, Blitar, East Java, is 1:1.25 (male:female) (Tjahjo et al., 2017).

Besides being a food source, it is also traded as an ornamental fish. Although the IUCN Red List of Threatened Species 2020 lists this species as least concern (LC) (Daniels, 2020), habitat stress due to environmental degradation and intensive exploitation threatens its population. Consequently, domestication efforts for aquaculture and conservation are essential to ensure the species' sustainability.

Conservation efforts have been initiated by the Animal Husbandry and Fisheries Service of the Government of Temanggung Regency, Central Java Province, in collaboration with the Research Center for Aquaculture through its implementing unit, the Research Institute for Freshwater Aquaculture in Bogor, West Java, Indonesia. The collaboration include domestication of native fish species, studies on barred loach hatchery operation and conservation support within their natural habitat. This chapter will discuss insights gained from the adaptation process, including reproduction, spawning, and seedling maintenance of barred loach in ex situ environments during domestication.

3.2 Taxonomy and Morphology

The subfamily Nemacheilinae was long classified as a part of the family Balitoridae, but after being recognized as a monophyletic lineage (Bohlen & Šlechtová, 2011), Kottelat (2012) designated Nemacheilidae as a distinct, valid family. According to the classification by Valenciennes (Cuvier & Valenciennes, 1846), *Nemacheilus fasciatus* is categorized as follows:

Kingdom:	Animalia
Phylum:	Chordata
Class:	Actinopterygii
Order:	Cypriniformes
Family:	Nemacheilidae
Genus:	*Nemacheilus*
Species:	*N. fasciatus Val. 1846*

The morphology of *N. fasciatus* in Sumatra and Java populations are based on color and shape characteristics and the number of black bands found in their bodies compared to other species. Hadiaty and Yamahira (2014), reported *N. fasciatus* is characterized as follows: Color pattern consisting of 14–18 dark blotches along lateral line with 11–12 dark saddles across the back, color pattern consisting of dark brown blotches and dorsal saddles, and anterior naris pierced on front side of a flap–no acuminate or enlarged scales on caudal peduncle (Fig. 3.1).

Fig. 3.1 Barred loach *Nemacheilus fasciatus*

3.3 Distribution

Species of the genus *Nemacheilus* are spread over the Southeast Asian, ranging from mainland Asia to the Sunda Islands. A total of 33 species have been recorded and validly identified by Kottelat et al. (1993), while Hadiaty and Yamahira (2014) identified 16 species distributed in the Sunda Islands (Sumatra, Kalimantan, and Java). These species are grouped based on location distribution where six species came from Sumatra, eight species were obtained from Kalimantan, and two species from Java. *Nemacheilus fasciatus* is a freshwater species distributed throughout Java and Sumatra (Kottelat et al., 1993). Husnah (2015) observed that this species occurs in the upper streams of the Musi River, Tanjung Raya, South Sumatera, while Daniels (2020) found this fish in the Progo River, stationed in Temanggung, Central Java. The distribution map of *Nemacheilus fasciatus* is presented in Fig. 3.2.

3.4 Reproduction

Reproductive behavior in barred loach has been observed in an ex situ environment by replicating conditions similar to their natural habitat. Spawning was conducted with six pairs of broodstock in a 100 L container. The spawning process occurred in pairs, with one female being followed by one male, and spawning took place 12 h after pairing (Kristanto et al., 2021). Barred loach eggs are nonadhesive and sink, with ovulated egg diameters ranging from 0.5 to 0.6 mm. In the spawning medium, the eggs expand to five times their initial diameter, reaching full size approximately 6 h after fertilization. During spawning, some broodstock were observed eating freshly released eggs, which negatively impacted productivity (Prakoso et al., 2021). After spawning, the broodstock are transferred to rearing containers, and the eggs typically hatch within 11–16 h.

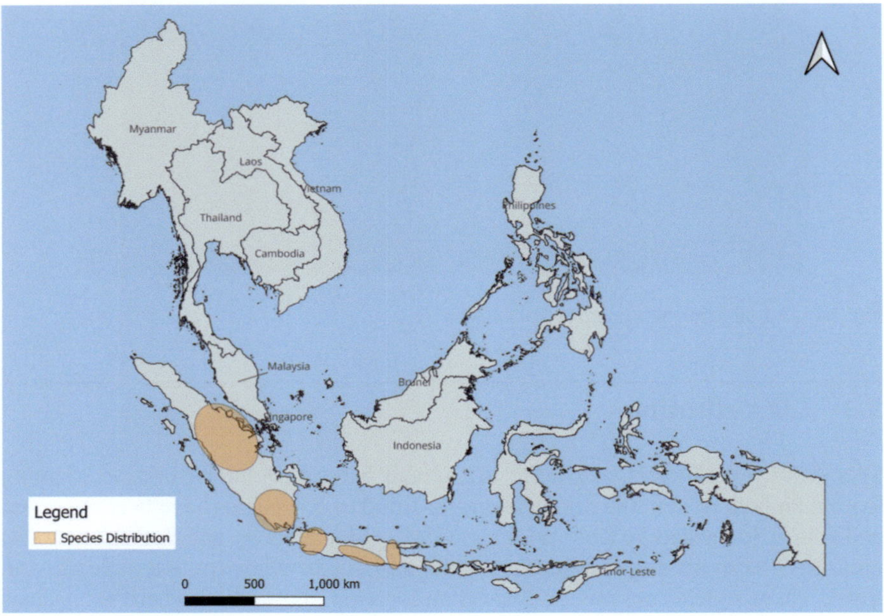

Fig. 3.2 Distribution map of *Nemacheilus fasciatus*

3.5 Domestication of Barred Loach

Conservation efforts through stock enhancement outside their natural habitat have been carried out since 2015. These efforts have contributed to improving natural populations through restocking, which aligns with the introduction of pioneering fish farming activities within local communities. Reproduction is a key factor in supporting successful domestication, and a well-researched understanding of barred loach hatchery practices is essential for producing aquaculture candidates that are adaptive to ex situ environments (Prakoso et al., 2017).

The domestication process for barred loach has included: collecting broodstock candidates from their natural habitat (specifically from the Progo River in Temanggung, Central Java, Indonesia), adapting them to ex situ environments (such as rearing containers) and artificial feed, implementing semi-artificial propagation methods, and rearing larvae.

3.5.1 Collection and Adaptation

Barred loach broodstock were collected from the Progo River, Temanggung, Central Java, Indonesia (Ath-thar et al., 2018). The collected fish were acclimated in an aquarium equipped with a gravel and sand substrate to simulate their natural habitat. This aquarium was connected to a recirculating aquaculture system

The shape and color of the thick solid black band on the adult female of barred loach *Nemacheilus fasciatus*

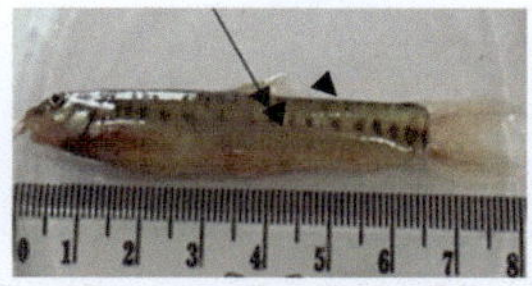

The shape and color of the thin black band on the adult male of barred loach *Nemacheilus fasciatus*. The body color is brighter

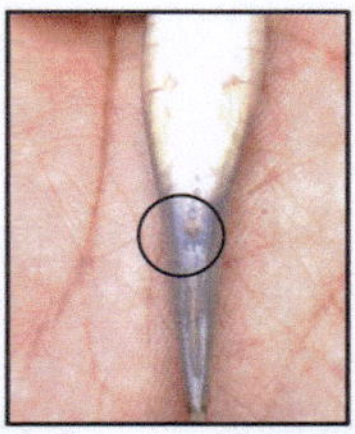

The abdomen part of adult female of barred loach *Nemacheilus fasciatus* and its genital shape

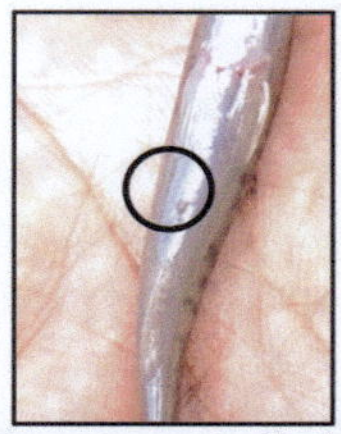

The abdomen part of adult male of barred loach *Nemacheilus fasciatus* and its genital shape

Fig. 3.3 Visual characteristics of adult male and female barred loach *Nemacheilus fasciatus* (Prakoso et al., 2017)

(RAS), which maintained a water temperature of 24–28 °C and was equipped with aeration. During the rearing period, the broodfish were gradually adapted to a commercial diet, initially fed natural food (*Tubifex*) until they accepted the commercial feed.

For gonad maturation, adult barred loach were first selected based on sexual characteristics (Fig. 3.3), then separated by sex into different aquariums. Maturation was supported by feeding commercial floating feed with 45% crude protein twice daily at a rate of 3–5% of fish biomass. Vitamin E supplements (60 mg/kg feed) were also provided by coating the feed with egg albumen (Subagja et al., 2019).

In their natural habitat, barred loach fecundity ranges from 1665 to 7567 eggs (Risyanto et al., 2012). In aquariums, fecundity ranged from 786 to 4198 eggs, with an ovary somatic index of 0.14–0.20 and egg diameters of 0.60–0.63 mm (Kristanto et al., 2021). Variation in fecundity among fish of similar size is common and can result from environmental factors like temperature, food availability, and genetic differences (Blaxter, 1969; Gibson & Ezzi, 1978; Hoda & Akhtar, 1985). In general, fecundity increases with increasing length or weight of the fish; however, in small fish, the gonadosomatic index values vary greatly.

3.5.2 Artificial Propagation

Broodstock selection for propagation is based on visual observation of the abdomen and genitals. Mature barred loach exhibit the following characteristics: adult females have more distinct black spots, a rounded abdomen that narrows near the genitals, and rounded genitalia. Adult males, as they begin to mature, show slightly faded black spots, a somewhat rounded abdomen, and elongated genitalia. Once selected, female and male barred loaches are separated for spawning.

Spawning techniques can be either semi-artificial or fully artificial (KKP News, 2016), using a male-to-female ratio of 2:1 or 3:1. Both methods start with a single hormone injection administered near the dorsal fin area. The hormone commonly used is Ovaprim, with doses ranging from 0.5 to 0.6 mL/kg for females and 0.3 mL/kg for males (Prakoso et al., 2021).

In semi-artificial spawning, following hormone injection, six pairs of broodfish are placed in a specially prepared aquarium equipped with a barrier net cage designed for spawning (Fig. 3.4). This netting allows ovulated eggs to fall to the aquarium floor while preventing adult fish from reaching and consuming the eggs. After spawning, the adult females are transferred to a separate aquarium. Fertilized eggs typically hatch within 10–14 h.

Meanwhile, artificial spawning can be conducted with external fertilization by mixing eggs and sperm. Mature broodfish ready to spawn are injected with a fish breeding hormone at doses of 0.5–0.6 mL/kg for females and 0.3 mL/kg for males. After injection, the male and female broodstock are placed in separate containers or aquariums. After 12 h, eggs are manually stripped from the female and immediately fertilized with a preprepared sperm solution. The artificial fertilization process involves pouring the sperm solution into a container with the eggs and stirring thoroughly with dry chicken feathers until the mixture is even (Prakoso et al., 2021).

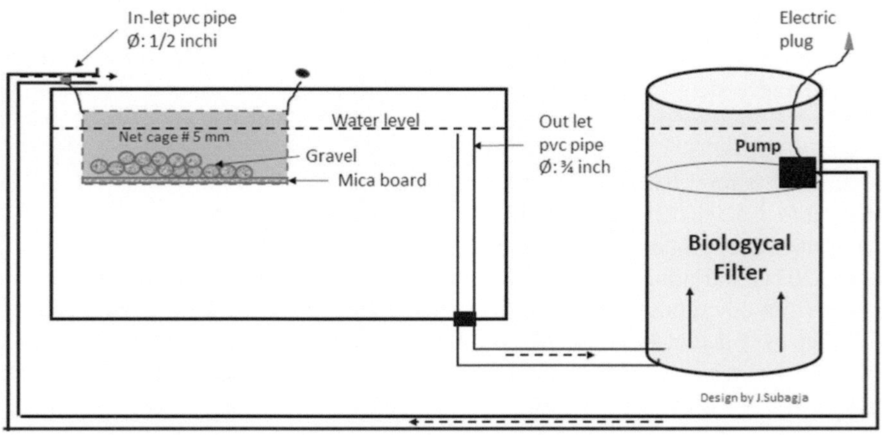

Fig. 3.4 Semi-artificial spawning method of barred loach *Nemacheilus fasciatus* using aquarium equipped with net cage in a recirculating water system. (Designed by Jojo Subagja)

After mixing, two to three parts of clean water are added to dilute the egg-sperm mixture, and it is stirred gently with a chicken feather or shaken for about 1 min to promote fertilization. The fertilized eggs are then transferred to hatching tanks, as barred loach eggs are nonadhesive (Kristanto et al., 2021; Prakoso et al., 2017).

The 8–10 h after fertilization are a critical period for the development of barred loach eggs. During this time, the eggs should not be disturbed, and water quality must be closely monitored. The hatching water temperature is maintained at a minimum of 25 °C, with dissolved oxygen levels kept above 5 mg/L for optimal hatching conditions (Prakoso et al., 2016). Regular observations of the eggs' development are essential to assess the success of the hatching process.

3.6 Aquaculture Technology of Barred Loach

Barred loach larvae begin to hatch about 14 h after fertilization. It is crucial to separate newly hatched larvae from dead or unfertilized eggs, as these decomposing eggs can release toxins that may harm the larvae. Separation is typically done by siphoning out the dead or unfertilized eggs. Approximately 2 days after hatching, once the yolk sac is fully absorbed, the larvae begin to require external food sources. Initially, larvae can be fed phytoplankton, such as *Chlorella* or *Nannochloropsis*, for 3 days. After this period, as their mouth opening widens, they can be fed *Artemia*. During larval rearing, the tank water should be kept clean through careful siphoning (Prakoso et al., 2021).

Once barred loach larvae reach 1–2 cm in length, they enter the nursery stage. This stage is conducted indoors in an aquarium, using a stocking density of five fish per liter for optimal growth (Iswantari et al., 2019; Subagja et al., 2019). The fish are fed *Artemia* ad libitum. When they reach 3 cm, they can be gradually transitioned to commercial floating pellets (30% protein), fed at 3% of their biomass twice daily.

Water quality is essential during the nursery stage and should be maintained with regular siphoning. Ideal water conditions include a temperature of 28 °C, pH of 6–7, and dissolved oxygen levels above 5 mg/L to support healthy physiological development (Wijaya et al., 2019; Prakoso & Kurniawan, 2020). Rearing continues until the fish reach 6–8 cm, at which point they can be used as broodstock for breeding purposes. Typically, it takes 6–7 months for barred loach reared in aquariums to mature as young broodfish (Cahyanti et al., 2020).

Barred loach hatcheries can be efficiently managed in indoor systems that require minimal space. However, further research is needed to advance hatchery methods for broader applications in conservation and commercial aquaculture. Additionally, ongoing monitoring of fishery resources in natural habitats is vital to ensure the sustainability of fish populations (Putranto, 2019).

References

Ath-thar, M. H. F., Ambarwati, A., Soelistyowati, D. T., & Kristanto, A. H. (2018). Keragaan genotipe dan fenotipe ikan uceng *Nemacheilus fasciatus* (Valenciennes, 1846) asal Bogor, Temanggung, dan Blitar [Genotypic and phenotypic performance of uceng (barred loach) *Nemacheilus fasciatus* (Valenciennes, 1846) from Bogor, Temanggung, and Blitar]. *Jurnal Riset Akuakultur, 13*(1), 1–10.

Blaxter, J. H. S. (1969). Development: Eggs and larvae. In W. S. Hoar & J. J. Randall (Eds.), *Fish physiology* (pp. 177–184). Academic Press.

Bohlen, J., & Šlechtová, V. (2011). A new genus and two new species of loaches (Teleostei: Nemacheilidae) from Myanmar. *Ichthyological Exploration of Freshwaters, 22*, 1–10.

Cahyanti, W., Putri, F. P., & Prakoso, V. A. (2020). Performa pertumbuhan dua generasi Barred loach (*Nemacheilus fasciatus* Val. 1846) dalam pemeliharaan di akuarium [Growth performance of two generations of barred loach (*Nemacheilus fasciatus* Val. 1846) in aquarium rearing]. *Media Akuakultur, 15*(1), 9–14.

Daniels, A. (2020). Nemacheilus fasciatus. *The IUCN Red List of Threatened Species, 2020*, e.T91174777A91174799. https://doi.org/10.2305/IUCN.UK.2020-1.RLTS.T91174777A91174799.en. Accessed on 3 March 2023.

Gibson, R. N., & Ezzi, I. A. (1978). The biology of a Scottish population of fries go by *Lesuerigobius friesil*. *Journal of Fish Biology, 12*, 371–389.

Hadiaty, R. K., & Yamahira, K. (2014). The loaches of the genus *Nemacheilus* (Teleostei: Nemacheilidae) in Sunda Islands, with an identification key. *Jurnal Iktiologi Indonesia, 14*(2), 83–100.

Hadie, L. E., Kristanto, A. H., & Hadie, W. (2011). Kajian komoditas prospektif ikan Uceng (*Nemacheilus fasciatus*) sebagai kandidat ikan budidaya [Study of the prospective commodity of uceng (barred loach) (*Nemacheilus fasciatus*) as a candidate for aquaculture]. *Prosiding Forum Inovasi Teknologi Akuakultur, 1*(1), 317–321.

Hoda, M. S., & Akhtar, Y. (1985). Maturation and fecundity of mudskipper *Boleophthalmus dentatus* in the northern Arabian Sea. *Indian Journal of Fisheries, 32*(1), 64–73.

Husnah, H. (2015). *Diversity, morphological characters and habitat of fish in Musi River drainage area, South Sumatra*. Center for Fisheries Research and Development.

Iswantari, A., Kurniawan, K., Priadi, B., Prakoso, V. A., & Kristanto, A. H. (2019). Konsumsi oksigen *Nemacheilus fasciatus* (Valenciennes, 1846) pada kondisi padat tebar yang berbeda [Oxygen consumption of *Nemacheilus fasciatus* (Valenciennes, 1846) under different stocking density conditions]. *OLDI (Oseanologi dan Limnologi di Indonesia), 4*(2), 79–87.

KKP News. (2016). *Pemijahan ikan uceng secara buatan berhasil dilakukan* [*Successful artificial spawning of uceng (barred loach) achieved*]. https://news.kkp.go.id/index.php/pemijahan-ikan-uceng-secara-buatan-berhasil-dilakukan/

Kottelat, M. (2012). Conspectus cobitidum*: An inventory of the loaches of the world (Teleostei: Cypriniformes: Cobitidae). *The Raffles Bulletin of Zoology, 26*, 1–199.

Kottelat, M. (2013). The fishes of the inland waters of Southeast Asia: A catalogue and core bibliography of the fishes known to occur in freshwaters, mangroves and estuaries. *Raffles Bulletin of Zoology, 27*, 1–663.

Kottelat, M., & Freyhof, J. (2007). *Handbook of European freshwater fishes* (646 p.). Publications Kottelat.

Kottelat, M., Whitten, A. J., Kartikasari, S. N., & Wirjoatmodjo, S. (1993). *Freshwater fishes of Western Indonesia and Sulawesi* (221 p.). Periplus Editions.

Kristanto, A. H., Subagja, J., Ath-thar, M. H. F., Arifin, O. Z., Prakoso, V. A., Ardi, I., & Samsudin, R. (2021). *Koleksi, identifikasi, karakterisasi, dan aspek ekobiologi ikan Uceng dan manggabai* [*Collection, identification, characterization, and ecobiological aspects of uceng (barred loach) and manggabai hish*] [Unpublished]. Laporan Teknis [Technical Report]. BRPBATPP.

Prakoso, V. A., & Kurniawan. (2020). Oxygen consumption of barred loach *Nemacheilus fasciatus* (Valenciennes, 1846) at different temperatures. *IOP Conference Series: Earth and Environmental Science, 457*(1), 012065.

Prakoso, V. A., Ath-thar, M. H. F., Subagja, J., & Kristanto, A. H. (2016). Pertumbuhan ikan uceng (*Nemacheilus fasciatus*) dengan padat tebar berbeda dalam lingkungan ex situ [Growth of uceng (barred loach) (*Nemacheilus fasciatus*) with different stocking densities in ex-situ environments]. *Jurnal Riset Akuakultur, 11*(4), 355–362.

Prakoso, V. A., Subagja, J., & Kristanto, A. H. (2017). Aspek biologi reproduksi dan pola pertumbuhan ikan uceng (*Nemacheilus fasciatus*) dalam pemeliharaan di akuarium [Reproductive biology aspects and growth patterns of uceng (barred loach) (*Nemacheilus fasciatus*) in aquarium rearing]. *Media Akuakultur, 12*(2), 67–74.

Prakoso, V. A., Subagja, J., Arifin, O. Z., Ath-thar, M. H. F., & Kristanto, A. H. (2021). Pembenihan ikan uceng *Nemacheilus fasciatus* [Seedlings production of barred loach *Nemacheilus fasciatus*]. In A. H. Kristanto, R. Gustiano, & K. Sugama (Eds.), *Pembenihan ikan air tawar asli perairan Indonesia [Breeding of native freshwater fish from Indonesian waters]* (pp. 87–96). Amafrad Press.

Putranto, Y. (2019). Upaya penyadaran masyarakat terhadap kelestarian spesies endemik Sungai Serayu ikan uceng (*Nemacheilus fasciatus* CV) di Kabupaten Banyumas [Community awareness efforts for the conservation of the endemic species of Serayu River, uceng (barred loach) (*Nemacheilus fasciatus* CV), in Banyumas Regency]. *Dinamika Journal: Pengabdian Masyarakat, 1*(3), 1–8.

Risyanto, S., Ardli, E. R., & Sulistiyo, I. (2012). Biologi ikan uceng (*Nemacheilus fasciatus* C.V.) di Sungai Banjaran Kabupaten Banyumas [Biology of uceng (barred loach) (*Nemacheilus fasciatus* C.V.) in the Banjaran River, Banyumas Regency]. *Biosfera, 29*(1), 51–58.

Subagja, J., Prakoso, V. A., Arifin, O. Z., & Kristanto, A. H. (2019). Pengaruh perbedaan padat tebar larva terhadap pertumbuhan dan sintasan pada ikan uceng (*Nemacheilus fasciatus*) [Effect of different larval stocking densities on growth and survival of uceng (barred loach) (*Nemacheilus fasciatus*)]. *Berita Biologi, 18*(2), 209–214.

Tjahjo, D. W. H., Purnama Ningtyas, S. E., & Purnomo, K. (2017). Aspek eko-biologi ikan uceng (*Nemacheilus fasciatus*) di sungai Lekso, Blitar [Eco-biology of uceng (barred loach) (*Nemacheilus fasciatus*) in Lekso River, Blitar]. *Jurnal Penelitian Perikanan Indonesia, 6*(2), 13–21.

Wibisono, S. A. (2014). *Kajian aspek eko-biologis ikan uceng (Nemacheilus sp.) di Desa Babatan, Kecamatan Wlingi, Kabupaten Blitar* [*Study on bio-ecological aspects of barred loach (Nemacheilus sp.) in Babatan Village, Wlingi District, Blitar Regency*] (Bachelor thesis, Universitas Brawijaya).

Wijaya, H., Widodo, M. S., & Soeprijanto, A. (2019). The effect of different temperature on cortisol, glucose and glycogen level of uceng fish (*Nemacheilus fasciatus*). *The Journal of Experimental Life Science, 9*(1), 1–6.

Domestication of the Striped Wallago Catfish (*Wallago leerii*) in Indonesia

4

Wahyulia Cahyanti, Otong Zenal Arifin, Kurniawan, and Bambang Gunadi

4.1 Introduction

The freshwater waters of the Sunda Shelf mainland are home to approximately 900 fish species, with over 190 of these being endemic (Prianto et al., 2015). Among these species is the striped wallago catfish (*Wallago leerii*), a prominent fish found in the Kampar River in Riau (Fig. 4.1). This species is highly valued by the local community due to its large size, reaching up to 1.5 m in length and weighing 35 kg (Yurisman et al., 2010), and its excellent taste. The striped wallago catfish, scientifically known as *Wallago leerii*, is also referred to by other names such as *Wallagonia leerii* Bleeker, 1851, *Wallago nebulosus* Vaillant, 1902, and *Wallago tweediei* Hora & Misra, 1941 (Roberts, 2014). It is one of the five species within the genus *Wallago*, which are all predatory catfish with significant economic importance.

W. Cahyanti (✉) · O. Z. Arifin
Research Center for Applied Zoology, Research Organization for Life Sciences and Environment, National Research and Innovation Agency, Cibinong, Indonesia
e-mail: wahy066@brin.go.id

Kurniawan
Research Center for Conservation of Marine and Inland Water Resources, Research Organization for Earth Sciences and Maritime, National Research and Innovation Agency, Cibinong, Indonesia

B. Gunadi
Research Center for Fishery, Research Organization for Earth Sciences and Maritime, National Research and Innovation Agency, Cibinong, Indonesia

E. Nugroho et al. (eds.), *Aquaculture Development in Archipelago*,
https://doi.org/10.1007/978-981-95-1084-9_4

Fig. 4.1 *Wallago leerii* Bleeker, 1851. (Personal picture)

4.2 Distribution

The striped wallago catfish, also known as the Great Tapah or helicopter catfish, is a fascinating species native to Southeast Asia (Fig. 4.2). This fish is predominantly found in countries such as Vietnam, Laos, Cambodia, Bangladesh, Thailand, Malaysia, northern Indonesia, and Singapore, especially within major river systems like the Mekong River (Ng, 1992; Roberts, 2014). As an adult, the striped wallago catfish prefers the waters of large rivers. However, during its larval stage, it is commonly found at the confluences of tributaries that flow into these large rivers. Its habitat typically features a muddy bottom with surface vegetation on the water's surface (Kottelat & Widjanarti, 2005). One notable characteristic of this species is its migratory behavior. The striped wallago catfish undertake long migrations influenced by seasonal changes. Additionally, this fish can adapt to migrate into forested areas when these regions become inundated during the rainy season. In terms of water conditions, the striped wallago catfish thrives in temperatures ranging from 72 to 78 F (22 to 26 °C). It prefers slightly acidic to neutral pH (6.0–7.0) and can tolerate water hardness up to 20°H.

The striped wallago catfish, specifically *Wallago leerii*, is a fascinating species with notable ecological and economic importance. Indigenous to Indonesia, particularly in the Sunda Shelf region encompassing Kalimantan and Sumatra, this fish thrives in the fast-flowing rivers of these islands. Its high market value, ranging from Rp. 120,000 to Rp. 250,000, underscores its economic significance for local fishermen (Rengi et al., 2013; Utomo, 2016). Despite its current classification by the IUCN as "Least Concern," indicating a lower risk of extinction, the striped wallago catfish faces significant challenges (Ng, 2019). Overfishing and habitat degradation have led to a noticeable decline in its population (Irawan et al., 2021). This decline highlights the urgency of conservation efforts and sustainable management practices. Given the pressures on wild populations, the potential for domestication and farming striped wallago catfish grows increasingly relevant.

Cultivating this species in controlled environments could alleviate some of the pressure on wild stocks, contribute to conservation efforts, and create new economic opportunities. Aquaculture also offers a more sustainable way to meet market demand and ensure the species' long-term viability. Efforts to domesticate and farm the striped wallago catfish should address various factors, including habitat requirements, breeding practices, and disease management, to ensure success. Collaboration

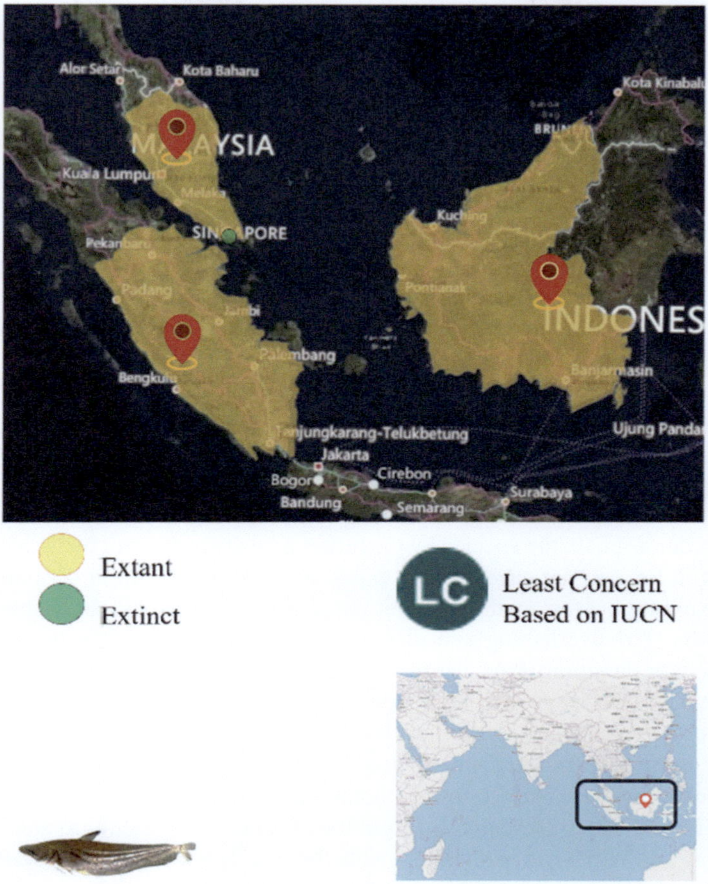

Fig. 4.2 Distribution map of *Wallago leerii*. (Google Earth with modification, based on IUCN data)

among local communities, researchers, and policymakers will be crucial in developing and implementing effective strategies for the sustainable management and conservation of this valuable fish species.

4.3 Classification and Morphology

According to Kottelat et al. (1993), striped wallago catfish are included in Kingdom: Animalia, Phylum: Chordata, Class: Actinopterygii, Oder: Siluriformes, Family: Siluridae, Genus: *Wallago*, Species: *Wallago leerii* (Blkr 1851). This species has a scaleless body with a variety of colors. It has a beautiful color pattern, with an olive-brass hue covering its back and head, and a white and black body highlighted by two brown bands running lengthwise (Ng, 1992). Its body is elongated and laterally compressed, with a very large head slightly longer than it is wide. The fish has a wide mouth with large teeth and a slightly protruding lower jaw. Its small eyes are positioned just above the mouth and are not covered by skin (Gupta, 2015). It has a short, spineless dorsal fin and a smaller ventral fin, while its anal fin is notably long. The pectoral fins are strong and finely serrated, and the gills are webbed. Both the anal and tail fins have a slight black coloration (Ng, 1992). This species can reach up to 2 m (6 feet 7 inches) in length and up to 95 kg (209 lb) in weight.

As a member of the Siluridae family, this fish is resistant to low-oxygen conditions, often referred to as a "blackfish." It is a freshwater species that still lives in the wild, classified as a carnivore and nocturnal fish, being active at night. The striped wallago catfish is also demersal, as it spends all its activities and hunts for food near the bottom of the waters (Muslih & Syari, 2016).

4.4 Domestication

Communities in the native regions of the striped wallago catfish have made early efforts to domesticate the species by capturing catfish seeds from the wild and raising them in floating net cages. These cages are typically set up in rivers near their natural habitats, but this method has not yet achieved optimal or sustainable production results. The overfishing and disrupted spawning migration patterns, exacerbated by river damming, have led to a marked decline in wild populations, making it increasingly challenging to find large fish and seeds. In Indonesia, various academics, practitioners, and researchers have begun efforts to domesticate the species, though current research remains focused on studying their behavior, seed maintenance technology, and growth in their natural environment.

In 2015, the Bogor Research Institute for Freshwater Aquaculture and Fisheries Extension (RIFAFE), in partnership with the Sei Tibun Fish Seed Center, Riau Province, conducted successful research on breeding striped wallago catfish using wild broodstock. They achieved this through artificial spawning technology, resulting in the production of larvae that grew and developed well (Kristanto et al., 2021). RIFAFE continues to advance its research into domestication, focusing on accelerating gonad maturation and enhancing breeding techniques for striped wallago catfish outside their natural habitat.

4.5 Development of Cultivation Technology

To date, striped wallago catfish have primarily been sourced from wild catches of various sizes. While community efforts have been made to cultivate the fish, these initiatives have mostly focused on rearing fry until they reach a size suitable for consumption. According to Muslih and Syari (2016) and Ulfa et al. (2020), many fish farmers in the Buluh Cina and Tanah Bawah Village areas of Kampar Regency engage in rearing activities. In this region, fish are raised in controlled system ponds with regular feeding, employing a combination of trash fish and pellets to optimize growth and help fish adapt to artificial feeds. Feeding occurs three times a day until the fish are satiated, and harvesting is done once the fish reaches the desired consumption size or according to market demand, typically after 3–4 months of growing period. Although the community has made efforts to use natural broodstock for breeding, these efforts have yet to fully address the challenges and achieve successful seed production.

In Indonesia, successful breeding of striped wallago catfish was first achieved by RIFAFE Bogor in collaboration with Sei Tibun Fish Seed Center in 2015, and this work continued in 2021 at the Freshwater Fisheries Germplasm Research Installation in Cijeruk, Bogor. Both broodstock and fry of striped wallago catfish have been observed to thrive in ponds with temperatures ranging from 24.8 to 28.0 °C, dissolved oxygen levels of 5.05–9.14 mg/L, and a pH between 7.6 and 9.9 (Prakoso et al., 2021). Although artificial spawning techniques have been developed and applied successfully, the success rates of spawning and larval survival remain suboptimal and require further research and refinement. Key factors influencing the success of striped wallago catfish breeding include the availability of mature gonad broodstock, adequate feed to stimulate spawning, optimal hormone dosage, and strategies to reduce larval cannibalism.

4.6 Reproduction

The reproductive patterns of *Wallago leerii* are similar to those of its relative *Wallago attu*, with both species exhibiting sexual dimorphism throughout all life stages. Males and females can be distinguished by their bone structure and morphological features. Secondary sexual characteristics, particularly the urogenital organs, provide a clear distinction: in males, the urogenital organ is slightly elongated and tapers toward the tail, while in females, it is more rounded (Fig. 4.3). In India, this species typically breeds once a year during the rainy season, from May to August, with peak spawning occurring in June to July in eastern regions and June to August in northwestern areas (Gupta, 2015). Research by Riza et al. (2021) indicates that striped wallago catfish in the Sebangau River, Kalimantan, reach peak gonadal maturity and spawning in December to January.

Broodstock fish captured from the wild, are generally distinguished based on physical characteristics (Fig. 4.3). Wild-cought broodstock with a minimum size of 1 kg, can achieve gonadal maturity when reared in ponds and fed a diet of chicken intestines and tilapia fry (Kurniawan et al., 2021). Research into gonad maturation has shown that although providing commercial feed (Yurisman et al., 2010) and adding Vitamin E (Nuraini & Nasution, 2015) can enhance gonadal development, these methods alone are not sufficient to induce spawning. However, the use of HCG hormones and Ovaprim injections has proven effective in stimulating gonadal maturation and enabling artificial spawning in fish averaging 1.60 ± 0.53 kg in weight and 55.59 ± 5.21 cm in length (Prakoso et al., 2021).

Broodstock selected for spawning should have reached gonadal maturity level IV, with egg diameters ranging from 1.9 to 2.2 mm (Kurniawan et al., 2021). The artificial spawning technique used for striped wallago catfish involves hormonal

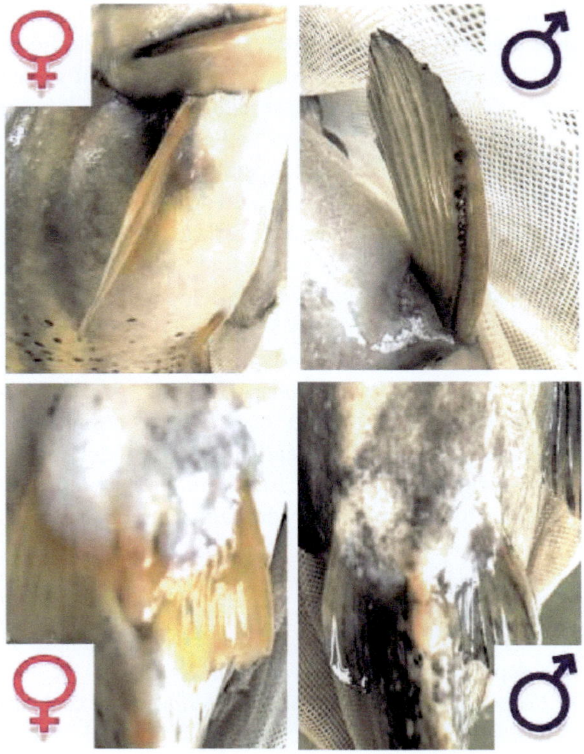

Fig. 4.3 Differences between female (left) and male (right) striped wallago catfish broodstock can be seen from the shape of the urogenital, pectoral fins, and ventral fins. (Modification of personal picture)

treatments, following the protocol established by Kristanto et al. (2021). This process includes administering two Ovaprim injections to female fish, at a dosage of 0.6 mL/kg. The first and second injections are given 8 h and 30 min apart. For male broodstock, a single Ovaprim injection of 0.2 mL/kg is administered concurrently with the second injection to the female. Ovulation in the females is monitored 9 h after the second injection. Fertilization is then carried out using the striping method.

Broodstock reared in ponds produce between 1281 and 8000 eggs/kg of fish (Kristanto et al., 2021). In contrast, wild-caught fish average around 43,334 eggs per fish, which can vary with fish size (Riza et al., 2021). Striped wallago catfish eggs typically hatch within 25–30 h, with a hatching rate of 65.00–72.50% (Kristanto et al., 2021). Newly hatched larvae remain in the hatching aquarium, which is equipped with aeration. Initial feeding begins with artemia, introduced 50 h post-hatching, at a frequency of six to seven times daily for 5 days. Once the larvae are 5 days old, they can be transferred to another aquarium or a fiberglass tub and fed with Tubifex worms (Kurniawan et al., 2021). During larval rearing, striped wallago catfish achieve optimal growth (32.30 g) and high survival rates (100%) when maintained at a water temperature of 29 °C (Tang et al., 2017).

References

Gupta, S. (2015). Wallago attu (Bloch and Schneider, 1801), a threatened catfish of Indian waters. *International Journal of Research in Fisheries and Aquaculture, 5*(4), 140–142. http://www.urpjournals.com

Irawan, D., Sundari, S., & Priadi, B. (2021). Pola pertumbuhan dan gambaran darah ikan tapah *Wallago leerii* pada tahap awal domestikasi [Growth pattern and blood images of striped wallago catfish in the early stage of domestication]. *Buletin Teknik Litkayasa Akuakultur, 19*(2), 67–71. https://doi.org/10.15578/btla.19.1.2021.67-71

Kottelat, M., & Widjanarti, E. (2005). The fishes of Danau Sentarum National Park and the Kapuas Lakes Area, Kalimantan Barat, Indonesia. *Raffles Bulletin of Zoology, 13*(Supplement), 139–173.

Kottelat, M., Whitten, A. J., Kartikasari, S. N., & Wirjoatmodjo, S. (1993). *Freshwater fishes of Western Indonesia and Sulawesi* (221p). Periplus Editions.

Kristanto, A. H., Subagja, J., Arifin, O. Z., Kurniawan, K., Zainal, S. A., Taryani, T., & Haspami, H. (2021). *Laporan awal pemijahan dan pemeliharaan larva ikan tapah* (*Wallago leerii*). [*The first record of artificial reproduction and larval rearing of striped wallago catfish* (*Wallago leerii*)]. Technical Report. Research Institute for Freshwater Aquaculture and Fisheries Extension (RIFAFE), Bogor.

Kurniawan, Subagja, J., Prakoso, V. A., Cahyanti, W., Arifin, O. Z., & Kristanto, A. H. (2021). Pembenihan ikan tapah *Wallago leerii* Bleker, 1851 [Hatchery of tapah fish *Wallago leerii* Bleker, 1851]. In A. H. Kristanto, R. Gustiano, & K. Sugama (Eds.), *Pembenihan ikan air tawar asli perairan Indonesia [Breeding of native freshwater fish from indonesian waters]* (Vol. 1, 1st ed., pp. 63–74). Amafrad Press.

Muslih, K., & Syari, I. A. (2016). Teknologi domestikasi ikan tapah di Desa Tanah Bawah Kabupaten Bangka [The striped wallago catfish domestication technology in Tanah Bawah Village, Bangka Regency]. *Jurnal Pengabdian Kepada Masyarakat Universitas Bangka Belitung, 3*(1), 48–57.

Ng, P. K. L. (1992). The giant malayan catfish, *Wallago leerii* Bleker, 1851, and identities of *Wallagonia tweediei* Hora & Misra, 1941 and *Wallago maculatus* Inger & Chin, 1959 (Teleostei: Siluridae). *Raffles Bulletin of Zoology, 40*(2), 245–263.

Ng, H. H. (2019). Wallagonia leerii. *The IUCN Red List of Threatened Species, 2019*, e.T187957A1839282. https://doi.org/10.2305/IUCN.UK.2019-3.RLTS.T187957A1839282.en. Downloaded on 28 February 2023.

Nuraini, & Nasution, S. (2015). Spawning technology and seed production of tapah fish (*Wallago leerii* B) by the injection of different doses of sGnRH + domperidone. *Journal of Biology, Agriculture and Healthcare, 5*(10), 126–129. www.iiste.org

Prakoso, V. A., Arifin, Z. A., Cahyanti, W., Kurniawan, K., Sundari, S., Putri, F. P., Subagja, J., & Kristanto, A. H. (2021). Evaluation of hormonal treatments on bio-reproductive and physiological performances of striped wallago catfish *Wallago leerii* during domestication. *Aquaculture International, 32*(2), 1801–1815.

Prianto, E., Pusptasari, R., Kartamihardja, E. S., Zulfia, N., Rachmawati, P., & Oktaviani, D. (2015). Kajian kebijakan konservasi sumber daya ikan di Paparan Sunda [Study of fish resource conservation policies on the Sunda Shelf]. *Prosiding Seminar Nasional Ikan Ke, 8*(1), 29–40.

Rengi, P., Bustari, & Sumarto. (2013). Kajian kelestarian ikan lokal (Ikan Tapah dan Kelemak) di wilayah Kabupaten Kampar Provinsi Riau [Study of local fish sustainability (Striped wallago catfish and Kelemak) in the Kampar Regency, Riau Province]. *Berkala Perikanan Terubuk, 41*(2), 82–91.

Riza, N., Butet, N. A., Kamal, M. M., Wibowo, A., Simon, O., & Ansori, M. (2021). Beberapa aspek biologi reproduksi catfish predator Indonesia (*Wallagonia leerii* Bleeker, 1851) di Sungai Sebangau Kalimantan Tengah [Some aspect biology reproduction of the Indonesian predatory catfish (*Wallagonia leerii* Bleeker, 1851) in Sebangau River Central Kalimantan].

Jurnal Pengelolaan Sumberdaya Alam dan Lingkungan [Journal of Natural Resources and Environmental Management], 11(2), 276–284. https://doi.org/10.29244/jpsl.11.2.276-284

Roberts, T. R. (2014). *Wallago* Bleeker, 1851, and *Wallagonia* Myers, 1938 (Ostariophysi, Siluridae), distinct genera of tropical Asian catfishes, with a description of wallago maemohensis from the Miocene of Thailand. *Bulletin of the Peabody Museum of Natural History, 55*(1), 35–47. https://doi.org/10.3374/014.055.0103

Tang, U. M., Muchlisin, Z. A., Syawal, H., & Masjudi, H. (2017). Effect of water temperature on the physiological stress and growth performance of tapah (*Wallago leeri*) during domestication. *Archives of Polish Fisheries, 25*(3), 165–171. https://doi.org/10.1515/aopf-2017-0016

Ulfa, N., Ridar, H., & Kusai. (2020). Persepsi pembudidaya ikan terhadap usaha budidaya ikan tapah (*Wallago* sp) dalam keramba jaring apung di Desa Buluh Cina Kabupaten Kampar Provinsi Riau [Perceptions of fish farmers on striped wallago catfish (*Wallago* sp.) farming business in floating net cages in Buluh Cina Village, Kampar District, Riau Province]. *Jurnal Sosial Ekonomi Pesisir [Coastal Socio-Economic Journal], 1*(3), 49–57. https://sep.ejournal.unri.ac.id/index.php/jsep/article/view/52

Utomo, A. D. (2016). Strategi pegelolaan suaka perikanan rawa banjiran di Sumatera dan Kalimantan [Management Strategies for Fisheries Reserve in Sumatera and Kalimantan Floodplain]. *Jurnal Kebijakan Perikanan Indonesia, 8*(1), 13–20. http://ejournal-balitbang.kkp.go.id/index.php/jkpi

Yurisman, Sukendi, & Putra, R. M. (2010). Domestikasi dan pematangan gonad ikan tapah (*Wallago* sp) dari perairan Sungai Kampar, Riau [Domestication and gonad maturation of striped wallago catfish (*Wallago* sp) from the Kampar River, Riau]. *Berkala Perikanan Terubuk, 38*(1), 107–117.

Hoven's Carp (*Leptobarbus hoevenii* Blkr) in Indonesia: Domestication and Cultivation

5

Anang Hari Kristanto, Wahyulia Cahyanti, Irin Iriana Kusmini, and Brata Pantjara

5.1 Introduction

The Thailand-Malaysia Peninsula, Sumatra, and Kalimantan are home to many different types of fish. One native fish from this area is Hoven's carp (*Leptobarbus hoevenii*), also known as the Sultan fish (Fig. 5.1). In Malaysia, Hoven's carp is a valuable commodity for consumption with high selling price reaching thousands of ringgits due to its tasty meat and soft scales. However, Hoven's carp is now overfished and extinct in the wild, further increasing its value and triggering fishermen to target it more aggressively.

Leptobarbus hoevenii is sport fish and are not suitable for aquarium keeping. The current state of the natural environment for the fish, also known as Jelawat, is concerning. Fish catches show that the size of individual fish is decreasing. Today, fish of various sizes are traded, without adherence to regulations regarding species size or age at maturity. This situation is alarming, as local communities have limited knowledge of effective cultivation techniques, and stakeholders lack cooperation in developing suitable techniques.

A. H. Kristanto · W. Cahyanti (✉) · I. I. Kusmini
Research Center for Applied Zoology, Research Organization for Life Sciences and Environment, National Research and Innovation Agency, Cibinong, Indonesia
e-mail: wahy066@brin.go.id

B. Pantjara
Research Center for Fishery, Research Organization for Earth Sciences and Maritime, National Research and Innovation Agency, Cibinong, Indonesia

E. Nugroho et al. (eds.), *Aquaculture Development in Archipelago*,
https://doi.org/10.1007/978-981-95-1084-9_5

Fig. 5.1 Hoven's carp *Leptobarbus hoeveni* Bleeker, 1851. (Personal documentation)

5.2 Distribution

Hoven's carp is a native fish found in Peninsular Malaysia and nearby areas. It is also found in countries in ASEAN, including Indonesia. In Malaysia, this species is only found in the states of Perak and Pahang. One was also caught in Sungai Bernam, Selangor (Mohsin & Ambak, 1991). In Indonesia, Hoven's carp inhabits two big islands of the Sunda Shelf, namely Sumatra and Kalimantan. According to Rainboth (1996), Hoven's carp is typically found in flowing rivers. While they do not frequently migrate between different bodies of water, they do travel short distances to and from wet forests or flood areas in search of food. According to Hardjamulia (1992), Hoven's carp are commonly found at river mouths and in small pools in the middle to lower reaches of rivers. They prefer streams in hilly areas surrounded by forest. Junenils live in still water adjacent to rivers, and when the water goes down, groups of Hoven's carp swim upstream to higher section of the river. They thrive in water with a pH between 5 and 7 and a temperature of 25 and 37 °C. The Hoven's carp is a "good swimmer" and capable of quickly catching food. This species of fish often feeds on the bottom of the water bodies (Aryani, 2018) (Fig. 5.2).

Hoven's carp is popular in many Asian countries and has various local names. It is called Jerjayar in Sumatra, Majuhan in Kalimantan, Sultan in Malaysia, and Praba or Riau Klemak in Thailand. In Sumatra, Hoven's carp is found in major rivers such as the Musi River, Batang Hari River, Kampar River, and Indragiri River (Aryani, 2018). The habitat of the Hoven's carp is deteriorating, making this fish even more difficult to find. Lumbantobing (2020) states that the IUCN (International Union for Conservation of Nature) has determined that the Hoven's carp is currently at low risk of extinction.

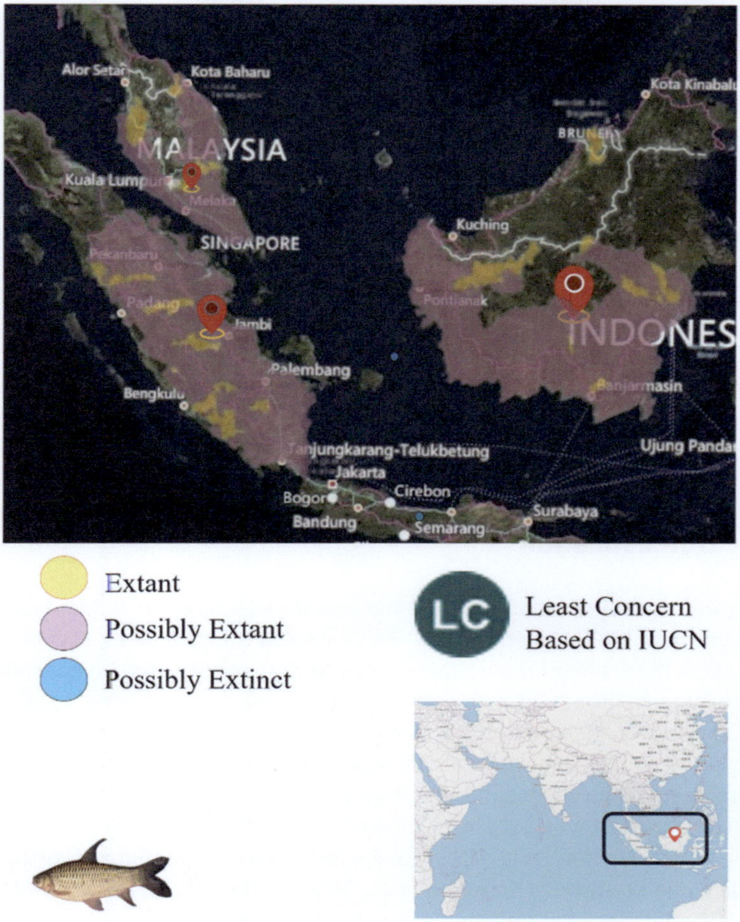

Fig. 5.2 Distribution map of *Leptobarbus hoevenii*. (Google Earth with modification based on IUCN data)

5.3 Classification and Morphology

According to the Bleeker system, Hoven's carp are classified as follows:

Kingdom: Animalia
Phylum: Chrodata
Class: Actinopterygii
Ordo: Cypriformes
Family: Cyprinidae
Genus: *Leptobarbus*
Species: *L. hoevenii* (Bleeker, 1851)
(Kottelat et al., 1993)

The Hoven's carp has a long body that is covered in shiny silver scales, with a slightly dark back and head. When they are young, the anal and pelvic fins are a reddish-orange color. This fish has 5–8.5 forked points on its anal fin, no sharp spines, and a line running along the underside of its tail. It has 7–8.5 split rays on the dorsal (top) fin, with the last rays being smooth. There are four to five scales between the sideline and the dorsal fin.

The top of the head is a bit flat, and the fish has a medium-sized mouth. The lateral line along the body is smooth, with red-colored pectoral and pelvic fins. The lateral line also curves downward along the body, and the fish features two pairs of barbels (Saputra et al., 2016). Hoven's carp has large scales. Its big mouth is positioned slightly downward at the front and can protrude outward, similar to the lips of other carp species (Rimalia, 2014).

The jelawat fish looks more unique than other carp. Juvenile fish have a black stripe running along the side of their bodies from the head to the tail fin, which fades as they grow up (Farida et al., 2015).

5.4 Domestication

The Hoven's carp is Indonesia's first indigenous freshwater fish and began to be domesticated in 1988 (Sunarno et al., 1988). At that time, the domestication of fish did not meet the standard requirements. Research is still focused on adaptability to the cultivation environment with compound feed and gonad maturation (Kristanto, 2022). These domestication efforts have increased public awareness to master the techniques of rearing naturally caught seed fish in ex situ environments. However, fish farming techniques are still in development.

Spawning of Hoven's carp was achieved using both hormones from the carp pituitary gland (Sunarno et al., 1988) and a hormone commercially available under the trade name Ovaprim® (Kristanto et al., 1994a). The success of artificial spawning of Hoven's carp has influenced the development of intertidal aquaculture in Sumatra and Kalimantan as well as fish farming in flooded marshes (Kristanto et al., 1993, 2000, 2007). Although breeding techniques for Hoven's carp are well mastered (Kristanto, 2022), technology transfer to fishing communities remains very limited. Public interest in breeding techniques is also relatively low. Many people remain unaware that natural fish supplies will decrease if efforts are not made to preserve or utilize aquaculture techniques.

5.5 Cultivation Technology Development

Hoven's carp were initially stocked in Lake Jambi Bay, followed by local breeding efforts led by the Research Institute for Freshwater Aquaculture (RIFA) Bogor in collaboration with the West Kalimantan Provincial Fisheries Department in 1992. Hoven's carp seeds were collected, dispersed, and cultured in floating net cages and ponds (Sunarno et al., 1988; Kristanto et al., 1994b, c, d).

The main challenge in developing Hoven's carp in West Kalimantan is the low frequency of gonad maturation, along with poor hatching and larval survival rates. These issues can be addressed by providing compound feed to promote gonad maturation and using artificial spawning techniques.

Hoven's carp (*Leptobarbus hoeveni*) is currently farmed in selected areas and offers breeders a profitable opportunity due to its straightforward farming techniques and low investment requirements. Although it has been cultivated for a long time and was once very popular locally, Hoven's carp farming still depends heavily on natural catches as the primary production source, with seed availability being seasonal. According to Abidin (2009), at the beginning of the rainy season, collected seeds measure about 0.5–1.0 cm (postlarvae). By mid-rainy season, they grow to 3–5 cm, and by the season's end, they reach 10–15 cm.

Government agencies are already collaborating with breeders to cultivate Hoven's carp. However, the number of seeds produced through breeding remains insufficient to meet the high demand. The Fisheries Office of Kampar Regency, Riau Province, reported that 32 million fish are farmed annually, driving significant demand for Hoven's carp seeds. Besides meeting local needs in Kampar, these seeds are sold to other regions. The price per seed ranges from 2000 to 2500 rupiahs (Taslim et al., 2015).

Hoven's carp can be reared in ponds, cages, or similar facilities. When raised in ponds, the fish reach a consumable size of approximately 125 g after 3–4 months. In cages, they can be harvested after 6–12 months, typically weighing between 1 and 2 kg/fish.

5.6 Reproduction

Hoven's carp (*Leptobarbus hoeveni*, Blkr) spawns naturally in the upper reaches of the river. Observations of spawning behavior have been made during the rainy season in the Jambi area, particularly around the Bungo Tebo waters and Batanghari River (Abidin, 2009). Hoven's carp are hermaphroditic species with a somatic gonad index of 14.40% and an egg diameter of 1.55 mm (Tan, 1980). Meanwhile, Aryani (2007) found that the egg diameters ranged from 0.88 to 0.92 mm, the mature gonads measuring 46–58 cm. The weight of mature females' range between 1400 and 3125 g, and their fecundity is estimated at 200,000–400,000 eggs per fish (for a fish average weighing 3 kg).

According to Saputra et al. (2016), Hoven's carp grow to about 26 cm at the start of gonadal maturity stage III (TKG III) and 32 cm at the start of stage IV (TKG IV). The distribution of these gonadal stages indicates that the spawning season coincides with the rainy season.

In the wild, Hoven's carp spawn two to three times during the rainy season when river water levels rise, and nearby areas flood. Juveniles migrate to the mouths of tributaries to spawn under these conditions (AARD, 1992). Spawning typically occurs in the morning during rainfall. Hoven's carp eggs are buoyant and float downstream, hatching along the way. The larvae settle in puddles or still bodies of water near the river (Aryani, 2018).

Spawning can occur naturally or through artificial (induced) breeding. Successful spawning depends on ensuring that the broodstock have mature gonads. Broodstocks typically reach maturity at 2.5 years of age and should be healthy, with no injuries or defects. In controlled breeding environments, fish can become ready to spawn after 3–6 months (Kristanto et al., 1994a). Artificial spawning uses hormonal stimulation to achieve optimal seed quantity and quality.

Female broodstock can be identified by a slightly bulging abdomen near the anus, which feels soft when pressed. Male broodstock have rough-textured pectoral fins, and sperm are released when the abdomen is gently massaged (Aryani, 2018). The artificial spawning process begins with selecting broodstock and conditioning them by feeding for a day. Ovaprim is used as a hormone for stimulation, administered at a dose of 0.5 mL/kg of broodstock. The first injection, consisting of one-third of the dose, is given to the female, followed by the remaining two-thirds in a second injection. Male broodstock receive a single injection at the same time as the female's second injection. Ovulation occurs 4–7 h after the second injection, at which point the eggs can be stripped.

Fertilization involves mixing the stripped eggs with sperm in a plastic basin. Once the eggs have grown sufficiently, they are transferred to the hatchery. Eggs are incubated in funnel-shaped incubators, each 60 cm in diameter and 50 cm high, with a stocking density of 400–500 eggs/L. During incubation, water quality must be maintained (dissolved oxygen: 4–8 ppm, pH: 7.0–8.0, temperature: 25–28 °C). At this temperature, eggs typically hatch 18–24 h after fertilization. Fecundity ranges from 29,000 to 44,000 eggs/kg of broodstock, with 80% fertilization success and a 70% hatching rate.

On the third day after hatching, the larvae are fed with enough *Artemia* nauplii three times a day (morning, afternoon, and evening). After 7 days, the juvenile fish are ready to be transferred to ponds for further rearing (Aryani, 2018). Some examples of stadia in Hoven's Carp fish can be seen in Fig. 5.3.

(a) (b) (c)

Fig. 5.3 20 days old Hoven's carp fry (**a**). Hoven's carp fry (**b**). Size ready for harvest (**c**). (Personal picture)

References

Abidin, M. (2009). Pengoperasian alat tangkap tangkul untuk menangkap benih ikan jelawat (*Leptobarbus hoeveni* Blkr) di perairan Batanghari, Jambi [Operation of the tangkul fishing gear to catch the seeds of the jelawat fish (*Leptobarbus hoeveni* Blkr) in the Batanghari Waters, Jambi]. *Buletin Teknisi Litkayasa Sumber Daya dan Penangkapan, 7*(1), 7–9. https://doi.org/10.15578/btl.7.1.2009.7-9

Agency for Agriculture Research and Development. (1992). *5 years of agricultural research and development in Indonesia 1987–1991*. Agency for Agricultural Research and Development.

Aryani, N. (2007). Penggunaan hormon LHRH dan vitamin E untuk meningkatkan kualitas telur ikan jelawat (*Leptobarbus hoeveni* Blkr) [Use of LHRH hormones and vitamin E to improve the quality of jelawat fish (*Leptobarbus hoeveni* Blkr) eggs]. *Sigmatek: Jurnal Sain dan Teknologi, 1*(1), 36–51.

Aryani, N. (2018). *Teknologi tepat guna budidaya ikan jelawat [Appropriate technology for jelawat fish farming]* (Vol. 1). Bung Hatta University Press.

Farida, Rachimi, & Ramadhan, J. (2015). Imotilisasi benih ikan jelawat (*Leptobarbus hoevani*) menggunakan konsentrasi larutan daun bandotan (*Ageratum conyzoides*) yang berbeda pada transportasi tertutup [Imotilisasi fish fry of jelawat (*Leptobarbus hoevani*) using different concentration bandotan leaf (*Ageratum conyzoides*) in closed transportation]. *Jurnal Ruaya, 5*, 22–28.

Hardjamulia, A. (1992). Pembenihan ikan Jelawat (*Leptobarbus hoeveni* Blkr) [Hoven's carp breeding]. *Warta Penelitian dan Pengembangan Pertanian, 10*(6), 10–11.

Kottelat, M., Whitten, A. J., Kartikasari, S. N., & Wirjoatmodjo, S. (1993). *Freshwater fishes of Western Indonesia and Sulawesi*. Periplus Editions.

Kristanto, A. H. (2007). Penguasaan teknologi budidaya untuk menghasilkan benih ikan air tawar [Mastery of cultivation technology to produce freshwater fish seeds]. In *Prosiding Nasional Hari Pangan Sedunia XXVII* (pp. 109–115). Badan Penelitian dan Pengembangan Pertanian, Departemen Pertanian.

Kristanto, A. H. (2022). *Domestikasi ikan air tawar asli Indonesia mendukung produksi perikanan [Domestication of native Indonesian freshwater fish to support fisheries production]*. Badan Riset dan Inovasi Nasional. https://doi.org/10.55981/brin.663

Kristanto, A.H., N. Suhenda, E. Wahyudi, A. Hardjamulia, M.T.D. Sunarno, F. Sukadi. (1993). *Penelitian pengembanganteknologi budidaya ikan jelawat (Leptobarbus hoevenii) di Kalimantan Barat [Research on the development of jelawat fish(Leptobarbus hoevenii) farming technology in West Kalimantan]*. Prosiding Seminar Hasil Penelitian Perikanan Air Tawar1991/1992, 20- 22 Oktober 1992, Cipayung.

Kristanto, A. H., Hardjamulia, A., & Dharma, L. (1994c). Produksi benih ikan jelawat dengan suntikan hormone ovaprim di Kalimantan Barat [Hoven's carp seeds production by injection of the hormone ovaprim in West Kalimantan]. In *Prosiding Seminar Hasil Penelitian Perikanan Balai Budidaya Air Tawar* (pp. 189–191). Bogor.

Kristanto, A. H., Hardjamulia, A., & Dharma, L. (1994d). Pembenihan ikan jelawat di keramba jaring apung (KJA) [Hoven's carp nursery in floating net cages]. In *Prosiding Seminar Hasil Penelitian Perikanan Air Tawar* (pp. 205–207). Bogor.

Kristanto, A. H., Hardjamulia, A., Suhenda, N., & Wahyudin, E. (1994a). Pembenihan ikan jelawat di keramba jaring apung (KJA) menggunakan jenis pakan dan padat tebar berbeda [Hoven's carp nursery in KJA uses different types of feed and stocking densities]. In *Prosiding Seminar Hasil Penelitian Perikanan Air Tawar* (pp. 208–211). Bogor.

Kristanto, A. H., Hardjamulia, A., Wahyudin, E., & Dharma, L. (1994b). Penelitian pembenihan ikan jelawat di kolam [Research on the Hoven's carp nursery in ponds]. In *Prosiding Seminar Hasil Penelitian Perikanan Air Tawar* (pp. 192–194). Bogor.

Kristanto, A. H., Jaenuddin, D., & Juarsah, I. (2000). Potensi dan pengembangan perikanan di lahan gambut (PLG) Kalimantan Tengah [Potential and development of aquaculture in peatland Central Kalimantan]. In *Prosiding Temu Pakar dan Lokakarya Nasional Diseminasi dan Optimasi Pemanfaatan Sumber Daya Lahan Rawa* (pp. 235–242). Jakarta.

Lumbantobing, D. (2020). *Leptobarbus hoevenii. The IUCN Red List of Threatened Species, 2020*, e.T181011A1688346. https://doi.org/10.2305/IUCN.UK.2020-3.RLTS.T181011A1688346.en

Mohsin, A. K. M., & Ambak, M. A. (1991). *Ikan air tawar di Semenanjung Malaysia [Freshwater fish in the Malaysian Peninsula]*. Dewan Bahasa dan Pustaka.

Rainboth, J. W. (1996). *FAO species identification field guide for fishery purposes: Fishes of the Cambodian Mekong*. Food and Agriculture Organization.

Rimalia, A. (2014). Perbandingan induk jantan dan betina terhadap keberhasilan pembuahan dan daya tetas telur ikan jelawat (*Leptobarbus hoevenii*) [Parental comparison of male and female and success fertilization egg hatching rate jelawat fish (*Leptobarbus hoevenii*)]. *Ziraa'ah, 39*(3), 114–118.

Saputra, Y. H., Syahrir, M., & Budiarsa, A. A. (2016). Biologi reproduksi ikan jelawat (*Leptobarbus hoevenii* Bleeker 1851) di rawa banjiran Sungai Mahakam, Kecamatan Muarawis, Kabupaten Kutai Kartanegara, Provinsi Kalimantan Timur [Reproduction biology of mud barb (*Leptobarbus hoevenii* Bleeker 1851) in floodplains of Mahakam Rivers, Muarawis Subdistrict, Kutai Kartanegara Regency, East Kalimantan Province]. *Jurnal Ilmu Perikanan Tropis, 21*(2), 48–56.

Sunarno, M. T. D., Reksalegora, O., & Nurdawaty, S. (1988). Penelitian pembenihan ikan Jelawat (*Leptobarbus hoevenii* Blkr.) dengan cara hipofisasi [Research on the Hoven's carp (*Leptobarbus hoevenii* Blkr.) breeding by hypophysation]. *Bulletin Penelitian Perikanan Darat, 7*(2), 102–108.

Tan, E. S. P. (1980). Some aspects of the biology of Malaysian riverine cyprinids. *Aquaculture, 20*, 281–289.

Taslim, R., Ramli, M., & Hendrik. (2015). The aquaculture business of jelawat (*Leptobarbus hoevanii*) at the cages in Ranah Kampar Sub-District, Kampar Regency of Riau Province. *Jom FAPERIKA, 2*(2), 1–12. https://media.neliti.com/media/publications/201266-none.pdf

Domestication and Farming of Kissing Gourami (*Helostoma temminckii*)

6

Otong Zenal Arifin, Wahyulia Cahyanti, and Nunak Nafiqoh

6.1 Introduction

The fish *Helostoma temminckii*, commonly known as the kissing gourami (Fig. 6.1), is recognized by various local names depending on the region in Indonesia. In West Java, it is called "terbakan," while in Central Java, it is known as "tambakan." Other names include "sapil" in South Sumatra, "tambakalang" in Jambi, and "poni" or "biawan" in Kalimantan. The kissing gourami is a significant local freshwater species with strong potential for aquaculture development. This species is particularly well-suited for aquaculture due to its ability to tolerate low-oxygen environments (Efriyeldi & Pulungan, 1995). As an herbivorous fish, it reduces reliance on commercial feed. It also has a high fecundity rate, making it valuable for both consumption and ornamental purposes. Despite this potential, the production of kissing gourami still relies heavily on wild populations (Kottelat et al., 1993). The population is facing a decline due to overfishing. According to Gustiano et al. (2015), the kissing gourami is one of 16 local freshwater species that holds significant promise for future aquaculture development.

O. Z. Arifin (✉) · W. Cahyanti
Research Center for Applied Zoology, Research Organization for Life Sciences and Environment, National Research and Innovation Agency, Cibinong, Indonesia
e-mail: oton003@brin.go.id

N. Nafiqoh
Research Center for Veterinary Science, Research Organization for Health, National Research and Innovation Agency, Cibinong, Indonesia

E. Nugroho et al. (eds.), *Aquaculture Development in Archipelago*,
https://doi.org/10.1007/978-981-95-1084-9_6

Fig. 6.1 Kissing gourami (*Helostoma temminckii*). (Personal documentation)

6.2 Taxonomy and Morphology

Taxonomically, the classification of kissing gourami based on Saanin (1984) is as follows:

Phylum: Chordata
Class: Actinopterygii
Ordo: Perciformes
Subordo: Anabantoidei
Family: Helostomatidae
Genus: *Helostoma*
Species: *Helostoma temminckii* (Cuvier 1829)

Kissing gourami scales are classified as stenoids, which feel rough to the touch due to the presence of spines along the edges. The scales on the back are greenish or gray, becoming lighter toward the belly, and a longitudinal line is present (Saanin, 1984). The dorsal fin has 16–18 spines and 13–16 soft rays, while the anal fin has 13–15 spines and 17–19 soft rays. Arifin et al. (2017a) suggested the meristics of kissing gourami fins as follows: D. XVI–XVIII.13–16; A. XIII–XV.17–19; P. 10–11; C. 13–16. The fish has 44–48 scales along the lateral line (linea lateralis). Cycloid scales are found on the top of the head, while ctenoid scales are present in other areas. The lips have rough teeth, and there are many gill rakers. The maximum length of the species is about 30 cm. Kissing gouramis are filter feeders and grazers of benthic algae.

The body of the kissing gourami is vertically compressed, with the dorsal and anal fins being similar in size and shape. The caudal fin is almost round or slightly convex outward. The pectoral fins, one on each side of the body, are nearly round. Along both sides of the body, there is a lateral line with 43–48 scales, extending from the base of the gill slits to the base of the caudal fin. The mouth is elongated and protrudes forward, allowing the fish to gather food. The lips are covered with horn-like teeth, which are absent from other parts of the mouth, such as the pharynx, premaxilla, dentary, and palate. The gill rakers function to filter food particles from the water as it passes through.

Male and female kissing gourami can be distinguished by observing their primary and secondary sexual characteristics. Primary sexual characteristics are defined by the shape and color of the genital organs. Female kissing gourami have a genital orifice located at the front of the genital papilla, which appears reddish, while male fish have a genital orifice at the back of the genital papilla, which is pale in color. Secondary sexual characteristics are determined by body shape. Female kissing gourami have a more rounded body shape, and when mature, their gonads cause the body to appear fatter and feel soft to the touch. Male fish, on the other hand, have a slimmer body shape. When their gonads are mature, pressing their abdomen may result in the release of white sperm (Fig. 6.2).

6.3 Distribution

The kissing gourami (*Helostoma temminckii*, Cuvier 1829) is found throughout the Indochina region and Southeast Asia, inhabiting lakes and rivers. These freshwater fish thrive in rivers or swamps and can live in warm temperatures ranging from 25 to 30 °C, at altitudes between 150 and 750 m above sea level. Kissing gourami have the ability to breathe air directly. They prefer slow-moving waters with dense vegetation and consume a variety of plants and animals, including green algae, zooplankton, and aquatic insects near the surface.

Kissing gourami are benthopelagic freshwater fish, meaning they live between the surface and deeper waters. Their natural habitat consists of tropical, shallow waters with calm currents and abundant aquatic vegetation. As omnivores, they feed on a wide range of food sources, from mosses and aquatic plants to zooplankton and aquatic insects. They typically spawn under floating vegetation. A map of the distribution of the kissing gourami is shown in Fig. 6.3.

6.4 Reproduction

The kissing gourami (*Helostoma temminckii*) is a heterosexual species, meaning that sperm and egg cells are produced by separate individuals. According to Lisna (2016), the development of gonads—ovaries in females and testes in males—occurs independently from the early stages of life, with each fish remaining distinctly male or female throughout its lifetime. Female kissing gourami typically reach sexual maturity at around 7 months of age, with an average weight of 28.5 ± 5.56 g and a standard length of 9.6 ± 0.86 cm. Males reach maturity slightly earlier, at around 6 months of age, with an average weight of 18.3 ± 3.45 g and a standard length of 8.2 ± 0.64 cm (Arifin et al., 2016).

The gonads of mature kissing gourami can make up as much as 12% of their body weight. Female ovaries are yellow, while male testes are transparent white (Rahardjo, 1980). The gonadosomatic index (GSI) for the species is 7.9 ± 2.86%, indicating the proportion of body weight occupied by the gonads. The gonads are positioned vertically near the spine and curve forward at their ends, located just

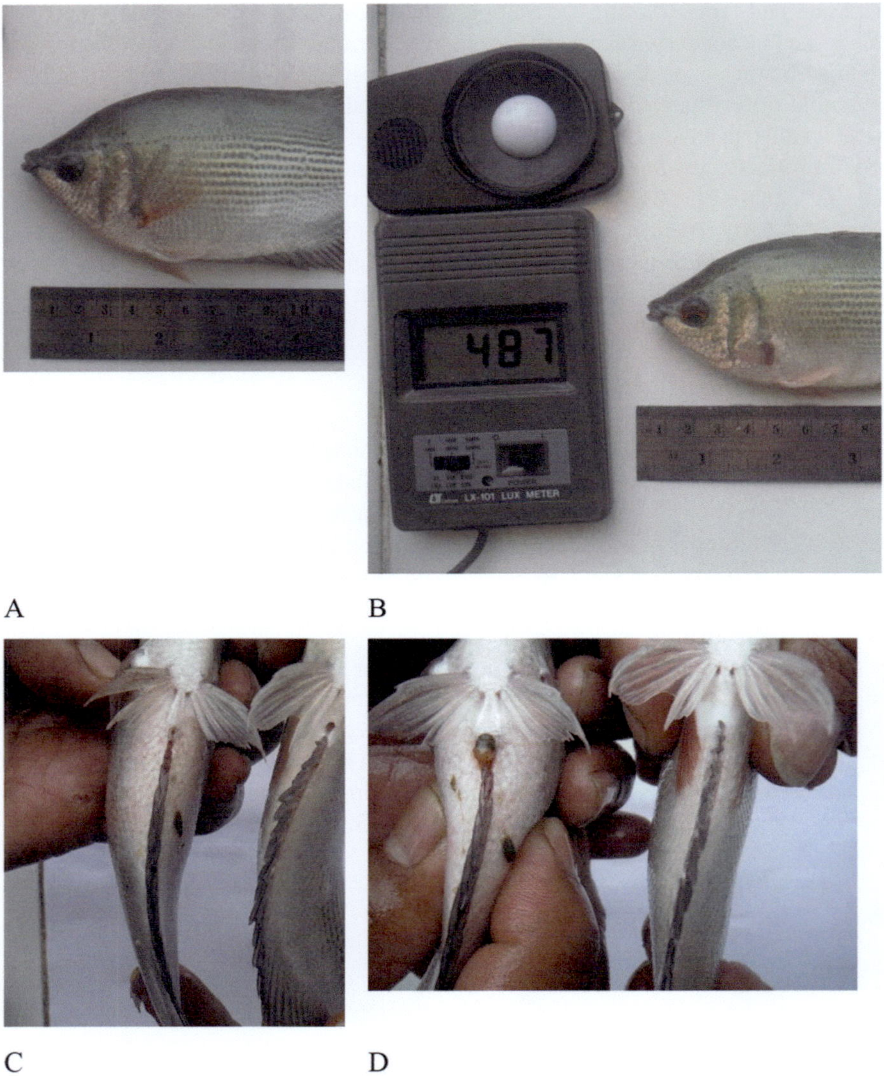

Fig. 6.2 Characteristics of female and male kissing gourami. (**a**) Female kissing gourami, (**b**) male kissing gourami, (**c**) the genital shape of female and male fish, (**d**) the broodstock body when the stomach is pressed will release eggs and sperm. (Source: Arifin et al. (2016))

behind the urogenital opening. Mature female gonads are bright yellow, while male gonads are milky white, as observed by Cahyanti et al. (2021) (Fig. 6.4).

Spawning in kissing gourami can be achieved using three methods: natural, semi-artificial, and artificial spawning. In natural spawning, mature male and female broodstock are paired in a container or pond, with the addition of environmental stimuli to encourage spawning. In semi-artificial spawning, the male and female

Fig. 6.3 Distribution map of kissing gourami (*Helostoma temminckii*). (Source: Google satellite with modification)

broodstock are hormonally stimulated to induce ovulation and spawning. The broodstock are then paired in a container to facilitate the spawning process. The ratio of males to females is typically 1:1 or 2:1. Female ovulation occurs approximately 8–12 h after the hormonal injection (Cahyanti et al., 2021).

Artificial spawning involves injecting gonadotropin hormones at a dose of 0.6 mL/kg for female broodstock and 0.4 mL/kg for male broodstock. Each fish receives a single injection. Once the female is ready to ovulate (8–12 h after injection), eggs are stripped and mixed with sperm. Fertilized eggs are then cleaned of excess sperm and placed in an aquarium for hatching. Kissing gourami eggs are buoyant (nonadhesive). Egg incubation occurs in an aquarium with an average stocking density of 10 eggs/cm^2 of water surface area.

The fertilized eggs undergo embryogenesis, the developmental process through which the egg matures into a larva. Embryogenesis includes stages such as cell division (cleavage), morulation, blastulation, gastrulation, and organogenesis, culminating in hatching. Cahyanti et al. (2021) found that temperature differences affect the duration of the embryonic phase and the hatching time of kissing gourami. At a temperature of 29–31 °C, the incubation period was approximately 14.87 h, with a hatching rate of 75 ± 2.64%. At 23–26 °C, the incubation period extended to 19.06 h, with a hatching rate of 71 ± 2.08%.

Fig. 6.4 Kissing gourami gonads after surgery. (**a**) The position of the ovary in the fish's body, (**b**) ovarian shape, (**c**) position of the testicles in the body, (**d**) testicular shape. (Personal documentation)

6.5 Domestication

As part of community cultivation efforts, domestication activities have been carried out, including the following:

(a) Characterizing and evaluating the performance of kissing gourami from various populations in Indonesia to identify potential biological advantages in the populations used as the initial generation of collections (G0).
(b) Observing phenotypic and genotypic characters using the PCR-RAPD method, along with reproductive and production traits such as growth, survival, feed efficiency, disease resistance, and environmental tolerance in the first (G1) and second (G2) generations.

The goal of domestication activities is to develop kissing gourami that are well-adapted to cultivation environments, with known characteristics related to reproduction, growth, survival, environmental conditions, disease resistance, cultivation technology (seeding, nursery, and growth phases), phenotype, genotype, and meat quality.

The results of domestication showed that kissing gourami produced an average of 10,445 ± 2171.6 eggs, with an average fecundity of 75.3 ± 12.74 eggs/g of broodstock weight. The percentage of fertilized eggs was 70.4 ± 1.40%, and 71.8 ± 7.26% of the eggs hatched.

The average growth of first-generation (G1) kissing gourami in the nursery phase, up to 60 days of age, resulted in a standard length of 2.85 ± 0.095 cm, a total length of 3.62 ± 0.217 cm, and a weight of 0.84 ± 0.159 g. The absolute growth in standard length was 0.47 ± 0.095 cm, in total length was 0.65 ± 0.198 cm, and in weight was 0.41 ± 0.159 g. The specific growth rate for standard length was 0.60 ± 0.112% per day, for total length was 0.65 ± 0.198% per day, and for weight was 2.14 ± 0.659% per day, with a survival rate of 84.6 ± 8.24%. During the enlargement phase, by 210 days from hatching, the gonads matured, and the fish achieved an average standard length of 8.6 ± 0.70 cm, a total length of 10.7 ± 1.03 cm, and a weight of 24.5 ± 7.67 g. The absolute growth in standard length was 5.7 ± 0.84 cm, in total length was 7.1 ± 0.79 cm, and in weight was 23.1 ± 5.59 g. The specific growth rate for standard length was 0.73 ± 0.087% per day, for total length was 0.72 ± 0.085% per day, and for weight was 2.25 ± 0.330% per day.

In the second generation (G2), growth performance varied based on stocking densities. According to Arifin et al. (2019), during the nursery phase, up to 60 days of age, the average standard length ranged from 3.52 ± 0.255 to 3.76 ± 0.179 cm, total length ranged from 4.72 ± 0.314 to 5.02 ± 0.220 cm, and weight ranged from 1.91 ± 0.357 to 2.27 ± 0.110 g. The specific growth rate for standard length was 0.84 ± 0.301% to 1.07 ± 0.216% per day, for total length was 1.00 ± 0.295% to 1.21 ± 0.218% per day, and for weight was 4.01 ± 0.878% to 4.62 ± 0.373% per day. In the enlargement phase, during the 150-day rearing period, the highest growth was observed at a stocking density of 50 individuals/m^2, with an average final standard length of 8.15 ± 0.206 cm, total length of 10.18 ± 0.282 cm, and weight of 20.16 ± 0.645 g. The specific growth rate for standard length was 0.56 ± 0.041% per day, for total length was 0.50 ± 0.037% per day, and for weight was 1.58 ± 0.127% per day, with a survival rate of 73 ± 8.45%.

Kissing gourami can survive in salinities of up to 10 g/L, live in pH levels ranging from 5 to 9, and tolerate water temperatures between 20 and 35 °C. The fish can survive for 30–38 min in closed water conditions with a biomass of 700 g in 10 L of water (Arifin et al., 2017c). However, kissing gourami are not resistant to *Aeromonas hydrophila* infection. In challenge tests, all test fish died within 2 days. Mortality began at the fourth hour, with 50% dying by the eighth hour, and 100% mortality occurring by the 49th hour.

The protein content in kissing gourami meat is relatively high, at 15.73% of the wet weight. The dressing percentage is also high, at 53.4 ± 7.8%, with a fillet yield of 39.4 ± 5.9%. Hatchery productivity is 0.136 ± 0.0033 fry/m^2, with a total of 67.1 ± 1.46 fry. During the enlargement phase, the productivity is 0.69 ± 0.055 kg/m^2, with a feed conversion ratio (FCR) of 1.81 ± 0.04.

Genetic diversity in the early generations was relatively high, with 10–12 fragments ranging in size from 200 to 1000 bp, a polymorphism percentage of 56.25%, and heterozygosity of 0.2413 (Kristanto et al., 2017). Observations over generations, using three primers in one selected population as a candidate for domestication, revealed a decrease in genetic diversity. The early generation (G0) had a polymorphism value of 35.48% and heterozygosity of 0.16. In G1, the polymorphism value dropped to 6.45%, with heterozygosity at 0.03. However, in G2, polymorphism increased to 16.13%, with heterozygosity at 0.08 (Arifin et al., 2017b).

6.6 Cultivation Technology

Kissing gourami has been cultivated for a long time, but is still maintained traditionally and has not yet been cultivated intensively (Cholik et al., 2005). As one of the adaptable wild fish genetic resources, kissing gourami production remains heavily reliant on natural catches. Controlled cultivation in ponds or other containers is still rarely practiced (Gustiano et al., 2015).

Kissing gourami can be cultivated in various types of ponds, such as concrete, earthen, or tarpaulin ponds. To grow kissing gourami to consumption size, the fry must be raised in ponds. Before use, the ponds must be dried to kill pests and diseases. The pond is then fertilized and filled with water to a depth of 40–70 cm. The pond is left for a week to allow natural food to grow. The fry are raised in the ponds for 4 months at a density of 30 fish/m^2, resulting in kissing gourami with an average weight of 50 g/fish.

Kissing gourami production is primarily reliant on natural catches, with the highest production occurring in Kalimantan and Sulawesi. According to the Central Bureau of Statistics for Tabalong Regency (2018), the production of kissing gourami has steadily declined, from 103 tons in 2009 to 21.2 tons in 2016. Similarly, the Central Bureau of Statistics for East Kotawaringin Regency recorded a production of 60.36 tons in 2016.

References

Arifin, O. Z., Subagja, J., Cahyanti, W., & Kristanto, A. H. (2016). *Kertas ilmiah ikan tambakan (Helostoma temminckii) hasil domestikasi* [*Academic paper of kissing gourami fish (Helostoma temminckii) results of domestication*] [Unpublished]. Balai Penelitian dan Pengembangan Budidaya Air Tawar.

Arifin, O. Z., Cahyanti, W., Subagja, J., & Kristanto, A. H. (2017a). Keragaan fenotipe ikan tambakan (*Helostoma temminckii*, Cuvier 1829) jantan dan betina generasi kedua hasil domestikasi [Phenotype performance on male and female from the second generation of domesticated kissing gourami (*Helostoma temminckii* Cuvier 1829)]. *Media Akuakultur, 12*(1), 1–9. http://ejournal-balitbang.kkp.go.id/index.php/ma

Arifin, O. Z., Cahyanti, W., & Kristanto, A. H. (2017b). Keragaman genetik tiga generasi ikan tambakan (*Helostoma temminckii*) dalam program domestikasi [Genetic diversity of three generations of kissing gourami (*Helostoma temminckii*) under a domestication program]. *Jurnal Riset Akuakultur, 12*(4), 295–305. http://ejournal-balitbang.kkp.go.id/index.php/jra

Arifin, O. Z., Prakoso, V. A., & Pantjara, B. (2017c). Ketahanan ikan tambakan (*Helostoma temminckii*) terhadap beberapa parameter kualitas air dalam lingkungan budidaya [The resilience of kissing gourami (*Helostoma temminckii*) on several water quality parameters in aquaculture environment]. *Jurnal Riset Akuakultur, 12*(3), 241–251. http://ejournal-balitbang.kkp.go.id/index.php/jra

Arifin, O. Z., Cahyanti, W., & Prakoso, V. A. (2019). Keragaan pertumbuhan ikan tambakan (*Helostoma temminckii* Cuvier, 1829) dengan kepadatan berbeda [Growth performance of kissing gourami (*Helostoma temminckii* Cuvier, 1829) at different stocking densities]. *Media Akuakultur, 14*(2), 83–87. http://ejournal-balitbang.kkp.go.id/index.php/ma

Cahyanti, W., Subagja, J., Kusdiarti, Irawan, D., & Arifin, O. Z. (2021). Keragaan bioreproduksi tiga generasi ikan tambakan (*Helostoma temminckii* Cuvier, 1829) [Bioreproductive characteristics of three generations of kissing gourami fish (*Helostoma temminckii* Cuvier, 1829)]. *Media Akuakultur, 16*(1), 1–6. https://doi.org/10.15578/ma.16.1.2021.1-6

Central Statistics Agency of Tabalong Regency. (2018). Fish catch production by fish type 2009-2016. Accessed on January 19, 2021, at 12:43 p.m. WIB, from https://tabalongkab.bps.go.id/statictable/2015/12/29/1405/fish-catch-production-by-fish-type-2009-2016.html.

Cholik, F., Jagatraya, A. G., Poernomo, P., & Jauzi, A. (2005). *Aquaculture the foundation of hope for the nation's future* (415 p.). Masyarakat Perikanan Nusantara.

Efriyeldi, & Pulungan, C. P. (1995). *Hubungan panjang, bobot dan fekunditas ikan tambakan (Helostoma temminckii C.V) dari perairan sekitar Taratak Buluh* [*Relationship between length, weight and fecundity of kissing gourami fish (Helostoma temminckii C.V) from the waters around Taratak Buluh*] [Unpublished]. Pusat Penelitian Universitas Riau, Pekanbaru.

Gustiano, R., Kusmini, I. I., & Ath-thar, M. H. F. (2015). *Mengenal sumber daya genetik ikan spesifik lokal air tawar Indonesia untuk pengembangan budidaya [Genetic resources of local freshwater fish in Indonesia for aquaculture development]* (51 p.). PT IPB Press.

Kottelat, M., Whitten, A. J., Kartikasari, S. N., & Wirjoatmodjo, S. (1993). *Freshwater fishes of Western Indonesia and Sulawesi* (221 p.). Periplus Editions.

Kristanto, A. H., Subagja, J., Cahyanti, W., & Arifin, O. Z. (2017). Evaluasi variasi fenotipe dan genotipe populasi ikan tambakan dari Kalimantan Tengah, Jawa Barat, dan Jambi dengan truss morfometrik dan Random Amplified Polymorphic DNA (RAPD) [Evaluation of phenotype and genotype variations of kissing gourami from Central Kalimantan, West Java, and Jambi through trust morphometric and RAPD]. *Jurnal Riset Akuakultur, 12*(3), 203–211. http://ejournal-balitbang.kkp.go.id/index.php/jra

Lisna. (2016). Aspek biologi reproduksi ikan tambakan (*Helostoma temminckii*) di perairan umum Kecamatan Kumpeh Ulu Kabupaten Muaro Jambi [Reproduction aspects of Kissing gourami (*Helostoma temminckii*) in public waters of Kumpeh Ulu Subdistrict, Muaro Jambi Regency]. *Biospecies, 9*(1), 15–22.

Rahardjo, M. F. (1980). *Iktiologi [Ichthyology]*. Departemen Biologi, Fakultas Perikanan, Institut Pertanian Bogor.

Saanin, H. (1984). *Taksonomi ikan dan kunci identifikasi [Fish taxonomy and key identification]* (360 p.). Bina Cipta.

The Status of Domestication and Aquaculture System of the Knifefish *Notopterus chitala* in Indonesia

7

Lies Setijaningsih and Bambang Gunadi

7.1 Introduction

Featherbacks, or knifefishes (*Notopterus chitala*), known locally in Indonesia as Belida (Fig. 7.1), have great potential as a local commodity for aquaculture development). Expanding the cultivation of local fish species in aquaculture is crucial in Indonesia, as it aligns with efforts to conserve the native fish germplasm in natural environments.

The knifefish is easily recognized by its physical morphology. In general, it has a concave head positioned close to the back, and its jaw extends with age, reaching far beyond the eye in large specimens. There are more than ten rows of preoperculum scales, 117–127 rays on the anal fin, and 43–49 pairs of small spines along the abdomen. According to Kottelat et al. (1993), knifefish exhibit different color patterns based on their size. In the maculosus phase, which occurs when they are between 150 and 270 mm in size, the body and fins are covered with small round spots. As they grow into the borneensis phase, between 300 and 600 mm, sloping lines with dots appear on the anal fin and the back of the body, along with a black spot at the base of the pectoral fin. When knifefish exceed 600 mm, they enter the hypselonotus phase, where only one black spot remains at the base of the pectoral fin. In India, knifefish can grow to over 1 m, while in Thailand, they typically reach 70–75 cm.

Knifefish prefer river waters with many fallen trees and forested floodplains. These ecosystems serve small fish as prey while also providing suitable surfaces for shrimp and aquatic invertebrates to thrive. Furthermore, the submerged wood, both

L. Setijaningsih (✉) · B. Gunadi
Research Center for Fishery, Research Organization for Earth Sciences and Maritime, National Research and Innovation Agency, Cibinong, Indonesia
e-mail: lies007@brin.go.id

E. Nugroho et al. (eds.), *Aquaculture Development in Archipelago*, https://doi.org/10.1007/978-981-95-1084-9_7

Fig. 7.1 The knifefish *Notopterus chitala*. (Personal documentation)

Fig. 7.2 Condition of the dead trees in the water body where knifefish broodstock lay fertilized eggs in Kampar Regency, Riau Province. (Source: Personal collection)

live and dead, serves as a substrate for fertilized knifefish eggs to attach (Figs. 7.2 and 7.3) (Adjie et al., 1999).

According to Giesen (1987), knifefish in Kalimantan are commonly found in the Kapuas River and its tributaries. They are usually found in lakes, swamps, flooded forests, and small rivers that flow into the swamps. Knifefish are top predators (carnivores), and their main food is fresh fish, with shrimp as an additional food source. The potential conditions for knifefish to produce eggs depend greatly on the breeding environment and the water level in the cultivation ponds, which should ideally be in shady regions to avoid extreme heat. Knifefish are nocturnal, active at night and during the day, they stay or hide among the vegetation. They hunt for prey at night (Wibowo & Sunarno, 2006).

Fig. 7.3 Condition of the dead trees in the water body where knifefish broodstock lay fertilized eggs in Riam Kanan Reservoir, South Kalimantan Province. (Source: Personal collection)

7.2 Taxonomy and Morphology

According to Kottelat et al. (1993), the taxonomy of knifefish is as follows:

Phyllum:	Chordata
Class:	Actinopterygii
Order:	Osteoglossiformes
Family:	Notopteridae
Genus:	*Chitala/Notopterus*
Species:	*Chitala lopis* (Kottelat et al., 1993)
Synonym:	*Notopterus chitala* (Weber & Beaufort, 1913)

According to Robert (1992) and Kottelat et al. (1993), the knifefish can reach the standard length of 150 cm. A well-known clown knifefish, *Chitala chitala* (Hamilton, 1822) can reach a length of 150 cm (Froese & Pauly, 2021), while *Chitala ornate* can reach a length of up to 100 cm (Froese & Pauly, 2021). Meanwhile, the so-called giant featherback, *Chitala lopis* (Bleeker, 1851), can reach a length of 75–122 cm (Froese & Pauly, 2021).

7.3 Distribution

As an endemic fish species, this knifefish are distributed across specific regions of Indonesia, and their presence limited to the islands of Kalimantan, Java, and Sumatra. Adjie et al. (1999) reported that there are three species of knifefish found in Indonesia, namely, *Notopterus notopterus*, *Notopterus borneensis*, and *Notopterus chitala*. However, the Decree of the Minister of Marine Affairs and Fisheries of the Republic of Indonesia Number 1 of 2021, concerning Protected Fish Species, has established four species of knifefish that receive full protection status to avoid their extinction, namely, *Chitala borneensis* (Bleeker, 1851) (*Notopterus borneensis*, Borneo knifefish), *Chitala hypselonotus* (Bleeker, 1851) (*Notopterus hypselonotus*, Sumatra knifefish), *Chitala lopis* (Bleeker, 1851) (*Notopterus lopis*, Lopis knifefish), and *Notopterus notopterus* (Pallas, 1769) (*Gymnotus notopterus*, Java knifefish). With full protection status, knifefish are prohibited from being caught and utilized at all stages of their life cycle, including their body parts and derivative products.

Outside Indonesia, knifefish have been reported in Asian countries such as Thailand, Cambodia, Vietnam, and Malaysia. Their presence in these areas demonstrates their broad dispersion throughout Southeast Asia (Wibowo, 2011; Kottelat et al., 1993).

7.4 Reproduction

Knifefish broodstocks that were fed with vitamin supplementation for 11 spawning cycles produced 18.492 million eggs with a fertility rate of 75–92% and SAI (Spawning Activity Index) of 3.8–5.68. In comparison, broodstocks fed without vitamin supplementation only produced 10.398 million eggs with a fertility rate of 50–75% and SAI of 0.28–4.90 (Kristanto et al., 2008).

The size of knifefish eggs is large enough to make them easier to count, as illustrated in Fig. 7.4. Figure 7.5 shows the knifefish egg cell growth phase.

Knifefish eggs from pond-cultured broodstocks go through a series of developmental stages, starting with fertilization and cell division. During the organogenesis phase, black eye spots were visible. The color of the eggs indicated their viability, where eggs with cloudy white eggs indicate they will not hatch, while the slightly red eggs will hatch into larvae phase (Fig. 7.6).

A trial of artificial spawning of knifefish was carried out using human chorionic gonadotropin (HCG) hormone induction at a dose of 500 IU/kg, followed by luteinizing hormone-releasing hormone (LHRH) analog at a dose of 0.5 mL/kg, administered 24 h after the HCG induction (Kristanto & Subagja, 2010). This hormone induction technique produced a high fertilization rate of 86.7% (Setijaningsih et al., 2018). In this study, a single female produced 386 eggs weighing a total of 20 g. Artificial spawning was successful because the broodstock had reached the required gonadal maturity, enabling effective hormonal stimulation.

Fig. 7.4 The size of the knifefish egg. (Source: Personal collection)

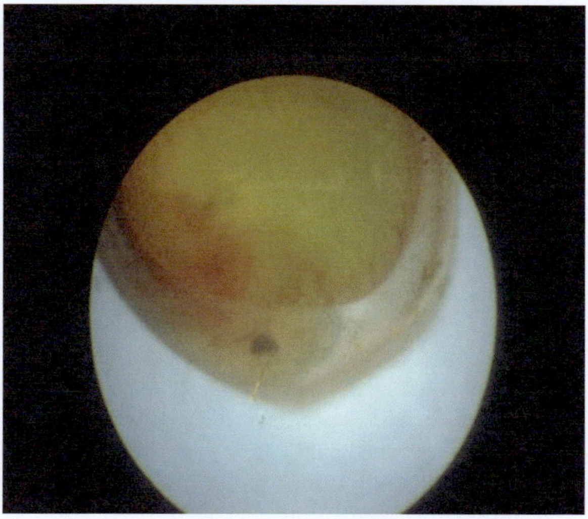

Fig. 7.5 The knifefish egg cell growth phase. (Source: Personal collection)

7.5 Domestication and Breeding Program

In 2004, the Mandiangin Freshwater Aquaculture Center, a Technical Implementation Unit under the Directorate General of Aquaculture, Ministry of Marine Affairs and Fisheries, began collecting knifefish broodstock from the waters of Riam Kanan Reservoir. The cultivation of knifefish (*Chitala lopis*) in Indonesia was first tested in 2004, with the application of gonad maturation in ponds (Wahyutomo & Ermayani, 2005). By 2005, gonadally mature broodstock had been successfully produced, yielding 324 knifefish fry following ten natural spawning cycles (ten females and ten males), with an average length of 15.5 cm and weight of 11.63 g.

Fig. 7.6 The different colors of knifefish eggs (pale whitish and reddish color). (Source: Personal research collection)

Wahyutomo and Ermayani (2005) conducted more research on the naturally controlled spawning of knifefish (*Chitala lopis*) with 47 egg attachment collector units and produced 1993 knifefish fry with an average length of 15.5 cm and an average weight of 11.63 g. The survival rate was 56.3%.

In a controlled rearing, knifefish were fed small fish with a length range of 5 cm such as carp, tilapia, and nilem fry. Shrimp can be fed in frozen form. The frequency of feeding was twice a day, in the morning and evening, totaling 5% of the weight of the fish biomass. Knifefish can spawn throughout the year with peak spawning occurring in the dry season (Kristanto et al., 2008).

Observations of natural spawning in controlled ponds utilizing wood collectors over 5 months showed that the highest spawning frequency, fecundity, fertility, and survival rates occurred between June and October (Setijaningsih, 2015). The spawning frequency reached 58.33%, with a fecundity of 62.64% (totaling 1955 eggs) and survival rates ranging from 45.97% to 81.95%.

Additionally, Setijaningsih et al. (2018) observed that knifefish cultured for 3 months in a concrete-wall pond with natural spawning showed the highest percentage of spawning frequency (50.00%) in the second month of broodstock rearing, whereas only 16.67% of the broodstock spawned in the third month. The highest fecundity was found in the second month of rearing, with a total egg of 932 from 1.5 to 2.0 kg of broodstock, compared to 225 eggs obtained in the third month. In a separate study, Gustomi et al. (2016) reported an average yield of 1000 eggs from knifefish broodstock weighing approximately 2.0 kg. Meanwhile, Hossain et al. (2006) discovered that broodstock weighing over 4.0 kg produced 5761 eggs.

7.6 Aquaculture

A study on knifefish (*Chitala lopis*) fry nursing in aquariums for 30 days reported a survival rate (SR) of 80.45–88.00%, starting with an initial stocking of 50–75 fish and harvesting 44–60 fish. In contrast, knifefish fingerlings reared in ponds had a lower survival rate of 41.67–52.75% from an initial stocking of 400–600 fish, yielding a harvest of 211–250 fish. Disease and predation by dragonfly larvae and tilapia prevented the 60% target survival rate from being achieved (Wahyutomo et al., 2007).

In a separate grow-out experiment, Wahyutomo et al. (2007) grew knifefish in a cage system within the Riam Kanan River, Banjar Regency, South Kalimantan Province for 10 months. The fry was stocked at a density of 30 fish/m^2, or 200 fish per cage, with an initial length of 19.58–19.62 cm and weight of 47.07–49.45 g. The fish were reared in two cages of 3 × 2 × 1 m in size. At the end of the experiment, the fish reached an average length of 22.40 cm and a weight of 556.01 g.

Wooden board collectors can be utilized to increase productivity in knifefish production since broodstock prefer smooth surfaces for egg attachment. Knifefish prefer smooth textures because their scaleless skin prevents egg attachment on rough surfaces.

7.7 Production

Knifefish is considered a nonmajor fishery commodity in Indonesia and is thus classified as "Other Fish" in statistics data. This classification makes it difficult to collect particular statistics on knifefish production at both the national and local levels from statistical reports produced by the National Statistics Agency and the One Data Program of the Ministry of Fisheries and Maritime Affairs of the Republic of Indonesia.

The main factor driving the increase in knifefish production is public interest in consuming knifefish. Apart from fresh consumption, knifefish are also used as ingredients for processed products (Kottelat et al., 1993). For the people of South Sumatra, knifefish, commonly known as belido fish, is commonly used as one of the main raw materials for Palembang's specialty food, called empek-empek. Meanwhile, in South and Central Kalimantan, knifefish is known as flat fish and is commonly used as raw material for Kalimantan fish crackers. Another business opportunity is also raised for the knifefish species *Notopterus notopterus* due to its beautiful abdominal pattern, which makes it popular as a freshwater ornamental fish even at the fry stage (Madang, 1999).

Knifefish is known for its delicious and distinctive palatability, attributed to its high fat, protein, and vitamin A content (Sunarno, 2002) and the knifefish bones can be used as a fortification source of calcium in culinary preparations or food products (Asikin et al., 2019). This makes knifefish a valuable raw material and ingredient of processed food with strong market potential. Therefore, aquaculture production is expected to assist alleviate knifefish overexploitation in the wild by supporting fry restocking efforts to conserve the population. This approach allows farmers to meet the demand for processed knifefish products while helping to conserve wild populations.

References

Adjie, S., Husnah, & Abdul. (1999). Studi biologi ikan belida (*Notopterus chitala*) di daerah aliran Sungai Batanghari, Provinsi Jambi [Biological study of knifefish (*Notopterus chitala*) in the Batanghari River basin, Jambi Province]. *Jurnal Penelitian Perikanan Indonesia, 5*(1), 38–43.

Asikin, A. N., Kusumaningrum, I., & Hidayat, T. (2019). Effect of knife-fish bone powder addition on characteristics of starch and seaweed kerupuk as calcium and crude fiber source. *Current Research in Nutrition and Food Science, 7*(2), 584–591. https://doi.org/10.12944/CRNFSJ.7.2.27

Froese, R., & Pauly, D. (2021). *FishBase*. World Wide Web Electronic Publication. Version (12/2019). https://www.fishbase.de

Giesen, W. (1987). *Danau Sentarum wildlife reserve – Inventory, ecology and management guidelines*. WWF Report for the Directorate of Forest Protection and Nature Conservation (PHPA). https://doi.org/10.13140/RG.2.2.31780.32642

Gustomi, A., Sulistiono, & Yonvitner. (2016). Biologi reproduksi ikan belida (*Notopterus notopterus* Pallas, 1769) di Kolong-Bendungan Simpur, Pulau Bangka [Reproductive biology of knifefish (*Notopterus notopterus* Pallas, 1769) in the Simpur Dam Basin, Bangka Island]. *Jurnal Ilmu Pertanian Indonesia (JIPI), 21*(1), 56–62. https://doi.org/10.18343/jipi.21.1.56

Hamilton, F. (1822). *An account of the fishes found in the river Ganges and its branches*. Archibald Constable and Company.

Hossain, M. Y., Ahmed, Z. F., Leunda, P. M., Jasmine, S., Oscoz, J., Miranda, R., & Ohtomi, J. (2006). Condition, length-weight, and length-length relationships of the Asian striped catfish (*Mystus vittatus* (Bloch, 1794)) (Siluriformes: Bagridae) in the Mathabhanga River, southwestern Bangladesh. *Journal of Applied Ichthyology, 22*(4), 328–333. https://doi.org/10.1111/j.1439-0426.2006.00803.x

Kottelat, M., Murdy, E. O., Whitten, A. J., Kartikasari, N., & Wirjoatmodjo, S. (1993). *Freshwater fishes of western Indonesia and Sulawesi (Edisi Dwi Bahasa)*. Periplus Editions.

Kristanto, A. H., & Subagja, J. (2010). Respons induk ikan belida terhadap hormon pemijahan [Response of knifefish broodstock to spawning hormones]. In *Prosiding Forum Inovasi Teknologi Akuakultur* (pp. 113–116). Pusat Riset Perikanan Budidaya.

Kristanto, A. H., Nuryadi, Yosmaniar, & Sutrisno. (2008). Perkembangan telur dan sperma ikan belida (*Notopterus chitala*) yang dipelihara di kolam [Development of eggs and sperm of knifefish (*Notopterus chitala*) in cultivation ponds]. *Jurnal Riset Akuakultur, 3*(1), 73–82.

Madang, K. (1999). *Morfologi habitat dan keragaman genetik kerabat ikan belida di perairan Sumatera Selatan* [*Morphogeology, habitat, and genetic diversity of knifefish (Maiacopterygii; Notopteridae) in South Sumatra*] [Dissertation, Institut Pertanian Bogor].

Robert, T. R. (1992). Systematic revision of the old-world freshwater fish family Notopteridae. *Ichthyological Exploration of Freshwaters, 2*(4), 361–383.

Setijaningsih, L. (2015). Pemijahan ikan belida (*Notopterus chitala*) dengan perbedaan substrat di kolam budidaya [Spawning of knifefish (*Notopterus chitala*) with different substrates in cultivation ponds]. In *Prosiding Seminar Nasional Perikanan Indonesia* (pp. 491–497). Sekolah Tinggi Perikanan.

Setijaningsih, L., Subagja, J., Radona, D., Pantjara, B., Kristanto, A. H., & Widyastuti, Y. R. (2018). Performa pemijahan ikan belida (*Notopterus chitala*) melalui induksi hormon HCG dan LHRH [Spawning performance of knifefish (*Notopterus chitala*) through HCG and LHRH hormone induction]. *Jurnal Riset Akuakultur, 13*(2), 115–122. https://doi.org/10.15578/jra.13.2.2018.115-122

Sunarno, M. T. D. (2002). Selamatkan plasma nutfah ikan belida [Save the germplasm of knifefish]. *Warta Penelitian Perikanan Indonesia, 8*(4), 2–6.

Wahyutomo, A. R., & Ermayani. (2005). *Pemijahan ikan belida (Chitala lopis) secara alami terkendali* [*Natural controlled spawning of knifefish (Chitala lopis)*]. Technical Report. Loka Budidaya Air Tawar Mandiangin, South Kalimantan, pp. 29.

Wahyutomo, A. R., Rochman, A., & Ermayani. (2007). *Pembesaran ikan belida (Chitala lopis) yang dipelihara dengan sistem karamba apung di perairan umum* [*Rearing of knifefish (Chitala lopis) using a floating cage system in public waters*]. Technical Report. Loka Budidaya Air Tawar Mandiangin, South Kalimantan.

Wibowo, A. (2011). *Kajian bioekologi dalam rangka menentukan arah pengelolaan ikan belida (Notopterus chitala) (Bleeker 1851) di Sungai Kampar, Provinsi Riau* [*Bioecological study in the framework of determining the direction of management of knifefish (Notopterus chitala) (Bleeker 1851) in the Kampar River, Riau Province*] [Dissertation, Institut Pertanian Bogor].

Wibowo, A., & Sunarno, M. T. D. (2006). *Karakteristik habitat ikan belida (Notopterus chitala)* [*Habitat characteristics of theknifefish*]. BAWAL, 1(1), 19–24.

Domestication of Native Climbing Perch, Kalimantan, Indonesia

8

Rahmat Hidayat and Bastiar Nur

8.1 Introduction

Climbing perch (*Anabas testudineus*) is a native freshwater fish that is cultured in Southeast Asia (Chotipuntu & Avakul, 2010). It is a locally specific freshwater fish in Indonesia that is spread across the islands of Sumatra, Java, and Kalimantan. Climbing perch is not an endangered category, as reported in the *Red List International Union for Conservation of Nature and Natural Resources* (Ahmad et al., 2019). This fish is known as Papuyu in South Kalimantan or called Puyu-puyu in West Sumatra. Uniquely, climbing perch has economic value as an exotic food menu of Banjar people in South Kalimantan (Syarifuddin et al., 1993; Candra et al., 2019; Widianti et al., 2021). As a source of daily food, the Banjar people get more supplies of climbing perch from natural catches. Despite the conservation status of the climbing perch as not an endangered fish, the climbing perch population is decreasing in areas where people have a high preference for this fish due to uncontrolled fishing activities as well as settlement and plantation development activities (Fig. 8.1).

Adaptive climbing perch fry produced by nurseries in South Kalimantan have started to be in demand by the farmers after domestication activities. Climbing perch domestication was conducted in 2009 and the result was then released in 2014 (Bunasir et al., 2014). Climbing perch culture in grow-out ponds, which is usually done with seed sources caught from nature, is now starting to rely on adaptive domesticated fry. Moreover, the climbing perch culture was advanced by the biofloc system that was introduced and developed by the farmer (Riswan et al., 2017; Ahmadi et al., 2021). Therefore, domestication became the initial milestone for the

R. Hidayat (✉) · B. Nur
Research Center for Fishery, Research Organization for Earth Science and Maritime, National Research and Innovation Agency, Jakarta, Indonesia
e-mail: rahm060@brin.go.id

E. Nugroho et al. (eds.), *Aquaculture Development in Archipelago*,
https://doi.org/10.1007/978-981-95-1084-9_8

Fig. 8.1 Climbing perch (*Anabas testudineus* Bloch, 1792). (Photo: R. Hidayat, personal collection)

fish farmer to start obtaining domesticated climbing perch as an alternative source of food without relying on nature.

Climbing perch are classified as follows: Kingdom: Animalia; Subkingdom: Bilateria; Infrakingdom: Deuterostomes; Phylum: Chordata; Subphylum: Vertebrata; Infraphylum: Gnathostomata; Superclass: Actinopterygii; Class: Teleostei; Superorder: Acanthopterygii; Order: Perciformes; Suborders: Anabantoidei; Family: Anabantidae; Genus: *Anabas* Cloquet, 1816; and Species: *Anabas testudineus* (Bloch, 1792) (www.itis.gov, 2023). It has hard scales, with a maximum body length of 25 cm, but most natural catches are less than 20 cm. The side of the body of the climbing perch is yellowish, and the back is dark black, brown, or greenish, depending on their living environment. In addition, there is a transverse dark line that is irregular and faint on the underside of the body. At the back end of the gill cover, there is a black spot that is vague or not clear and has a hard and sharp texture resembling a thorn. Figure 8.2 shows the morphology of the climbing perch.

In low tide conditions or dry land, climbing perch use their hard gills to walk on land and migrate in search of water sources. Under these conditions, the fish, which has a labyrinth in its respiratory system, is still able to survive in low-oxygen water conditions, even in mud if its gills are wet.

8.2 Distribution

The climbing perch's habitat is in swamps, rice fields, small rivers, and ditches, as well as in ponds that get flood water or are connected to open water. Its distribution is very wide, starting from Bangladesh, Brunei Darussalam, Cambodia, China, India, Indonesia, Lao People's Democratic Republic, Malaysia, Burma, Nepal,

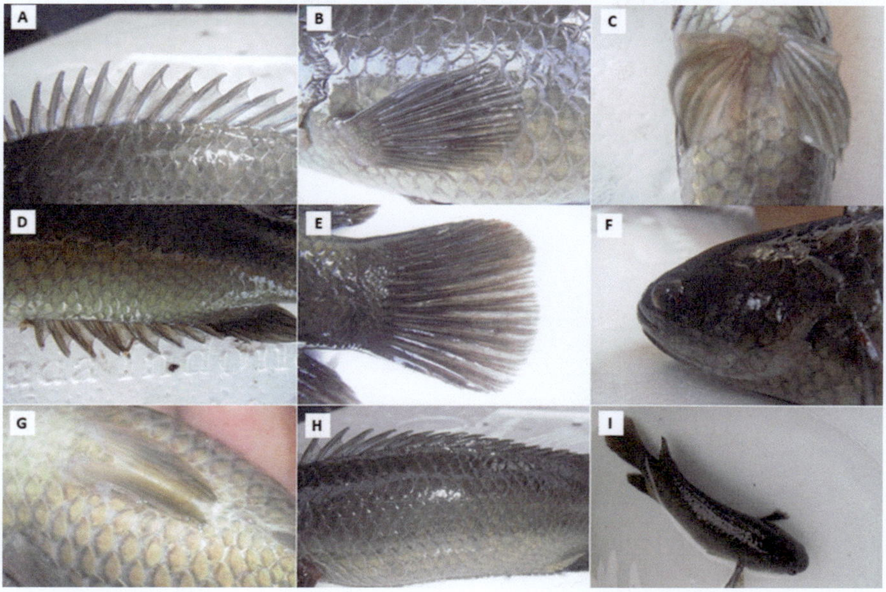

Fig. 8.2 Morphology of climbing perch (*Anabas testudineus* Bloch.). (**a**) dorsal fin, XVII.8–9; (**b**) pectoral fin, 14–15; (**c**) ventral fin, I.5; (**d**) anal fin, X.9–10; (**e**) caudal fin, 14; (**f**) operculum, silvery yellow black; (**g**) ventral, yellowish white; (**h**) lateral line scales; (**i**) dorsal, dark greenish brown. (Source: Bunasir et al. (2014))

Pakistan, Singapore, Sri Lanka, Thailand, and Vietnam (Axelrod, 1983; Shinsuke, 2009; Ahmad et al., 2019; Ndobe et al., 2022). In Indonesian waters, the climbing perch fish are widely distributed, covering the islands of Sumatra, Nias, Bintan, Sulawesi, Bangka, Sumbawa, Pati, Ambon, Java, Bacau, Halmahera, Kalimantan, and Madura (Djuhanda, 1981). Climbing perch fish can live in fresh and brackish water environments and have limited migratory properties in freshwater (potamodromous) (Riede, 2004). Figure 8.3 shows the spreading of climbing perch in the world.

8.3 Reproduction

Climbing perch is a type of swamp fish that spawns throughout the rainy season, and its spawning peak occurs when the rainfall is highest. The growth pattern of this fish is negative allometric (Hasnidar et al., 2022; Fauzi et al., 2022), and the condition factor varies and fluctuates. Although Zworykin (2012) stated that the climbing perch does not have sexual dimorphism, other research has proven that the growth of females is faster than males (Hidayat et al., 2016). The size of the female brood stock is larger than the male fish at the same age. The morphology of climbing perches distinguishes them sexually by body shape. The male tends to be slender

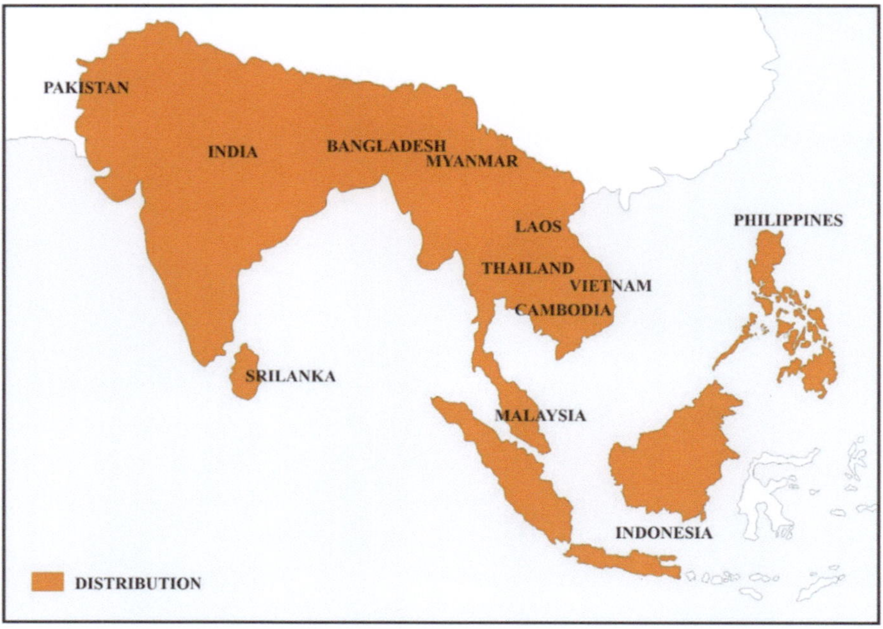

Fig. 8.3 The map of climbing perch spreading in the world (Ahmad et al., 2019). (Source: Google maps with modification, based on IUCN data)

Fig. 8.4 Climbing perch broodstock (*Anabas testudineus* Bloch.). (**a**) Female, TL: 18.5 cm; SL: 14.0 cm; weight: 87.3 g, (**b**) Male, TL: 13.5 cm; SL: 10.3 cm; weight: 36.8 g TL total length, SL standard length. (Photo: R. Hidayat, private collection)

and elongated, while the characteristics of the female are larger and wider in body size. The difference in sex characteristics is easy to observe during spawning season. Matured males are characterized by the release of milky white sperm from the genital pores after massaging the abdomen toward the genitals. The mature females show an enlarged soft abdomen, and the genitalia are reddish. Figure 8.4 shows the morphological difference of male and female climbing perch.

Climbing perch does not make nests, do not prepare spawning substrates, and do not take care of their eggs. The spawning system is also an unfaithful fish in one pair. The climbing perch eggs are floating. This characteristic is rarely found in freshwater fish in general (Zworykin, 2012). Pre-spawning behavior shows the male chasing the female before ovulation occurs. After the constant chase, the male suddenly bends around the female and the male and female attempt to hug each other. Both the males and females, at that time, bring their genital openings closer together for about 10–15 s (Behera et al., 2016).

8.4 Domestication

The domestication of climbing perch in Indonesia, especially in the province of South Kalimantan, is something special for the native farmer. This is due to the high public interest in fish consumption. In this case, the climbing perch in the South Kalimantan region consisted of green and yellow (galam) variants (Rohansyah, 2010). Furthermore, climbing perch domestication conducted on the green variant has reached three generations, starting from collections and spawning of the basic broodstock, producing the first generation in 2009–2010 to the second generation producing third-generation seeds from 2011 to 2012. Typically, a series of domestication activities include the collection of broodstock from nature, broodstock reproduction, fry production, rearing climbing perch in controlled grow-out ponds to produce broodstock and fish for consumption in each generation.

The first stage of the domestication of the climbing perch was by collecting natural broodstock from three different habitats in one district: rivers, swamps, and paddy fields. Male and female brooders were reared separately in concrete tanks with a size of $1.5 \times 1\ m^2$ and a water depth of 0.8–1.0 m with a stocking density of 50–100 individuals.m^{-2}, fed with commercial pellets (protein >30%) and a feeding rate of 3–5% of biomass per day.

The broodstock maturity is seasonal, which starts at the beginning of the rainy season and begins to decrease when the dry season begins. At this early stage of domestication, most G0 broodstocks were ready to spawn during the rainy season (December–May), with the number of broodstock ready to spawn 71–89 per one hundred broodstock reared in controlled tanks of $1.5 \times 1 \times 1\ m^3$. During the spawning season, broodfish can spawn again up to 2 months after spawning. With controlled rearing of the broodstock, the climbing perch can still mature and spawn even outside the spawning season.

Climbing perch spawned by induced analog LHRH hormones. Hormone at a dose of 0.2 mL.kg^{-1} brood weight injected intramuscularly to males (mean weight 38.35 ± 9.62 g) and females (mean weight 74.04 ± 15.14 g) and spawned in a fiberglass $1 \times 2\ m^2$, water level 0.4 m with a latency of 6–8 h. The spawning of a male and female weight ratio was 1:1. Because the weight of the female is greater than the male, spawning is usually conducted by mating of 2–4 males and one female. After spawning, the broodstocks are immediately separated from the eggs and returned to the brood-rearing container. Brood spawning during domestication

Table 8.1 Broodstock performance of parents, first generation, and second generation during domestication

Broodstock performance	Parents	First generation	Second generation
Male average weight (g)	38.35 ± 9.62	51.97 ± 3.17	55.07 ± 3.78
Female average weight (g)	74.04 ± 15.14	72.54 ± 5.97	81.12 ± 7.63
Fecundity (eggs/g)	183 ± 5	192 ± 17	201 ± 26
Hatching rate (%)	83.36 ± 8.10	79.52 ± 9.16	89.04 ± 5.49

Source: Bunasir et al. (2014)

produced broodstock until the second generation, with the production of the third generation of hatchlings. The broodstock performance of climbing perch all generations, during domestication shown in Table 8.1.

The hatching climbing perch larvae are reared in nursery ponds that have previously been prepared by the application of lime at a dose of 250 $g.m^{-2}$ and fertilizer at a dose of 500 $g.m^{-2}$, then filled with water to reach a minimum level of 60 cm. Furthermore, 2 days after hatching, larvae were reared on nursery ponds with a stocking density of 200–250 $individuals.m^{-3}$. Monitoring the abundance of aquatic plankton is carried out during the fish nursery until the fifth day, and re-fertilization is needed immediately if plankton deficiency occurs, characterized by a water brightness level of more than 20 cm. The pond should maintain a maximum water level of 80 cm. Climbing perch larvae on the fifth day of nursery were fed with powdered feed with a minimum protein content of 40% using a blind feeding method until the larvae were 10 days old. Then, they were fed with a feeding rate of 3–6% biomass based on tenth-day-old fish sample measurement results.

Climbing perch are reared in fixed nets in earthen ponds for 240 days for each generation. Fish stocked with a density of fifty $individuals.m^{-2}$. The juveniles that were stocked at the grow-out stage had an average weight of 1.56 ± 0.33 g. $individuals^{-1}$ with a total length of 3–5 cm. During culture, the fish are fed with a pellet (protein content of 30%) two times day^{-1}, a dose of 5–8% of body weight.

The results of climbing perch domestication through three generations provide a significant positive impact to the native fish farmer, namely climbing perch cultured on the controlled ponds, especially as they are adaptive to commercial feed. In terms of basic feeding habits, the climbing perch belongs to the omnivorous fish group, with the types of food found consisting of insects, fish, crustaceans, litter, Bacillariophyceae, Chlorophyceae, Cyanophyceae, and unidentified organisms, and their main food is insects (Haloho, 2008). The research results of Amornsakun et al. (2005) stated that climbing perch larvae aged 3–10 days consumed rotifer plankton, and then larvae aged 8–15 days consumed *Moina*. The type of food for larvae aged 5–10 days after hatching at the nursery stage consists of Brachionus sp., Scenedesmus, Botrydiopsis, Synedra, Acanthocephala, and Dictyosphaerium. Climbing perch aged above 10 days start to consume pollen or commercial flour with a protein content of at least 40%, then young fish until they reach consumption size, and the broodstock fully consume commercial pellet feed with a protein content of at least 30%, which is available in the market.

Based on the results of the climbing perch domestication, which have been successfully obtained up to the third generation, the level of domestication of this fish is categorized at level 4. At this level, the entire life cycle of fish in the cultivation process can be carried out in full in a controlled environment without using wild individuals from nature anymore, but a selective breeding program has not yet been carried out for this fish in Indonesia. Teletchea and Fontaine (2014) classify the level of domestication into five levels. Level 0 is the lowest, still in the process of catching fish from nature. Level 1 is still in the stage of adaptation and acclimatization to a controlled environment. Level 2 is the level in which some of the fish life cycles have been successfully carried out in a controlled environment, but there are still obstacles encountered, for example, the reproductive stage or the rearing of the larvae. Level 3 is the state where the entire life cycle of fish in the cultivation process can be carried out in a controlled environment but still using wild individuals from nature; level 4: the entire life cycle of fish in the cultivation process can be carried out completely in a controlled environment without using wild individuals from nature anymore. Finally, level 5 is where selective breeding programs have been carried out for specific purposes such as growth, fillet yield, and meat quality.

References

Ahmad, A. B., Hadiaty, R. K., de Alwis Goonatilake, S., Fernado, M., & Kotagama, O. (2019). *Anabas testudineus (errata version published in 2020).* The IUCN red list of threatened species 2019: e.T166543A174787197. https://doi.org/10.2305/IUCN.UK.2019-3.RLTS.T166543A174787197.en. Accessed 8 March 2023.

Ahmadi, A., Lilimantik, E., & Sri Mahreda, E. (2021). Marketing efficiency of the climbing perch (*Anabas testudineus*) cultured with biofloc system. *Egyptian Journal of Aquatic Biology and Fisheries, 25*(2), 561–572.

Amornsakun, T., Sriwatana, W., & Promkaew, P. (2005). Some aspects in early life stage of climbing perch, *Anabas testudineus* larvae. *Songklanakarin Journal of Science and Technology, 27*(1), 403–418.

Axelrod, H. R. (1983). *Exotic tropical fishes expanded edition*. T. F. H. Publications. ii + 1249 h.

Behera, S., Ahmed, A. S., Kumar, S., Gogoi, R., Jomang, O., & Baksi, S. (2016). Courtship behaviour and breeding success of climbing perch, *Anabas Testudineus* (Bloch) in three different breeding sets with the application of a synthetic hormone (WOVA-FH). *International Journal of Fisheries and Aquaculture Sciences, 6*(1), 1–6. ISSN 2248-9975.

Bunasir, H. A., Hidayat, R., Ilmi, A., Mudjiutami, E., Rifa'i, A., Susanti, W., Tulus, W., Webby, et al. (2014). *Domestication and culture of Papuyu*. Freshwater Aquaculture and Development Center Mandiangin. Directorate General of Aquaculture, Marine and Fishery Ministry. 59 pp.

Candra, P., Ikrima, N., & Adawyah, R. (2019). Organoleptic Papuyu (*Anabas testudineus* Bloch) presto with how different culture. *Fish Scientiae, 9*(2), 80–86.

Chotipuntu, P., & Avakul, P. (2010). Aquaculture potential of climbing perch Anabas testudineus in brackish water. *Walailak Journal of Science and Technology, 7*, 15–21.

Djuhanda, T. (1981). *Fish world*. Armico.

Fauzi, M., Fajri, N. E., Saputra, W., & Hendrizal A. (2022). *Biological aspects of Betok fish (Anabas testudineus) on Lake Tuok Tonga (oxbow Lake) Riau Province*. Proceedings of the National Fish Seminar XI "sustainable blue economy challenges in aquaculture, management and conservation of fish resources" Bogor, 21 June 2022. https://doi.org/10.32491/Semnasikan-MII-2022-p.135-144

Halolo, L. M. (2008). *Food habits of Betok fish (Anabas testudineus) in the flood swamp area of the Mahakam River, Kec. Kota Bangun, Kab. Kutai Kartanegara, East Kalimantan*. Department of Aquatic Resources Management, Faculty of Fisheries and Marine Sciences, Bogor Agricultural University. https://repository.ipb.ac.id/handle/123456789/5246. Accessed 10 March 2023.

Hasnidar, T. A., Ernaningsih, H., & Akram, A. M. (2022). Reproductive biology climbing perch *Anabas testudineus* (Bloch 1792) at Danau Tempe Kabupaten Wajo South Sulawesi. *Jurnal Iktiologi Indonesia, 22*(1), 17–34. https://doi.org/10.32491/jii.v22i1.567

Hidayat, R., Carman, O., & Alimuddin. (2016). Sexual dimorphism related to growth in climbing perch *Anabas testudineus*. *Jurnal Akuakultur Indonesia, 15*(1), 8–14. https://doi.org/10.19027/jai.15.8-14

Ndobe, S., Rusaini, Masyahoro, A., Serdiati, N., Madinawati, & Moore, A. M. (2022). Reproductive and morphometric characteristics of climbing perch *Anabas testudineus* in Sigi, Central Sulawesi, Indonesia. *Aquaculture, Aquarium, Conservation & Legislation, 13*(1), 167–182.

Riede, K. (2004). *Global register of migratory species: From global to regional scales: Final report of the R&D-Projekt 808 05 081*. Federal Agency for Nature Conservation. 330 pp.

Riswan, Sumoharjo, & Isriansyah. (2017). Performance of betok fish (*Anabas testudineus* Bloch) reared on biofloc media with intercropping system with tilapia (*Oreochromis niloticus*). *Aquawarman Journal, 3*(1), 20–26.

Rohansyah, E. d. M. R. (2010). Comparing analysis of morphology character of two varians climbing perch (*Anabas testudineus* Bloch). *Media Sains Proceedings, 2*(1), 77–81.

Shinsuke, M. S. (2009). Growth and morphological development of laboratory-reared larval and juvenile climbing perch *Anabas testudineus*. *Ichthyological Research, 56*, 162–171. The Ichthyological Society of Japan.

Syarifuddin, R., Kasuma, A., & Hermantedo, S. (1993). *Food: Form, variety, and function as well as how to serve it in South Kalimantan* (pp. 1–146). Department of Education and Culture. https://repositori.kemdikbud.go.id/12692/1/MAKANAN%20WUJUD%20VARIASI%20DAN%20FUNGSINYA%20SERTA%20CARA%20PENYAJIANNYA%20DAERAH%20KALIMANTAN%20SELATAN.pdf. Accessed 10 March 2023.

Teletchea, F., & Fontaine, P. (2014). Levels of domestication in fish: Implications for the sustainable future of aquaculture. *Fish and Fisheries, 15*(2), 181–195. https://doi.org/10.1111/faf.12006

Widianti, I., Ilhamiyah, & Hasiani, Y. (2021). *Study of fish processing business in Haruyan Seberang Village, Hulu Sungai Tengah Regency* (Case study on Hj. Ati's business) [Diploma thesis. Kalimantan Islamic University]. http://eprints.uniska-bjm.ac.id/5766

www.itis.gov. (2023). *Anabas testudineus (Bloch, 1792), Taxonomic Serial No.: 172585*. Integrated Taxonomic Information System – Report. https://www.itis.gov/servlet/SingleRpt/SingleRpt?search_topic=TSN&search_value=172585#null

Zworykin, D. D. (2012). Reproduction and spawning behavior of the climbing perch *Anabas testudineus* (Perciformes, Anabantidae) in an aquarium. *Journal of Ichthyology, 52*(6), 379–388.

Striped Snakehead (*Channa striata* Bloch 1793) Farming in Indonesia

9

Adang Saputra, Ruby Vidia Kusumah, and Siti Murniasih

9.1 Introduction

Snakehead is an important functional food commodity in aquaculture and capture in Southern and Southeastern Asia. This product, in addition to being consumed as a source of protein in fresh or processed form, can also be used as a raw material for medication. Snakehead contains albumin, which can be used to substitute albumin in humans for wound tissue recovery after surgery (Haniffa et al., 2014; Rosyidi et al., 2019; Lee et al., 2022). According to Asfar et al. (2014), the albumin concentration in snakehead flesh ranged from 14% to 21%. Economically consumption-sized fresh snakehead sells for around Rp. 55,000–75,000 per kg (Saputra et al., 2022), with prices increasing to more than Rp. 100,000 per kg on the online market (Tokopedia, 2023), therefore snakeheads are potentially for aquaculture.

Between 1758 and 2020, 56 snakehead species were discovered worldwide, with 53 species in the genus *Channa* located in Asia and more than three of the genus *Parachanna* found in tropical Africa (Wang et al., 2023). There are 12 species of snakehead in Indonesia, including *C. bankanensis*, *C. cyanospilos*, *C. gachua*, *C. lucius*, *C. marulioides*, *C. marulius*, *C. melanoptera*, *C. melasoma*, *C. micropeltes*, *C. orientalis*, *C. pleurophthalma*, and *C. striata* (Froese & Pauly, 2022). Among these species, *C. striata* Bloch 1973 has been successfully cultivated in Indonesia (Saputra et al., 2022; Fig. 9.1).

A. Saputra (✉) · R. V. Kusumah · S. Murniasih
Research Center for Fishery, National Research and Innovation Agency (BRIN), Kawasan Sains dan Teknologi (KST) Soekarno, Jl. Raya Jakarta - Bogor KM 46, Cibinong, Indonesia
e-mail: adan004@brin.go.id

E. Nugroho et al. (eds.), *Aquaculture Development in Archipelago*,
https://doi.org/10.1007/978-981-95-1084-9_9

Fig. 9.1 *Channa striata* Bloch 1793. (Source: personal documentation)

9.2 Classification and Morphology

Fowler classified snakeheads as Channidae in 1934. Furthermore, the fish of the genus *Channa* (Scopoli 1777) are grouped into the phylum Chordata, class Teleostei, and order Perciformes. In 1793, Bloch described the striped snakehead, *C. striata*. Morphologically, *C. striata* has a subcylindrical body with a depressed head and a rounded caudal fin (Allen, 1991). The dorsal region and sides are dark and mottled with black and ochre, with white on the abdomen, a large cranium resembling the head of a snake, a deeply open, fully toothed mouth, and large scales (Yamamoto & Tagawa, 2000). There are no spines in the fins and no lower jaw scale. Dorsal fin rays 42–45, anal fin rays 25–29, lateral line scale 55–65, v-shaped sharp isthmus, caudal fin base lacking ocellus, and body with chevron-shaped protruding bars upward (Musikasinthorn, 2000; Vishwanath & Geetakumari, 2009). *C. striata* has a total length of 61 cm, a highest standard length of 100 cm, and a maximum weight of up to 3 kg (Froese & Pauly, 2022). Snakeheads have extra respiratory organs called diverticula that allow them to take oxygen directly from the air (Ishimatsu, 2012). During the dry season, snakeheads can survive by burrowing themselves in the mud as long as their additional organs and skin stay moist, and they live off the fat reserves in their bodies.

9.3 Distribution and Habitat

C. striata is distributed between South Asia and Southeast Asia, including Pakistan, India, and Sri Lanka, east to Vietnam, Yunnan, and southeastern China, and from Thailand and Malaysia to Borneo and Indonesia. Now, this species has been introduced elsewhere (Fricke & Van der Laan, 2021). In Indonesia, snakeheads were discovered in western Sumatra, Java, and Kalimantan. Nowadays, striped snakeheads have been introduced into Eastern Indonesia. Their habitat is primarily freshwater, slightly salty, sluggish, and muddy water, depths 1–2 m, in swamps, reservoirs, flooded fields, paddy fields, ponds, brooks, streams, canals, rivers, and lakes. In nature, the pH of the waters that become a snakehead's habitat varies from 4.5 to 7.8, and the temperature is 23–31.7 °C (Saputra et al., 2021; Kusmini et al., 2022). The snakehead can live in marshy water with low dissolved oxygen levels (Saputra et al., 2021).

9.4 Reproduction and Spawning Season

Female fish have rounder heads and paler bodies, whereas male fish have elongated heads and darker bodies. Male fish have reddish-colored points in their genital holes that, when pressed, release a transparent fluid when their gonads are mature. The female parent's mature gonad shows an enlarged, soft, and reddish stomach (Kusmini et al., 2016).

The striped snakehead's first gonad maturity measurement was 18 cm (Froese & Pauly, 2022); 32.2 cm in male and female 27.8 cm in the Sebangau-Palangkaraya River Swamp (Selviana et al., 2017); 19.3–30.8 cm in male and female 22.7–33.3 cm in the Musi Basin (DAS) of South Sumatra (Karmon, 2011). The first gonad development in the Barito River was 14 cm and 350 g in females (Makmur & dan Prasetyo, 2006). Snakehead fecundity reaches 576–58,219 eggs with a total length of 23.3–47.2 cm and weight of 117.4–838.4 g (Saputra & Samsudin, 2017). Snakehead eggs have a diameter of 1.20 to 1.55 mm, are transparent, pale yellow, have a large oil globule, are floating, and are not adhesive (Saputra et al., 2021). Striped snakeheads reproduce during the period of rainfall (August to February), and the spawning peak lasts from the beginning to the middle of the rainy season (Utomo et al., 1992; Saputra & Samsudin, 2017). Male and female snakeheads build nests to protect fertilized eggs. Male guards fertilized eggs, which hatch in 24–35 h at 28–30 °C (Tongsanga, 1960).

9.5 Domestication

Domestication activities are expected to be able to replace Snakehead's production dependence on wild catches, which are declining in stocks with sizes that are getting smaller (Nasution, 2012). In Indonesia, snakehead domestication studies were conducted in Kalimantan between 2007 and 2012 (Kusmini et al., 2022). In general, efforts to domesticate snakehead fish are carried out starting from the adaptation process until the fish can stay alive, consume the feed provided, grow and develop, mature in the gonads, and can reproduce well in farming conditions. According to Saputra & Samsudin (2017), snakeheads have adapted well to the cultivating environment. The next challenge, though, is that snakeheads find it difficult to mature their gonads to the point where they can reproduce out of season. Mylonas et al.

(2010) state that this situation can happen because snakehead fish do not fully receive signals from the environment, resulting in reproductive dysfunction. Numerous studies have used enhanced reproductive dysfunction as a solution to these problems.

9.5.1 Gonad Maturation

Snakehead maturation of gonads can be facilitated by hormonal application and parental control of nutrition. Hormonally, using 2.0 mL of PMSG hormone, antidopamine (AD), and their combination per kilogram of snakehead fish may stimulate gonadal maturity to the vitellogenesis phase and improve egg diameter up to 0.38 ± 0.015 mm (Ath-thar et al., 2017). The gonadal level of maturity of snakehead can be facilitated by PMSG and AD hormones (2.50 mL/kg). In snakehead broodstock, rising water levels, along with injections of sGnRHa (salmon gonadotropin-releasing hormone analog) and AD, may stimulate gonad development until ovulation (Bijaksana, 2010). Ovulation can be stimulated after 9–12 h with an intramuscular injection of pituitary extract at a dosage of 45 mg/kg for the first injection and 80 mg/kg for the second injection (Hossain et al., 2008). The ovulation rate can be accelerated by delivering HCG hormones 500 IU + LHRHa 0.6 mL/kg, with an average ovulation rate of 21 h 8 min (Radona et al., 2019). Rakhmawati (2015) reported similar findings, showing that the morphological characteristics of oocytes stimulated by HCG at a dosage of 500 IU were effective at stimulating hormonal performance, resulting in an increased oocyte diameter.

9.5.2 Breeding

Snakehead breeding methods involve natural, semi-artificial, and artificial ways. Natural breeding occurs by allowing the snakehead in the water to reproduce in response to environmental cues. In general, spawning happens once a year, at the start of the wet season. Semi-artificial breeding is carried out through hormonal injections at a dose of 0.6 mL/kg broodstock, then allowed to spawn naturally in ponds/spawning containers. Artificial breeding uses hormonal injections, and spawning is done through stripping techniques. However, semi-artificial breeding is suggested to produce snakehead fish in Indonesia.

9.5.3 Larval Development

The final digestion of snakeheads begins 12 days after hatching (dah), according to studies on the organ development in larvae (Saputra, 2018). The snakehead larvae can be gradually adapted to artificial feed after 12 days and can be fed artificial feed beginning at 15 days. Thus, it is possible to create large-scale, intensive snakehead farming.

9.6 Aquaculture Technology

9.6.1 Broodstock Feeding Management

The development of farming in the nursery and growing parts is restricted by snakehead spawning, which occurs just once a year. Feeding complete nutrition to broodstock with artificial feed can increase spawning as much as three times per year (Saputra, 2018). The composition of the broodstock feed used is water content (6.76%), protein (36.12%), fat (3.17%), ash (10.27%), crude fiber (4.79%), and nitrogen-free extract (45.65%) (Saputra, 2018). The snakehead broodstocks are kept separately between males and females in order to speed up the development of the gonads. To prevent the fish from escaping, the broodstock maintenance pond should be a ground pond with a net covering the top. The broodstock maintenance pond has water that is at least one meter deep and provides cover for aquatic vegetation. Feeding is provided three times per day at a rate of 10% of the total biomass.

9.6.2 Larva Rearing

Rearing larvae can be done using indoor or outdoor hatcheries. The best rearing of snakehead larvae is provided in a recirculation system-equipped indoor hatchery.

Indoor hatchery The container used for larval rearing is an aquarium measuring 40 × 60 × 100 cm with a water level of 15–25 cm. The stocking density of larvae maintenance is 50 individuals/L (Saputra, 2018). Each container is heated to maintain temperature stability and aeration for oxygen supply. The key parameters of larval rearing are temperature, alkalinity, and pH. The good growth performance of snakehead fish larvae takes place at a water temperature of 28–30 °C, an alkalinity of 50–100 mg L^{-1} $CaCO_3$, and a water pH of 5.5–6.5. During rearing, snakehead fish larvae are given enough natural food (*Moina* sp.) until they are 15 days old, and from 10 days old, they have been adapted to artificial feeds with a crude fiber content of not more than 0.3%. The larvae rearing time is around 30 days until the size is 1–2 cm, then they are transferred to a nursery container.

Outdoor hatchery Apart from indoor hatcheries, larval rearing can also be carried out in outdoor hatcheries. The key to success in rearing larvae in indoor hatcheries is the provision of suitable natural feed. Natural food for snakehead larvae includes zooplankton *Moina* sp., because of the appropriate nutrition and size. To maintain zooplankton in prime condition, it is necessary to prepare a food source in the form of microalgae such as *Chlorella* sp. The availability of *Chlorella* sp. in the media for rearing snakehead larvae will ensure the availability of *Moina* sp. as a source of nutrition in prime conditions. One of the characteristics of *Chlorella* sp. growth and development is a change in the media for rearing fish larvae to become green. This condition is often called the green water system. The technique of rearing snakehead larvae with a green water system is an alternative solution in providing natural feed that is available all the time. The results of research by Saputra et al. (2018) showed the best density of *Chlorella* sp. for rearing snakehead larvae is 4.3×10^6 cells/mL. Proximate test results for *Moina* sp. reared in green water (in % wet weight) containing 92.62% water, 4.53% crude protein, 0.66% crude fat, 0.48% ash, 0.23% crude fiber, and 1.48% BETN (Saputra et al., 2018). With this system, snakehead seed production from outdoor hatcheries can be increased.

9.6.3 Aquatic Plant Application

In their natural habitat, snakeheads like to hide in weathered wood or aquatic plants from seed to maturity with the aim of avoiding and carrying out predation. In snakehead fish cultivation, it is necessary to provide shelter as a place to hide. This is in accordance with the results of Sunarno & Syamsunarno (2015), the main function of shade is to prevent predation. The type of shade that fish farmers mostly use is aquatic plants. According to Saputra & Samsudin (2017), the best type of shade for raising snakehead fish seeds is water hyacinth. One of the basics of using water hyacinth as a shelter is that it can act as a shelter and is able to degrade nutrients so that the water becomes nontoxic to fish. For water hyacinth to function optimally as a shelter, it needs to be kept at a certain population size. According to Saputra & Samsudin (2017) the amount of water hyacinth cover as the best shade is 50% of the surface area of the water.

9.6.4 Probiotic Application

Intensive aquaculture impacts high-pollutant sources, including nitrogen. In general, nitrogen accumulates in water and sediment (Rodehutscord & Pfeffer, 1995; Barak & van Rijn, 2000; Voslarova et al., 2008). High concentrations of nitrogen have a negative impact on the aquatic environment and contain toxic compounds to fish and other aquatic organisms (Lazzari & Baldisserotto, 2008). For this reason, intensive aquaculture management is focused on water quality management so that during the cultivation activity process, it is not constrained by changes in water quality, which can have a detrimental effect on productivity. Efforts to maintain water quality have been carried out by utilizing the work of bacteria (bioremediation), which aims to break down or degrade toxic compounds to become nontoxic. Giving probiotics will stabilize water quality and increase growth. The application of nitrogen-degrading bacteria is one of the most profitable alternative solutions to the problem of aquaculture waste because it can reduce inorganic nitrogen waste (Crab et al., 2007). *Pseudomonas* sp. can reduce nitrogen compounds in shrimp culture media in ponds (Widiyanto, 2006). These bacteria include heterotrophic bacteria. Based on the energy source, heterotrophic bacteria are divided into two groups, namely, photoheterotrophs, which use sunlight as an energy source, and chemoheterotrophs, which use organic and inorganic compounds as their energy source (Hogg & Stuart, 2013). Through the application of probiotics, snakehead seeds are proven to grow and develop optimally. Because in ponds that are given probiotics, the water quality conditions are very supportive of the growth performance of the fish being cultivated. The decrease in total nitrogen and *Channa striata* total phosphorus greatly affects the growth rate of the snakehead fish that are kept.

9.6.5 Cultivation in the Tidal Area

Snakehead can be cultivated in the lowlands to the highlands. One area that is very good for the development of snakehead fish farming is the tidal area. The area of swamps in Indonesia is estimated at 33,393,570 ha, consisting of 20,096,800 ha (60.2%) tidal areas and 13,296,770 ha (39.8%) nontidal swamps (*lebak*). The low productivity of fish farming in the tidal area is caused by the fact that the tidal area has many problems, especially water quality, such as H_2S toxin and acidity, turbidity, and ingress of seawater so that the water becomes brackish. This condition has an impact on low growth performance. To optimize the productivity of the tidal area, the utilization of secondary and tertiary channels has the potential to be used for snakehead cultivation. The cultivation system used in secondary/tertiary channels is more focused on fixed net cages. In addition, ponds in tidal areas are generally dug ponds with the characteristics of calm and stagnant water. The components of a still-water pond include the pond levee, pond bottom, inlet, outlet, water inlet, and drain. The snakehead cultivation system that is applied to still-water ponds is the cage pond system, namely, the placement of cages in the pond. Tarpaulin tubs are an alternative fish farming system when the area for fish farming is increasingly limited. In tidal areas, fish farming in tarpaulin tanks is one solution to overcome fluctuations in water quality parameters. The advantage of tarpaulin ponds is that they can be built in various places, not necessarily on ideal land like conventional pond construction.

References

Allen, G. R. (1991). *Field guide to the freshwater fishes of New Guinea*. Publication, no. 9. (p. 268). Christensen Research Institute.

Asfar, M., Tawali, A. B., Abdullah, N., & Mahendradatta, M. (2014). Extraction of albumin of snakehead fish (*Channa striatus*) in producing the fish protein concentrate (FPC). *International Journal of Scientific & Technology Research, 3*(4), 85–88.

Ath-thar, M. F. H., Gustiano, R., Kusmini, I. I., Prakoso, V. A., & Putri, F. P. (2017). Induksi hormonal maturasi gonad ikan gabus *Channa striata*. [Hormonal induction of gonad maturation in snakehead fish, *Channa striata*]. *Jurnal Riset Akuakultur, 12*(1), 9–20.

Barak, Y., & van Rijn, J. (2000). Biological phosphate removal in a prototype recirculating aquaculture treatment system. *Aquaculture Engineering, 22*, 121–136.

Bijaksana, U. (2010). *Kajian fisiologi reproduksi ikan gabus, di dalam wadah dan perairan rawa sebagai upaya domestikasi*. [Study of reproductive physiology of snakehead fish, in containers and swamp waters as a domestication effort] [Disertasi. Institut Pertanian Bogor].

Crab, R., Kochva, M., Verstraete, W., & Avnimelech, Y. (2007). Bio-flocs technology application in over-wintering of tilapia. *Aquacultural Engineering, 40*(2009), 105–112.

Fricke, R., & Van der Laan, R. (2021). *ECoF, Eschmeyer's catalog of fishes catalog of fishes*. http://researcharchive.calacademy.org/research/ichthyology/catalog/fishcatmain.asp.

Froese, R., & Pauly, D. (Eds.). (2022). *FishBase*. World Wide Web electronic publication. www.fishbase.org, (08/2022).

Haniffa, M. A. K., Jeya Sheela, P. A., Kavitha, K., & Jais, A. M. M. (2014). Salutary value of haruan, the striped snakehead *Channa striatus* – A review. *Asian Pacific Journal of Tropical Biomedicine, 4*(Suppl 1), S8–S15. https://doi.org/10.12980/APJTB.4.2014C1015

Hogg, & Stuart. (2013). *Essential microbiology* (2nd ed., p. 86). Wiley-Blackwell.

Hossain, M. K., Latifa, G. A., & Rahman, M. M. (2008). Observations on induced breeding of snakehead murrel, *Channa striatus* (Bloch, 1793). *International Journal of Sustainable Crop Production, 3*, 65–68.

Ishimatsu, A. (2012). Evolution of the cardiorespiratory system in air-breathing fishes. *Aqua-BioScience Monographs, 5*(1), 1–28. https://doi.org/10.5047/absm.2012.00501.0001

Karmon. (2011). *Pertumbuhan dan Reproduksi Ikan Gabus (Channa striata); Kaitannya dengan Kondisi Habitat Rawa Banjiran Daerah Aliran Sungai (DAS) Musi Sumatera Selatan*. [Growth and reproduction of snakehead fish (*Channa striata*); its relation to flood swamp habitat conditions in the Musi River basin area (DAS), South Sumatra] [Tesis, Institut Pertanian Bogor].

Kusmini, I. I., Gustiano, R., Subagja, J., Ath-thar, M. H. F., Prakoso, V. A., & Radona, D. (2016). *Pembenihan ikan gabus secara alami dan semi alami. Rekomendasi Teknologi Kelautan dan Perikanan. [Natural and semi-natural seeding of snakehead fish. Recommendations for Marine and Fisheries Technology]* (pp. 323–335). Balitbang KP.

Kusmini, I. I., Prakoso, V. A., & Radona, D. (2022). Status Terkini Domestikasi Induk Ikan Gabus. [Current status of domestication of snakehead fish broodstock]. In A. Saputra, M. T. D. Sunarno, R. Gustiano, & A. M. Lusiastuti (Eds.), *Ikan Gabus (Channa Striata Bloch, 1793) Pembenihan dan Pendederan. [Snakehead fish (Channa Striata Bloch, 1793) breeding and nursery]* (p. 128). PT Penerbit IPB Press.

Lazzari, R., & Baldisserotto, B. (2008). Nitrogen and phosphorus waste in fish farming. *Boletim do Instituto de Pesca, 34*(4), 591–600.

Lee, V. L. L., Norazit, A., Noor, S. M., & Shaikh, M. F. (2022). *Channa Striatus* protects against PTZ-induced seizures in LPS pre-conditioned zebrafish model. *Frontiers in Pharmacology, 13*(April), 1–9. https://doi.org/10.3389/fphar.2022.821618

Makmur, S., & dan Prasetyo, D. (2006). Kebiasaan makan, tingkat kematangan gonad dan fekunditas ikan haruan (*Channa striata* bloch) di suaka perikanan sungai Sambujur DAS Barito Kalimantan Selatan. [Feeding habits, gonad maturity level and fecundity of snakehead fish (*Channa striata* bloch) in the Sambujur River fisheries sanctuary, Barito Watershed, South Kalimantan]. *Jurnal Ilmu-ilmu Perairan dan Perikanan Indonesia, 13*(1), 27–31.

Musikasinthorn, P. (2000). *Channa aurantimaculata*, a new channid fish from Assam (Brahmaputra River basin), India, with designation of a neotype for C. amphibeus (McClelland, 1845). *Ichthyological Research, 47*(1), 27–37. https://doi.org/10.1007/BF02674310

Mylonas, C. C., Duncan, N. J., & Asturiano, J. F. (2010). Hormonal manipulations for the enhancement of sperm production in cultured fish and evaluation of sperm quality. *Aquaculture, 472*, 21–44. https://doi.org/10.1016/j.aquaculture.2016.04.021

Nasution, Z. (2012). *Kelembagaan pengelolaan sumberdaya perikanan "lelang lebak lebung" dan kemiskinan masyarakat nelayan (studi kasus di Kabupaten Ogan Ilir-Sumatera Selatan)*. [Institutional management of fisheries resources "lebak lebung auction" and poverty of fishing communities (case study in Ogan Ilir Regency, South Sumatra)] [Tesis, Sekolah Pascasarjana IPB] (p. 162).

Radona, D., Kusmini, I. I., Prihadi, T. H., Khasani, I., & Astuti, D. N. (2019). Hormonal induction for maturation, ovulation and successful level of spawning on striped snakehead (*Channa striata*). *Aquaculture, Aquarium, Conservation & Legislation, 13*(1), 118–124.

Rakhmawati, E. (2015). *Induksi perkembangan gonad betina ikan gabus (Channa striata, Bloch) dengan penyuntikan hormon HCG dalam wadah budi daya*. [Induction of female gonad development in snakehead fish (*Channa striata*, Bloch) by injecting HCG hormone in the cultivation container]. Skripsi. Departemen Budi Daya Perairan, Fakultas Perikanan dan Ilmu Kelautan. IPB.

Rodehutscord, M., & Pfeffer, E. (1995). Effects of supplemental microbial phytase on phosphorus digestibility and utilization in rainbow trout (Oncorhynchus mykiss). *Water Science and Technology, 31*, 143–147.

Rosyidi, R. M., Januarman, J., Priyanto, B., Islam, A. A., Hatta, M., & Bukhari, A. (2019). The effect of snakehead fish (*Channa striata*) extracts capsule on the albumin serum level of post-operative neurosurgery patients. *Biomedical and Pharmacology Journal, 12*(2), 893–899. https://doi.org/10.13005/bpj/1714

Saputra, A. (2018). *Teknologi produksi benih ikan gabus Channa striata melalui pendekatan fisiologi, manajemen pakan, dan rekayasa lingkungan*. [Production technology of snakehead seed, *Channa striata* through physiological, feed management and environmental engineering approaches] [Doctoral dissertation, IPB Graduate School, Bogor].

Saputra, & Samsudin. (2017). Penentuan jenis gulma air sebagai naungan (shelter) pada pendederan ikan gabus *Channa striata* di kolam. [Determination of the type of water weeds as shelter for the nursery of snakehead, *Channa striata*, in ponds]. *Jurnal Perikanan dan Kelautan, 7*(2), 100–111.

Saputra, A., Ath-thar, M. H. F., Samsudin, R., Putri, F. P., & Prakoso, V. A. (2018). Reproductive biology of striped snakehead (*Channa striata* Bloch, 1973) in Bogor and Bekasi, West Java. *Berita Biologi, 16*(3), 3015. https://doi.org/10.14203/beritabiologi.v16i3

Saputra, A., Syamsunarno, M. B., & Sunarno, M. T. D. (2021). Development of seed mass production of snakehead (*Channa striata*) in Indonesia. *IOP Conference Series: Earth and Environmental Science, 715*(1). https://doi.org/10.1088/1755-1315/715/1/012060

Saputra, A., Sunarno, M. T. D., Gustiano, R., & Lusiastuti, A. M. (2022). In A. Saputra, M. T. D. Sunarno, R. Gustiano, & A. M. Lusiastuti (Eds.), *Ikan Gabus (Channa Striata Bloch, 1793) Pembenihan dan Pendederan. [Snakehead fish (Channa Striata Bloch, 1793) breeding and nursery]*. Penerbit IPB Press.

Selviana, E., Affandi, R., & dan Kamal, M. M. (2017). Aspek reproduksi ikan gabus (*Channa striata*) di rawa banjiran aliran Sungai Sebangau. [Reproductive aspects of snakehead fish (*Channa striata*) in the flood swamps of the Sebangau River, Palangkaraya]. *Jurnal Ilmu Pertanian Indonesia, 25*(1), 10–18.

Sunarno, M. T. D., & Syamsunarno, M. B. (2015). Pengaruh naungan terhadap pematangan gonad dan pemijahan ikan belida (*Chitala lopis*) di kolam rawa. [The effect of shade on gonad maturation and spawning of belida fish (*Chitala lopis*) in swamp ponds]. *Jurnal Ilmu Pertanian dan Perikanan., 4*(1), 71–77.

Tokopedia. (2023). *Ikan gabus hidup 1kg fresh*. [Live snakehead fish 1kg fresh]. https://www.tokopedia.com/ajat17/ikan-gabus-hidup- 1kg-fresh?extParam=ivf%3Dfalse&src=topads. Accessed 10 March 2023.

Tongsanga, S. (1950). Life history of *Ophiocephalus Striatus*. *Thai Fisheries Gassette, 13*(1), 63–67.

Utomo, A. D., Nasution, Z., & dan Adjie, S. (1992). *Kondisi ekologi dan potensi sumberdaya perikanan sungai dan rawa*. [Ecological conditions and potential of river and swamp fishery resources]. Temu Karya Ilmiah Perikanan Perairan Umum Puslitbang Perikanan, Badan Litbang Pertanian, Departemen Pertanian.

Vishwanath, W., & Geetakumari, K. (2009). Diagnosis and interrelationships of fishes of the genus Channa Scopoli (Teleostei: Channidae) of northeastern India. *Journal of Threatened Taxa, 1*(2), 97–105. https://doi.org/10.11609/jott.o1788.97-105

Voslarova, V., Pistecova, V., Svobodova, Z., & Bedanova, I. (2008). Nitrite toxicity to danio rerio; effects of subchronic exposure on fish growth. *Acta Veterinaria, 77*, 445–460.

Wang, J., Xu, W., Liu, Y., Bai, Y., & Liu, H. (2023). Comparative mitochondrial genomics and phylogenetics for species of the snakehead genus Channa Scopoli, 1777 (Perciformes: Channidae). *Gene, 857*, 147186. https://doi.org/10.1016/j.gene.2023.147186. Epub 2023 Jan 7.

Widiyanto, T. (2006). Seleksi bakteri nitrifikasi dan denitrifikasi untuk bioremediasi di tambak udang. [Selection of nitrifying and denitrifying bacteria for bioremediation in shrimp ponds] [Doctoral dissertation, IPB Graduate School, Bogor]

Yamamoto, M. N., & Tagawa, A. W. (2000). *Hawaii's native and exotic freshwater animals* (p. 200). Mutual Publishing.

Domestication and Aquaculture Development of Tinfoil Barb *Barbonymus schwanenfeldii* in Indonesia

10

Vitas Atmadi Prakoso and Rasidi

10.1 Introduction

Tinfoil barb, *Barbonymus schwanenfeldii* (Bleeker, 1853), is one of the domesticated freshwater native fish species in Indonesia. This species has the potential to be developed as an aquaculture commodity. Tinfoil barb has market prices ranging from IDR 25,000 to 40,000 per kilogram. It is favored for consumption because of its savory taste. In addition, tinfoil barb can be used as an ornamental fish, partly because of its unique shape with a silvery-golden body and orange or blood-red dorsal and tail fins. This species is classified into the Cyprinid group, with omnivorous and benthopelagic types (Gustiano et al., 2015). Generally, the habitat of tinfoil barb is in flooded rivers and swamps, with a depth range of 1.0–4.0 m, a temperature range of 25–30 °C, a brightness range of 40–120 cm, a pH range of 5–7, and a weak current state (Kottelat et al., 1993). Morphologically, this species is similar to *Barbonymus gonionotus* and *Barbonymus balleroides,* which have been successfully farmed (Fig. 10.1).

This species has different local names in some areas, such as "lampam," "kapiek," and "lemeduk" in Sumatra (Gunawan et al., 2017; Desrita et al., 2021), and "tengadak lampam" in Borneo (Santoso & Wahyudewantoro, 2019). The trade name for this species is "tinfoil barb." Tinfoil barb has the potential to be farmed for both food

V. A. Prakoso (✉)
Research Center for Conservation of Marine and Inland Water Resources, National Research and Innovation Agency (BRIN),
Kawasan Sains dan Teknologi (KST) Soekarno, Jl. Raya Jakarta - Bogor KM 46, Cibinong, West Java 16911, Indonesia
e-mail: vita007@brin.go.id

Rasidi
Research Center for Fishery, National Research and Innovation Agency (BRIN),
Kawasan Sains dan Teknologi (KST) Soekarno, Jl. Raya Jakarta - Bogor KM 46, Cibinong, West Java 16911, Indonesia

E. Nugroho et al. (eds.), *Aquaculture Development in Archipelago*,
https://doi.org/10.1007/978-981-95-1084-9_10

Fig. 10.1 Tinfoil barb Barbonymus schwanenfeldii (Photo credit: Author)

and ornamental purposes (Kusmini et al., 2021; Oktavia et al., 2024). Tinfoil barb's habitat can be found in several aquatic environments, such as rivers, peat, and open water, which all provide suitable environments for this species. An inventory of fish biodiversity along different stretches of river flow in Arut-Kumai, Central Kalimantan, revealed that tinfoil barbs are restricted to swamp rivers (Santoso & Wahyudewantoro, 2019). This demonstrates how the tinfoil barb's natural habitat is becoming increasingly hostile. The number of wild populations can be expected to decline as a result of persistent fishing efforts. Land conversion for mining, plantations, cities, and factories is another human-made phenomenon. Their population is predicted to decline due to these activities, despite currently being classified as Least Concern (Desrita et al., 2022; Lumbantobing & Allen, 2020). Therefore, tinfoil barb farming is necessary as part of conservation efforts.

10.2 Taxonomy and Morphology

This species is included in Kingdom Animalia, Phylum Chordata, Class Actinopterygii, Order Cypriniformes, Family Cyprinidae, and Genus *Barbonymus*. The scientific name of this species is *Barbonymus schwanenfeldii* (Bleeker, 1853), with *Barbus schwanenfeldii* (Bleeker, 1853) and *Puntius schwanenfeldii* (Smith, 1945) as synonyms. The common name of this species in English is tinfoil barb, while in Malaysia, this species is called "Lampam Sungai" (Fricke et al., 2019). The classification of tinfoil barbs is as follows:

Kingdom:	Animalia
Phylum:	Chordata
Class:	Actinopterygii
Order:	Cypriniformes
Family:	Cyprinidae
Genus:	*Barbonymus*
Species:	*B. schwanenfeldii*

Barbonymus schwanenfeldii is a deep-bodied cyprinid with a silvery-golden body and red fins, the dorsal bearing a black tip blotch and the caudal deeply forked with a submarginal black stripe (Kottelat et al., 1993). The meristic features of this species, according to Radona et al. (2017), are as follows: dorsal fin (I.9), pectoral fin (11–13), ventral fin (I.8), anal fin (I.6), caudal fin between 15 and 17, and linea lateralis ranging from 31 to 36.

10.3 Distribution

This species is distributed in Southeast Asia (Dewantoro, 2015; Lumbantobing & Allen, 2020), and specifically in Indonesia, it can be found in Sumatra and Borneo (Gustiano et al., 2015; Kusmini et al., 2016; Radona, 2016; Radona et al., 2017). In Southeast Asia, except Indonesia, tinfoil barbs are distributed in many areas, such as the Chao Phraya Basin in Thailand, the Mekong Basin in Thailand, Cambodia, Lao PDR, Viet Nam, Malaysia (the peninsula, Sabah, and Sarawak), and Brunei (Lumbantobing & Allen, 2020). The distribution map of tinfoil barb is presented in Fig. 10.2.

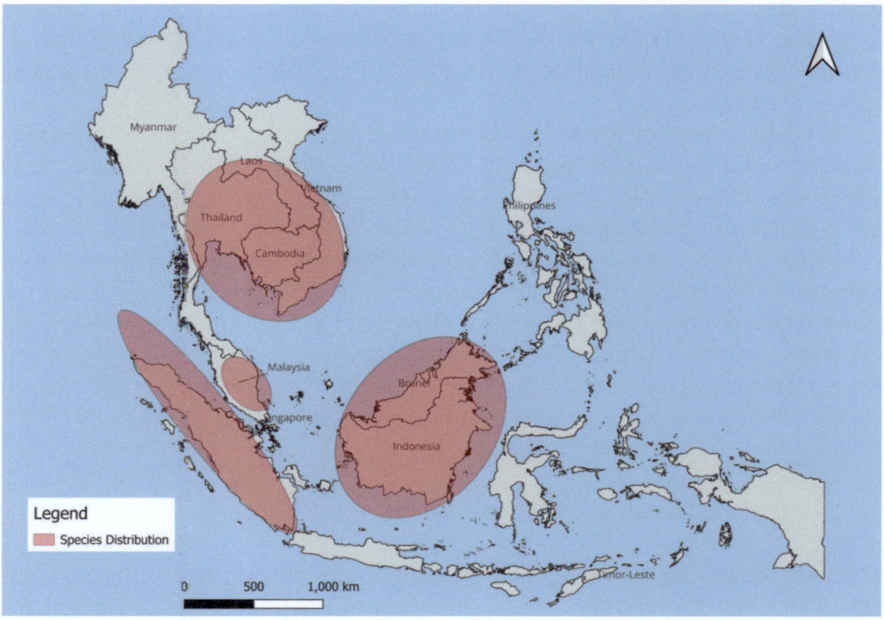

Fig. 10.2 Distribution map of tinfoil barb *Barbonymus schwanenfeldii*

Fig. 10.3 Rearing media for tinfoil barb *Barbonymus schwanenfeldii* broodstocks. Description: (**a**) concrete pond, (**b**) Semi-permanent pond, and (**c**) Net cages pond (Radona et al., 2021; Photo credit: Irin Iriana Kusmini)

10.4 Reproduction

The maintenance of tinfoil barb's broodstocks can be carried out in earthen, concrete, semi-permanent ponds, or floating nets (Fig. 10.3). For pond maintenance, a pond measuring 5 × 2 m can be used for broodstock. The pond needs to be drained first for 3–4 days, then refilled with 80–100 cm of water, which should be continuously flowed. The maintenance of the broodstocks is carried out separately between males and females in order to facilitate gonad maturity observation. The optimal density of broodstock maintenance is four fish per m^3. The pond is equipped with an inlet and outlet.

Broodstock selection aims to select broodfish that are ready for spawning and to anticipate failures during spawning. The ideal size of female broodstock is 150–300 g, and the male broodstock ranges from 100 to 200 g, with a body that is not deformed, agile movements, an age >1 year, and rapid growth. Initial selection is carried out based on the physical completeness and health of the broodstocks. Subsequent selection of gonadal maturity level is performed by examining the diameter and position of the cell nucleus on the oocyte. This aims to ensure the level of readiness of the fish brood that has passed the selection invisibly. The egg diameter for a mature tinfoil barb's gonad is 0.7–1.0 mm. The broodfish that have passed the selection process are separated from the broodstock ponds for 1–2 days.

10.5 Domestication of Tinfoil Barbs

Tinfoil barb farming is carried out to meet the consumption needs of the community, which has been relying on catches from nature. At the Research Institute for Freshwater Aquaculture, Indonesia, the domestication of tinfoil barbs has been carried out since 2014. Currently, the breeding technology and larval rearing have been successfully implemented.

10.5.1 Natural Spawning

The natural spawning of tinfoil barb is similar to that of other cyprinid fish. The ratio between male and female broodfish is 1:1. Natural spawning can be carried out in earthen or concrete ponds that have previously been dried and cleaned from sediments, small fish, and other predators. The broodfish can be put into the spawning ponds, which females precede in the morning and then males in the afternoon. The ovulation process will usually occur in the early hours of the morning (2–5 a.m.). Considering that the eggs of tinfoil barb are benthopelagic, if spawning is carried out in an earthen pond, it is recommended to install the nets (hapa) in the pond (Fig. 10.4). This aims to ensure that sediments do not cover the fertilized eggs. If the fish eggs are covered with sediment, it will adversely affect the hatching rates.

10.5.2 Artificial Spawning

Artificial spawning is carried out using hormonal stimuli. Basically, the ratio of males and females for artificial spawning is 1:1. In artificial spawning, female broodfish will be stimulated using fish reproductive hormone (Ovaprim). The hormone is applied at a dose of 0.5 mL/kg of broodfish weight. To determine the amount (mL) of the hormone to be injected, a weighing is first carried out on the weight of the female broodstock to be spawned so that the amount of hormone injected is in accordance with the requirements of each broodfish.

Fig. 10.4 Ponds for natural spawning of tinfoil barb *Barbonymus schwanenfeldii* (Radona et al., 2021; Photo credit: Fera Permata Putri)

The injection is done intramuscularly on the lower back of the dorsal fin with a slope of 45°. Injections can be done twice or only once. If the injection is done twice, the dose of the first injection is 40%, and the second is 60% of the amount of hormones. After 6 h from the first injection, the second injection was carried out. The initial injection is administered in the afternoon, followed by a second injection in the evening. Injected broodfish are then separated into males and females. Usually, 4–6 h after the injection, the fish will ovulate, and broodfish stripping can be performed.

The stripping process should be carried out in a dry and clean container. This is followed by mixing male sperm into the egg container while being given water for the activation of sperm and eggs. Stirring is carried out for 1 min using clean chicken feathers, and then the eggs are rinsed with clean water to remove the remaining unfertilized sperm. After the eggs are free from the remaining sperm, they can be stocked in hatching containers such as aquariums or funnels. Radona (2016) explained that one female broodfish (total length 14–16 cm; body weight 150–200 g) can produce an average fecundity of 70,000 eggs with a gonad maturity index of 12.26%. The egg diameter ranges from 0.7 to 0.9 mm.

10.5.3 Larval Rearing

In aquaculture, several important factors must be considered after spawning, including how to handle fish eggs properly and correctly so that the eggs can hatch optimally and produce healthy and perfect larvae, and handling larvae after the depletion of yolk reserves (yolk sac). The assessment of the success rate of hatchery activities is assessed based on the percentage of hatchability of fertilized eggs and the survival of larvae until they become ready-to-stock fry. Tinfoil barb eggs hatch optimally at 28 °C. Fertilized eggs are clear and transparent, whereas unfertilized eggs appear pale white. According to Radona (2016), the fertilization rate of tinfoil barb is 90–95%, with hatching rates of 70–80%.

10.5.4 Hatching

The methods of hatching tinfoil barb eggs can be carried out intensively or traditionally using several hatching media, such as a funnel system (Fig. 10.5) with rack maintenance, aquarium-based hatching and maintenance, or pond-based hatching and maintenance.

10.5.5 Nursery

The nursery of tinfoil barb can be carried out in ponds, nets, or aquariums. The nursery of tinfoil barbs in aquariums can begin at the postlarval stage. The optimum stocking density in the nursery of tinfoil barbs in an aquarium measuring

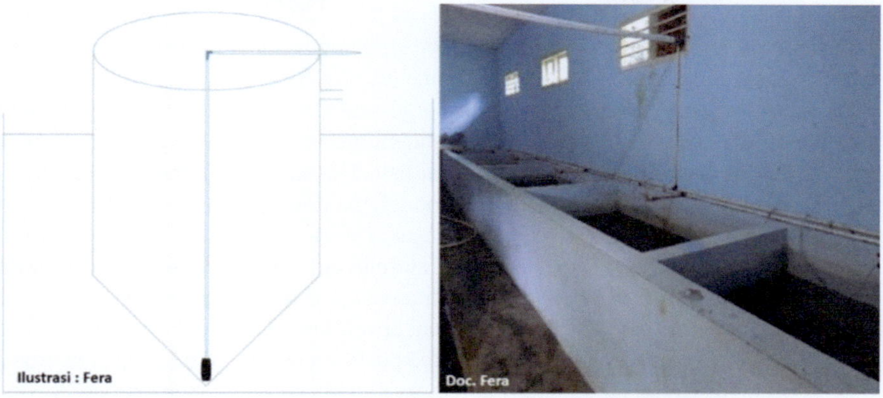

Fig. 10.5 (**a**) Illustration of the egg hatching funnel system (Illustration by Fera Permata Putri, based on Radona et al., 2021); (**b**) Larval rearing in indoor concrete ponds (Radona et al., 2021; Photo credit: Fera Permata Putri)

60 × 30 × 30 cm is five fish per liter. Postlarval maintenance is carried out in a 34-L aquarium with adequate aeration. During 90 days of postlarval rearing, they are fed commercial feed in the form of crumbs containing 40% protein at satiation with a frequency of three times a day (morning, afternoon, and evening). Water quality is maintained by replacing 30–50% of the aquarium water every 3 days, as needed.

For the nursery of tinfoil barbs in earthen ponds, it should be conducted after proper pond preparation. Pond management includes drying, fertilization, and filling with water. Pond drying aims to eliminate microbes and pests that may interfere with rearing. Fertilization is carried out using urea and TSP fertilizer at a 2:1 ratio, applied at 10 g/m^2. The pond is then gradually filled with water to promote optimal growth of natural feed. Once the pond is ready, larvae are stocked at a density of 50 fish/m^3 in the morning or evening to avoid stress from direct sunlight. During rearing, fish are fed commercial pellets in the form of crumble with a protein content of 28–32%, which are given *ad libitum* with a frequency of three times a day (morning, afternoon, and evening). Maintenance at Nursery Phase I is carried out for 30 days.

10.6 Aquaculture Technology of Tinfoil Barbs

Since 2010, research on the potential of tinfoil barb farming has been conducted. Nevertheless, the adoption of this technology by local communities remains limited. According to the findings of a previous study (Kusmini et al., 2020), it is suggested that female broodstock from Kalimantan and male broodstock from Java could serve as prospective sources of broodstock that have the potential to be developed for the production of tinfoil barb. The spawning stage of tinfoil barb farming faces many challenges, including slow gonadal development (which can take from 6 to 8 months), poor reproductive performance, and low larval survival. The value of recovering life and growth, as well as feed efficiency in relatively low-intensity culture conditions, are some of the challenges that have been encountered in the process of seed production (Kusmini et al., 2015).

The main obstacle to the expansion of large-scale fish farming is the lack of seed due to local fish spawning issues like infrequent broodstock spawning, poor quality eggs, and a shortage of mature gonad broodfish. Tinfoil barbs in their natural environment can spawn at any time of year, although the rainy seasons of September and October are the most likely to occur (Siregar, 1989). As a result, fewer seeds are likely to be produced. Nevertheless, both internal and external factors in its natural environment have an impact on the reproductive cycle of this species. Examples of internal components that contribute to sexual development include the expression of gonadotropins, testicular maturation, and the presence of steroid hormones in blood plasma, whereas examples of external factors include feed and environmental conditions such as season.

Ensuring fish eggs are available year-round is essential for the successful development of aquaculture (Migaud et al., 2013). Therefore, as part of the strategy, improving broodstock nutrition and administering hormones can enhance both the number and quality of broodstocks with mature gonads, as well as increase their spawning frequency. Using nutritional supplements in broodstock feed helps ensure that the gonads developed by the broodfish are of sufficient quality to produce healthy eggs (Singh et al., 2021). The feed processed by the endocrine system influences its ability to produce and release hormones to target organs (Liley & Stacey, 1983; Volkoff et al., 2009). Feed intake utilized during broodfish rearing has a significant effect on egg quality, with higher-quality feed resulting in better-quality eggs. In shrimp broodstock hatcheries, marine worms for shrimp feed (Laining et al., 2019), microalgae for freshwater and marine fish feed, and a combination of shellfish and sea worms for tiger shrimp feed are all examples of broodstock feeds that have been studied (Lante et al., 2014).

Water quality (temperature, dissolved oxygen, and pH), feed inputs, predators, and disease all influence reproductive success in fish (Schreck, 2010). The fish's skin, eyes, and nose detect environmental cues and relay them to the brain's hypothalamus. Furthermore, the pituitary gland is stimulated by gonadotropin-releasing hormone (GnRH), which is produced in the hypothalamus (GTH). The steroid hormones released as a result of GTH's circulation to the gonads induce reproductive behavior (Demski & Hornby, 1982). The aim of aquaculture environment management is to provide the optimal conditions for domesticated broodfish to thrive and maintain sustainable reproduction.

The nutritional approach for feeding the domesticated fish is adapted to meet the particular requirements of the broodstocks. Broodstock for freshwater fish generally require a protein diet of 30–35% (NRC, 1993). During the reproductive phase, broodfish demand high energy inputs for gonad development, steroid hormone synthesis, vitellogenesis, energy reserves for hatching larvae (endogenous feeding), fecundity (egg production), and embryogenesis. When these nutritional requirements are met, both basal metabolic needs and the energy demands of reproduction can be fulfilled. During reproduction, essential fatty acids are crucial. According to Watanabe et al. (1984), the low productivity of the fry produced from the broodfish that are spawned is a direct result of poor feed intake, which results in the production of eggs with low hatching rates and larvae that are frequently abnormal. The precursors of prostaglandins and cell membranes both require fatty acids.

Research on tinfoil barb has been increasingly conducted in the fields of environment, feed nutrition, genetics and reproduction, and resource management. Spawning and hatchery activities have been carried out effectively outside of their natural habitat. Hormonal induction (Sutriana et al., 2022; Kusmini et al., 2020) and hormonal combinations with enhanced broodstock nutrition have both been tested in laboratory-scale hatcheries for seed production (Lestari et al., 2016). Hormonal induction, in conjunction with a nutritional strategy for broodstock feed that includes *Spirulina* and turmeric, has been shown to increase growth rates by 2.2 times faster outside of the spawning season (Lestari et al., 2016). These advances highlight future opportunities for scaling seed production, improving broodstock performance, and integrating sustainable feed innovations into commercial aquaculture systems, thereby strengthening the role of tinfoil barb in Indonesia's aquaculture development.

During the juvenile rearing stage, tinfoil barb fish need feed that is 32–35% protein (Dewantoro et al., 2018; Mansour et al., 2017) and approximately 22.89% carbohydrates (Yanto et al., 2019). Farmed fish have already adapted to the commercial diets. The majority of commercial feed comes from fish meal and fish oil as protein and lipid sources. To reduce feed costs, alternative ingredients such as green leaves can be utilized. Tinfoil barb can be fed a wide variety of fresh foodstuffs in the floating cages; however, some of the foods they are fed cannot be consumed intact. For instance, cassava leaves can be used as a plant-based feed source. It was reported that tinfoil barb consumed up to 70% of banana bunches and 96% of the tubers and leaves of the cassava plant (*Manihot esculenta*). Less usefully, only 49% and 34% of *Passiflora* sp. and *Ipomoea aquatica* leaves were eaten, respectively (Christensen, 1993). Juvenile tinfoil barbs had a higher efficiency with corn and tapioca starches than taro flour (Nafees et al., 2022). These results have implications for the sustainable use of starch in aquafeeds and the future of aquaculture. For tinfoil barb, plant-based ingredients such as cassava leaves, banana waste, and starch, alongside animal-derived proteins, can be used as raw material sources in feed formulation to improve growth performance and lower feed costs.

Moreover, research on seed transportation has been conducted, including the use of a closed transportation system with anesthetic derived from tekelan (*Chromolaena odorata*) leaves (Ayuzar et al., 2021). However, this work remains at the laboratory scale, with limited development and application in the field. Future research should focus on farm-level validation of these methods to enhance seed survival and facilitate broader distribution of tinfoil barb seed.

References

Ayuzar, E., Rusydi, R., Muna, H., & Muliani, M. (2021). Anesthetic effectiveness of tekelan leaves on tinfoil barb fish in closed transportation. *Depik, 10*(2), 143–150. https://doi.org/10.13170/depik.10.2.18519

Bleeker, P. (1853). Nieuwe tientallen diagnostische beschrijvingen van nieuwe of weinig bekende vischsoorten van Sumatra. *Natuurkundig Tijdschrift voor Nederlandsch-Indië, 5*(3), 495–534.

Christensen, M. S. (1993). An economic analysis of floating cage culture of tinfoil barb, *Puntius schwanenfeldii*, in East Kalimantan, Indonesia, using chicken manure and other fresh feeds. *Asian Fisheries Science, 6*(3), 271–281. https://doi.org/10.33997/j.afs.1993.6.3.003

Demski, L. S., & Hornby, P. J. (1982). Hormonal control of fish reproductive behavior: Brain–gonadal steroid interactions. *Canadian Journal of Fisheries and Aquatic Sciences, 39*(1), 36–47. https://doi.org/10.1139/F82-006

Desrita, D., Hasugian, F. K., Yusni, E., Manurung, V. R., & Rambey, R. (2021). Feeding habits of tinfoil barb, *Barbonymus schwanenfeldii* in the Tasik River, South Labuhanbatu, North Sumatra, Indonesia. *Biodiversitas Journal of Biological Diversity, 22*(4), 2131–2135.

Desrita, D., Manik, B. R., & Rambey, R. (2022). Abundance and growth of tinfoil barb (*Barbonymus schwanenfeldii*) in Tasik River, South Labuhanbatu, North Sumatera, Indonesia. *Aquaculture, Aquarium, Conservation & Legislation, 15*(6), 3152–3158.

Dewantoro, E. (2015). Gonad performance of the barbfish (*Barbonymus schwanenfeldii*) after periodic injection of HCG hormone, [Keragaan gonad ikan tengadak (*Barbonymus schwanenfeldii*) setelah diinjeksi hormon HCG secara berkala]. *Jurnal Akuatika, 6*(1), 1–10.

Dewantoro, E., Dhahiyat, Y., Rostika, R., Zahidah, & Iskandar. (2018). Growth performance of tinfoil barb (*Barbonymus schwanenfeldii*) fed with different protein levels and energy/protein ratios on diet. *Aquaculture, Aquarium, Conservation & Legislation, 11*(4), 1300–1310.

Fricke, R., Eschmeyer, W. N., & Van der Laan, R. (Eds). (2019). *Eschmeyer's catalog of fishes: Genera, species, references*. Updated 5 August 2019. Available at: http://researcharchive.calacademy.org/research/ichthyology/catalog/fishcatmain.asp

Gunawan, R. H., Muchlisin, Z. A., & Mellisa, S. (2017). Kebiasaan makan ikan Lemeduk (*Barbonymus schwanenfeldii*) di Sungai Tamiang, Kecamatan Sekerak, Kabupaten Aceh Tamiang, Provinsi Aceh. [Feeding habits of Lemeduk fish (*Barbonymus schwanenfeldii*) in the Tamiang River, Sekerak District, Aceh Tamiang Regency, Aceh Province]. *Jurnal Ilmiah Mahasiswa Kelautan dan Perikanan Unsyiah, 2*(3), 379–388.

Gustiano, R., Kusmini, I. I., & Ath-thar, M. F. H. (2015). *Mengenal sumber daya genetik ikan spesifik lokal air tawar Indonesia untuk pengembangan budidaya. [Understanding the genetic resources of specific local freshwater fish in Indonesia for aquaculture development]* (p. 51). IPB Press.

Kottelat, M., Whitten, A. J., Kartikasari, S. N., & Wirjoatmodjo, S. (1993). *Freshwater fishes of Western Indonesia and Sulawesi* (p. 221). Periplus Editions.

Kusmini, I. I., Gustiano, R., Mulyasari, Iskandariah, & Huwoyon, G. H. (2015). Ikan lokal tengadak (*Barbonymus schwanenfeldii*) asal Kalimantan sebagai andalan untuk ikan budi daya. [The native tinfoil barb (*Barbonymus schwanenfeldii*) from Kalimantan as a potential candidate species for aquaculture development], in M. F. Rahardjo et al. (eds.) *Prosiding Seminar Nasional Ikan, 8*(2), pp. 177–187. Masyarakat Iktiologi Indonesia.

Kusmini, I. I., Gustiano, R., Radona, D., Prakoso, V. A., Putri, F. P., & Prihadi, T. H. (2016). Karakterisasi fenotipe dan genotipe tiga populasi ikan tengadak, *Barbonymus schwanenfeldii*. [Phenotypic and genotypic characterization of three populations of tengadak, *Barbonymus schwanenfeldii*]. *Jurnal Riset Akuakultur, 11*(3), 207–216.

Kusmini, I. I., Radona, D., Prakoso, V. A., Gustiano, R., & Dinar, T. (2020). *Outbreeding performance of tinfoil barb Barbonymus schwanenfeldii from Java and Kalimantan for Aquaculture Development* (p. 01009). EDP Sciences.

Kusmini, I. I., Gustiano, R., Radona, D., & Kurniawan, K. (2021). Domestication strategies of tinfoil barb *Barbonymus schwanenfeldii* (Bleeker, 1854): A potential candidate for freshwater aquaculture development. *IOP Conference Series: Earth and Environmental Science, 934*(1), 012003. IOP Publishing.

Laining, A., Trismawanti, I., Usman, U., & Masruri, M. H. (2019). Influence of squid liver meal in maturation diet on gonadosomatic index and gonadal amino acid content of golden spotted rabbitfish. *Siganus guttatus. Indonesian Aquaculture Journal, 14*(1), 31–38.

Lante, S., Laining, A., & Parenrengi, A. (2014). Performa reproduksi induk udang windu (*Penaeus monodon* Fab.) jantan alam dan domestikasi tambak. [Reproductive performance of wild and domesticated male tiger prawn (*Penaeus monodon* Fab.) broodstock in ponds]. *Prosiding Forum Inovasi Teknologi Akuakultur* (pp. 693–700).

Lestari, T. P., Sudrajat, A. O., & Budiardi, T. (2016). Combination of additional *Spirulina platensis* and *Curcuma longa* supplements in diet and hormonal induction to improve reproduction performance of tinfoil barb *Barbonymus schwanenfeldii* (Bleeker, 1854). *Jurnal iktiologi indonesia, 16*(3), 299–308.

Liley, N. R., & Stacey, N. E. (1983). Hormones, pheromones, and reproductive behavior in fish. In *Fish physiology* (Vol. 9, pp. 1–63). Academic Press.

Lumbantobing, D., & Allen, D. J. (2020). *Barbonymus schwanenfeldii*. The IUCN red list of threatened species 2020:ee. T181160A89800163. https://doi.org/10.2305/IUCN.UK.2020-RLTS.T181160A89800163.en. Downloaded on 15 July 2021.

Mansour, O., Idris, M., Noor, N. M., & Das, S. K. (2017). Growth performance of tinfoil barb (*Barbonymus schwanenfeldii*) fry feeding with different protein content diets. *Aquaculture, Aquarium, Conservation & Legislation, 10*(3), 475–479.

Migaud, H., Bell, G., Cabrita, E., McAndrew, B., Davie, A., Bobe, J., Herráez, M. P., & Carrillo, M. (2013). Gamete quality and broodstock management in temperate fish. *Reviews in Aquaculture, 5*, S194–S223.

Nafees, M. S. M., Kamarudin, M. S., Karim, M., Hassan, M. Z., & de Cruz, C. R. (2022). Effects of dietary starch sources on growth, nutrient utilization and liver histology of juvenile tinfoil barb (*Barbonymus schwanenfeldii*, Bleeker 1853). *Aquaculture Reports, 23*(February), 101069. https://doi.org/10.1016/j.aqrep.2022.101069

NRC. (1993). *Nutrient requirements of fish*. National Research Council (NRC).

Oktavia, N., Prakoso, V. A., Kamal, M. M., Arifin, O. Z., Subagja, J., Kurniawan, K., Iswantari, A., Kusmini, I. I., Irawan, D., Cahyanti, W., Radona, D., Putri, F. P., Bosman, O., Wibowo, A., Kristanto, A. H., & Gustiano, R. (2024). A study on the growth and physiological response of juvenile tinfoil barb *Barbonymus schwanenfeldii* (Bleeker, 1854) under the influence of pH changes. *Indonesian Aquaculture Journal, 19*(1), 35–44.

Radona, D. (2016). *Karakterisasi Genotipe, Biometrika dan Potensi Performa Ikan Barbonymus schwanenfeldii (Bleeker 1854) dari Sumatera, Jawa dan Kalimantan*. [Genotype characterization, biometrics and potential performance outbreed of tinfoil barb *B* (Bleeker 1854) from Sumatera, Java and Kalimantan] [Master Thesis, IPB (Bogor Agricultural University), Bogor] (in Bahasa Indonesia).

Radona, D., Kusmini, I. I., & Ath-thar, M. H. F. (2017). Karakterisasi meristik dan morfometrik tiga generasi ikan tengadak *Barbonymus schwanenfeldii* asal Kalimantan Barat, Indonesia. [Meristic and morphometric characterization of three generation of tengadak, *Barbonymus schwanenfeldii* from West Kalimantan, Indonesia]. *Jurnal Riset Akuakultur, 12*(1), 1–8.

Radona, D., Kusmini, I. I., Putri, F. P., Dewi, R. R. S. P. S., & Gustiano, R. (2021). Tinfoil barb *Barbonymus schwanenfeldii* farming. In A. H. Kristanto, R. Gustiano, & K. Sugama (Eds.), *Seedlings production of Indonesian freshwater native fish species* (pp. 75–85). Amafrad Press. (in Bahasa Indonesia).

Santoso, E., & Wahyudewantoro, G. (2019). Biodiversitas spesies ikan perairan gambut Arut-Kumai, Kabupaten Kotawaringin Barat, Kalimantan Tengah. [Biodiversity of fish species in the Arut-Kumai peat waters, West Kotawaringin Regency, Central Kalimantan]. *Jurnal Iktiologi Indonesia, 19*(2), 315–335.

Schreck, C. B. (2010). Stress and fish reproduction: The roles of allostasis and hormesis. *General and Comparative Endocrinology, 165*(3), 549–556.

Singh, S. K., Baidya, S., Das, P., & Biswas, P. (2021). Functional role of dietary supplements on reproductive physiology of fishes. In J. K. Sundaray, M. A. Rather, S. Kumar, & D. Agarwal (Eds.), *Recent updates in molecular endocrinology and reproductive physiology of fish*. Springer.

Siregar, S. (1989). *Kemungkinan budidaya ikan tinfoil barb (Puntius schwanenfeldii Blkr.) dari Sungai Kampar, Riau. Tesis Magister. IPB (Institut Pertanian Bogor), Bogor* [Possibility of culturing tinfoil barb (*Puntius schwanenfeldii* Blkr.) from Kampar River, Riau] [Master Thesis, IPB (Bogor Agricultural University), Bogor].

Smith, H. M. (1945). *The freshwater fishes of Siam or Thailand. Bulletin of the United States National Museum, 188*, xi+622 pp.

Sutriana, A., Baihaqi, A., Hasri, I., & Hafizuddin, H. (2022). Hormonal induction of gonad maturation in female tinfoil barb fish (*Barbonymus schwanenfeldii*) using spawnprim hormone. *Jurnal Akuakultur Indonesia, 21*(2), 118–124.

Volkoff, H., Xu, M., MacDonald, E., & Hoskins, L. (2009). Aspects of the hormonal regulation of appetite in fish with emphasis on goldfish, Atlantic cod and winter flounder: Notes on actions and responses to nutritional, environmental and reproductive changes. *Comparative Biochemistry and Physiology Part A: Molecular & Integrative Physiology, 153*(1), 8–12.

Watanabe, T., Ohhashi, S., Itoh, A., Kitajima, C., & Fujita, S. (1984). Nutritional studies in the seed production of fish – XV. Effect of nutritional composition of diets on chemical components of red sea bream broodstock and eggs produced. *Nippon Suisan Gakkaishi, 50*(3), 503–515.

Yanto, H., Setiadi, A. E., & Kurniasih, D. (2019). Pengaruh tingkat karbohidrat berbeda dalam pakan terhadap kinerja pertumbuhan ikan tengadak *(Barbonymus schwanenfeldii),* [The effect of different carbohydrate levels in feed on the growth performance of barbfish *(Barbonymus schwanenfeldii)*]. *Jurnal Ruaya, 7*(2), 39–46.

Development of Freshwater *Tor* sp. in Indonesia: Domestication and Culture

11

Yogi Himawan, Otong Zenal Arifin, and Jojo Subagja

11.1 Introduction

Tor sp. is a freshwater fish species found in various regions of Indonesia, including Sumatra, Java, and Kalimantan, and holds significant economic value. In Indonesia, the *Tor* sp. population comprises four species: *Tor tambroides*, *T. douronensis*, *T. tambra*, and *T. soro* (Kottelat et al., 1993; Roberts, 1999). A key challenge in culturing this species is its slow growth rate, with the development from larva to broodstock taking approximately 4 years (Fig. 11.1).

This finding is consistent with a study by Dinesh et al. (2015) on the aquaculture of Mahseer fish, specifically *Tor khudree*, in the lowlands of Kerala, South India. Their results indicated low survival rates and specific growth rate for *Tor khudree* in lowland aquaculture. Similarly, Raina et al. (1999) observed that the growth performance of *Tor putitora* in manured ponds with artificial supplementary feed resulted in a survival rate of less than 55%. For successful Mahseer aquaculture, both seed quality and rearing conditions are critical factors for the success of freshwater fish farming in rivers or reservoirs.

Y. Himawan (✉)
Research Center for Fishery, National Research and Innovation Agency, Jakarta, Indonesia
e-mail: yogi011@brin.go.id

O. Z. Arifin · J. Subagja
Research Center for Applied Zoology, National Research and Innovation Agency, Jakarta, Indonesia

E. Nugroho et al. (eds.), *Aquaculture Development in Archipelago*,
https://doi.org/10.1007/978-981-95-1084-9_11

Fig. 11.1 *Tor soro*. (Source: Personal documentation)

11.2 Taxonomy

For many years, most large-sized carps with big scales, barbels, and similar morphological traits were classified as mahseers. Carps characterized by large scales, fleshy lips that meet at the corners of the mouth with an interrupted fold or groove across the lower jaw, two pairs of large barbels, lateral-line scales ranging from 22 to 28, and a head length that is either equal to, greater than, or less than the body depth are taxonomically considered true mahseers, belonging to the genus *Tor*.

Misra (1959) classified this genus as follows:

Phylum Vertebrata
- Subphylum Craniata
  - Superclass Gnathostomata
    - Series Pisces
      - Class Teleostomi
        - Subclass Actinopterygii
          - Order Cypriniformes
            - Division Cyprini
              - Suborder Cyprinoidei
                - Family Cyprinidae
                  - Genus *Tor*
                    - Species *Tor tor* (Hamilton) 1822

In 1822, Hamilton first categorized mahseers, placing them in the genus *Cyprinus* and identifying three species: *Cyprinus tor*, *Cyprinus putitora*, and *Cyprinus mosal*. Later, researchers reassigned these species to two distinct genera: *Barbus* by Cuvier in 1817 and *Tor* by Gray in 1834. Hora (1943) suggested that dividing *Barbus* into additional genera would be premature until a comprehensive revision of the *Barbus* complex is conducted. However, he acknowledged that mahseers are likely subgenerically distinct from *Barbus*, and *Tor* could be used as a provisional subgeneric classification. Misra (1959) later classified all mahseers under the separate genus *Tor*.

Internationally, these fish are commonly referred to as mahseer. Several species from the genus *Neolissochilus* and a single species from the genus *Naziritor* are also referred to as mahseer due to their large scales and similar characteristics. The taxonomy of these genera remains unresolved, with no expert consensus. Based primarily on FishBase, with some revisions following Kottelat (2013), at least 21 species belong to the genus *Tor*, distributed across the Middle East, South and Southeast Asia, including countries like India, Pakistan, Bangladesh, China, Vietnam, Thailand, Malaysia, and Indonesia. These species include *Tor ater*, *Tor barakae* (Mahseer barakae), *Tor douronensis* (Semah, Semah daun, Semah mahseer, Mahseer kelah), *Tor hemispinus*, *Tor kulkarnii* (Dwarf mahseer, only one specimen identified so far), *Tor khudree* (black mahseer, Mahseer Dean), *Tor laterivittatus*, *Tor macrolepis* (species unclear), *Tor malabaricus* (Malabar mahseer), *Tor mosal*, *Tor Polylepis*, *Tor progeneius* (Jungha mahseer), *Tor putitora* (Himalayan or Golden mahseer), *Tor remadevii*, *Tor sinensis* (Chinese mahseer), *Tor tambra* (Tambera), *Tor tambroides* (Thai mahseer, Malayan mahseer), *Tor tor* (redfin or short gill mahseer), *Tor yingjiangensis*, *Tor yunnanensis*, and *Tor zhobensis* (Balochi mahseer).

11.3 Morphology

The head of *Tor* species is slightly shorter than the body depth, with the dorsal profile more sharply arched than the ventral profile. These fish possess thick, fleshy lips with a continuous labial fold across the lower jaw. The mouth is small, with the gape not extending below the eyes, and the snout is pointed, with both jaws being roughly equal in length. *Tor* species have two pairs of barbels, with the maxillary barbels being slightly longer than the rostral barbels but shorter than the eye. The interorbital space is flat. The dorsal fin is located either opposite or slightly ahead of the ventral fin, and the dorsal spine is smooth and shorter than the body depth. The pectoral fin reaches the pelvic fin and is as long as the head (excluding the snout), while the pelvic fin is shorter and, when laid flat, does not reach the base of the deeply forked caudal fin. The lateral line is complete, with nine predorsal scales, and each pelvic fin has a well-developed scaly appendage in the axil.

In terms of measurements, the head length is four times, and the body depth is 3.0–3.7 times the standard length. The eye diameter is 3.0–4.0 times the interorbital distance, and the minimum height of the caudal peduncle is 1.2–1.4 times its length. The coloration of *Tor* species is typically silvery or grayish green dorsally, with pinkish sides transitioning to greenish-gold above and light olive green below. The lower fins are reddish yellow, and these species can grow to a maximum size of 152.0 cm in the wild. In Indonesia, *Tor* fish exhibit variations in characteristics such as color, body shape, and scale count across different species (Table 11.1).

11.4 Distribution

There are 24 species of fish within the *Tor* genus globally (Kiat, 2004), distributed from the Himalayas to Southeast Asia. In Indonesia, four species have been identified: *Tor soro* Val. 1842, *Tor douronensis* Val., *Tor tambra* Val., and *Tor tambroides* Blkr. (Kottelat et al., 1993). In Lake Toba, North Sumatra, and nearby areas such as Aek Sirambe Site, Panahatan, Asahan River, Bahorok River, and Aek Sarula River in Tarutung Regency, *Tor soro* is locally known as the Batak ihan and holds cultural significance for the Batak people.

Across Indonesia, *Tor soro* is known by various local names. In Kuningan, Sumedang, and Majalengka (West Java), it is called Dewa Fish or Kancra Bodas,

Table 11.1 Morphological description of *Tor* species from Pasawahan-Kuningan (West Java), Sumedang (West Java), and Ambarita (North Sumatra)

No	Character	Location origin		
		Pasawahan (Kuningan, West Java)	Sumedang (West Java)	Ambarita (North Sumatra)
1	Dorsal fin	III,9	III,8–9	III,9
2	Pec*Tor*al fin	I,11–16	I,13–16	I,12–16
3	Ventral fin	III,8–9	III,8–9	III,8
4	Anal fin	III,5–6	III,6	III,5
5	Caudal fin	22	22	20–22
6	Grunt	2 pairs	2 pairs	2 pairs
7	Operculum color	Silver-gold (TC 061–624)	Green (TC 061–061)	Green (TC 056–068)
8	Ventral color	White (TC 622–624)	White (TC 622–624)	White (TC 624)
9	Amount of scale linea lateralis (LL)	22–28	22–27	24–28
10	Dorsal color	Silver (TC 598)	Silver-gold (TC 598)	Green-silver (TC 449–469)
11	Spine section	38	35–37	37
12	Gill sieve bones	19	18–19	18–19
13	Fecundity eggs/kg	572–1310	593–1204	646–1109
14	Early maturity:			
	Male Female	98 g 820 g	110 g 870 g	122 g 770 g
15	Carcass:			
	Flesh, bones, scales, intestines, and gills	61.2% 38.8%	66.72% 33.28%	71% 29%
16	Dressing			
	Percentage head Body Fin Scale Gonads Gill gut	11.10% 75.10% 3.10% 4.10% 4.30% 2.30%	13.10% 70.00% 3.00% 5.20% 6.20% 2.40%	13.20% 70.00% 3.10% 5.10% 6.30% 2.30%

Source: Arifin et al. (2019)

while in Bogor (Cisadane River), it is referred to as soro. In Blitar (East Java), it is known as Senggaring. In South, Central, and West Kalimantan, it is called Lomi fish, while *Tor tambroides* is known as sapan in South and Central Kalimantan. In West Sumatra, *Tor douronensis* is named Ikan Garing or Jurung. In Jambi, Riau, West Kalimantan, and South Sumatra, all *Tor* species are collectively called semah, while in Malaysia, they are known as kelah. Additionally, *Tor soro* is synonymous with *Labeobarbus soro*, with the common name being *Soro*.

11.5 Reproduction

Broodstock rearing for *Tor* species is conducted in running water ponds with a minimum depth of 80 cm. The pond bottom is lined with gravel, and water clarity is maintained (Secchi disk ≥50 cm), with dissolved oxygen levels of at least 4 mg/L and a pH range of 6.5–7.5. Alternatively, broodstock can be raised in floating net cages at a stocking density of 3–5 individuals per cubic meter, with a male-to-female ratio of 1:2. The broodstock are fed artificial feed containing 28–30% protein and 7% fat, with a daily ration of 2–3% of the biomass, and feeding is done twice daily. Visual signs of gonadal maturity in *Tor* species include changes in the anal fin color from dark to bright bluish-green, enlarged genital papillae, rough spots on the gill covers in females, and a distended abdomen.

To confirm gonadal maturity, egg samples can be collected using a catheter inserted through the genital opening. Mature eggs, ready for spawning, are typically uniform in size, large in diameter (2.7–3.0 mm), firm in texture, and bright in color. The seasonal cycle of female gonadal maturation is evaluated over a year by measuring the average monthly oocyte diameter. This is done using a micrometer attached to a binocular microscope (12× magnification) on 30–50 oocyte samples, periodically collected from 12 broodstock individuals using a plastic cannula (Subagja & Gustiano, 2006; Subagja et al., 2006). The largest average oocyte diameter (1.4 mm) was recorded in July, just before the dry season, with another peak in early January (1.35 mm) (Fig. 11.2). Outside these months, egg development is minimal, with some individuals showing signs of egg atresia.

Fig. 11.2 Seasonal observation on the reproduction of *Tor soro*. (Source: Personal documentation)

11.6 Domestication of *Tor* sp. in Indonesia

Domestication is the process of taming wild animals to make them more suitable and beneficial for human use. In the case of fish, domestication involves adapting species that originally lived in natural environments (in situ) to controlled or cultured environments (ex situ). Through this process, fish populations become more adaptable to culture environments over successive generations. One example of a domesticated *Tor* species is the Dewa Fish, specifically *Tor soro*, which was developed from populations in Sumedang and Pasawahan-Kuningan (West Java) and Ambarita (North Sumatra) between 1996 and 2011 at the Research Institute for Freshwater Fisheries Culture and Instructor in Cijeruk, Bogor. The domestication process began with the characterization of the fish based on DNA and morphometric profiles. The first generation (G0) was produced through selective breeding from the original population, which was further selected to establish a second generation (G2) with stable traits suited to culture. Testing from G0 to G2 was conducted to evaluate genetic diversity, growth, and other factors. Domestication efforts for *Tor* species began in 1996 with fish collected from North Sumatra and adapted to floating net cages in Jatiluhur Dam. However, an incident led to the death of the collected fish, prompting the continuation of the collection and maintenance efforts from 1997 to 2006.

Tor species have several unique advantages related to their name, mythology, texture, nutritional content, price, and also rich historical significance. During the Dutch East Indies era, they were known as Java Salmon and have also been referred to by various names such as Dewa Fish, Kancra (West Java), Semah (South Sumatra and Jambi), Garing (West Sumatra), and Jurung (Aceh). In West Java, a myth links this fish to King Siliwangi's army, while in North Sumatra, it is often featured in traditional ceremonies and weddings as an offering. The Chinese believe that keeping this fish brings luck and longevity. The Dewa Fish is popular for its tender meat and high albumin content, which is beneficial as a raw material for postoperative healing medications. This fish also commands a high market price, making it a premium, exclusive choice for consumption. As part of the local fish conservation program launched by the Ministry of Maritime and Fisheries Affairs, domestication efforts for *Tor soro* begin with data collection and research, including identification by phenotype and genotype, adaptation to the ex situ environment, and observation of egg development variations. These efforts are followed by the development of technological components such as spawning with different hormone concentrations, larval rearing, and seed nurseries. A technology package is being prepared in collaboration with local governments in Sumatra and Java to support these activities.

Tor species are known for their high protein albumin content. According to Andreeva (2011), serum albumin is a crucial protein in the fish metabolic system. Fish serum albumin (FSA) plays a vital role in transporting body metabolites such as fatty acids, hormones, and bilirubin, regulating blood colloid osmotic pressure, aiding in osmoregulation, and filtering fluids in body tissues. Additionally, FSA has bioactive properties beneficial for pharmacological purposes and has been developed into commercial nutraceutical products. FSA-based nutraceuticals, commonly derived from snakehead fish (*Channa striata*), are already produced by several small and medium enterprises (SMEs) in Indonesia and are available in the market. FSA extracts from fish are also in demand for treating hypoalbuminemia in hospitals, offering a more affordable alternative to synthetic FSA, which is relatively expensive. Further pharmacological research has found that protein extracts from snakehead fish, containing FSA and essential amino acids, exhibit bioactive properties that promote skin health and show antibacterial, antifungal, antinociceptive, and platelet aggregation balancing effects (Jais Mat, 2007).

11.7 Culture Technology Development

Broodstock rearing for *Tor* species is conducted in running water ponds with a minimum depth of 80 cm, where the pond bottom is lined with gravel and water clarity is maintained (Secchi disk ≥50 cm). Dissolved oxygen levels should be at least 4 mg/L, and the pH should range between 6.5 and 7.5. Alternatively, broodstock can be kept in floating net cages (KJA) with a stocking density of 3–5 individuals per cubic meter and a male-to-female ratio of 1:2. Feeding consists of artificial feed containing 28–30% protein and 7% fat, with daily rations amounting to 2–3% of the biomass, provided twice daily. Signs of gonadal maturity in Dewa fish include a color change in the anal fin from dark to bright bluish green, enlarged genital papillae, rough spots on the gill covers of females, and a distended abdomen. To confirm gonadal maturity, egg samples can be measured using a catheter inserted through the genital opening. Mature eggs are uniform in size, large in diameter (2.7–3.0 mm), firm in texture, and brightly colored, indicating readiness for spawning.

Artificial spawning involves several steps: selecting broodstock ready for hormone injection, injecting hormones, collecting sperm, fertilizing the eggs, activating the eggs, and washing them. Broodstock selection is based on the observation of egg oocytes with a diameter of 2.9–3.1 mm, with an 80% distribution. If the egg condition only reaches 60% at this diameter, a preliminary injection of 500 IU/kg of HCG, based on broodstock weight, is administered 24 h before the ovulation injection. Ovulation is induced by injecting the gonadotropin hormone (*ovaprim*) at a dose of 0.6 ml/kg of the female's weight, divided into two injections: the first at 0.2 ml/kg and the second, 8 h later, at 0.4 ml/kg. The hormone is injected into the dorsal area behind the fin. Ovulation occurs 14–16 h after the second injection at a water temperature of 23–25 °C (with a degree hour of 350–368°H). Sperm is prepared before fertilizing the eggs, diluted with physiological NaCl (0.09% concentration) at a 1:4 ratio of sperm to NaCl, and store at a refrigerator temperature of 5 °C.

Fertilized eggs become translucent 8 h after fertilization, while unfertilized or dead eggs turn white and must be removed by siphoning. The eggs hatch after 4–5 days, and water quality is carefully maintained to ensure optimal hatching conditions. Once the larvae have absorbed their yolk sacs, they are transferred to an aquarium at a stocking density of 10,000 individuals per square meter. Live food, such as *Artemia nauplii*, is provided for 5–7 days, starting on days 9–11 after hatching. The rearing process continues for 20–21 days, until the larvae reach 1.5–2 cm in size, at which point the fry are ready for the next stage of culture.

The grow-out phase for *Tor* species aims to produce fish 20–25 cm in length or 200–250 g in weight, typically carried out in stagnant ponds. The ideal grow-out pond should have walls or a sandy base, with a minimum area of 100 m^2, a height of at least 80 cm, and a water depth of at least 50 cm. The pond should be free of pests and leaks, with a water flow rate of at least 60 l per minute. Water sources can include tertiary irrigation canals, reservoirs, or lake outlets. Water quality requirements include dissolved oxygen >4 mg/L, pH between 6 and 8, temperature between 22 °C and 32 °C, CO_2 levels <4.5 mg/L, water hardness <12 mg/L, and a Jackson turbidity unit (JTU) value of <30 for water clarity. The optimal location for *Tor* species breeding is in areas 50–400 m above sea level.

References

Andreeva, A. M. (2011). Mechanisms of the plurality of Scorpaena porcus L. serum albumin. *Open Journal of Marine Science, 1*, 31–35.

Arifin, O. Z., Subagja, J., Asih, S., & Kristanto, A. H. (2019). *The culture of Dewa fish*. PT IPB Press.

Dinesh, K., Daisy, C. K., Akhil, P. B., Geeji, M. T., & Nair, C. M. (2015). Aquaculture indices of Deccan Mahseer, Tor khudree transplanted to the lowlands of Kerala while stocked with Indian major carps. *International Journal of Science, Environment and Technology, 4*(4), 1080–1088.

Hora, S. L. (1943). The game fishes of India XVI: The mahseer or the large-scaled barbels of India. Further observations on mahseer from the Deccan. *Journal of the Bombay Natural History Society, 44*, 1–8.

Jais Mat, A. M. (2007). Pharmacognosy and pharmacology of Haruan (Channa striatus), a medicinal fish with wound healing properties. *Boletín Latinoamericano y del Caribe de Plantas Medicinales y Aromáticas, 6*(3), 52–60.

Kiat, N. C. (2004). *The king of rivers: Masheer in Malayan and the region*. Inter Sea Fishery.

Kottelat, M., Whitten, A. J., Kartikasari, S. N., & Wirjoatmodjo, S. (1993). *Freshwater fishes of Western Indonesia and Sulawesi*. Periplus Editions (HK) Ltd.

Kottelat, M. (2013). The fishes of the inland waters of Southeast Asia: A catalogue and core bibliography of the fishesknown to occur in freshwaters, mangroves and estuaries. *The Raffles Bulletin of Zoology Suppl*. 27:1–663.

Misra, K. S. (1959). An aid to the identification of the common commercial fishes of India and Pakistan. *Records of the Zoological Survey of India, 57*, 156–157.

Raina, H. S., Sunder, S., Joshi, C. B., & Mohan, M. (1999). *Himalayan Mahseer*. National Research Centre on Coldwater Fisheries.

Roberts, T. R. (1999). Fishes of the cyprinid genus tor in the Nam Theun watershed (Mekong basin) of Laos, with description of a new species. *The Raffles Bulletin of Zoology, 47*(1), 225–236.

Subagja, J., Asih, S., & Gustiano, R. (2006). Management in the breeding of Tor soro fish. *Media Aquaculture, 1*(1), 84–89.

Subagja, J., & Gustiano, R. (2006). The effect of HCG hormone implantation on egg development, final gonadal maturation, and spawning of Tor soro fish. *Journal of Aquaculture Research, 1*(2), 219–225.

Domestication of Kelabau Fish (*Osteochilus melanopleurus*) from Indonesia

12

Sulasy Rohmy, Irin Iriana Kusmini, Rahmat Hidayat, and Mulyasari Mulyasari

12.1 Introduction

Kelabau fish (*Osteochilus* sp.) is one of the local fish from Indonesia. There are about seventeen (17) species of *Osteochilus* sp., spread in Asia, including East Kalimantan, North Kalimantan, Sumatera (Kampar River, Riau), Mekong dan Chao Praya River, and Malaysian Peninsula (Saanin, 1984; Roberts, 1989; Kottelat et al., 1993). Mulyasari et al. (2010) revealed that kelabau in Kalimantan Indonesia consists of two species, *Osteochillus kelabau* and *Osteochillus melanopleurus* but kelabau fish from Riau (Kampar, Rokan and Siak River) were from one species *Osteochilus melanopleurus,* Bleeker 1852 (Asiah et al., 2019).

Kelabau fish are found in little rivers or in little lakes with lush aquatic plants. The analysis of kelabau stomachs showed they consumed phytoplankton, algae, insect detritus and fruits along the river (Nasution & Nuraini, 2004). The presence of microflora, especially bacteria in the digestive tract, was very instrumental in supporting fish health both in terms of nutrient utilization and enhancement of the immune system (Nayak, 2010; Merrifield & Ringo, 2014).

Kelabau fish has economic value; the price was so high, but the availability of this fish decreased because of illegal fishing by electricity shock (Kusmini et al.,

S. Rohmy (✉) · R. Hidayat
Research Center for Fishery, National Research and Innovation Agency (BRIN), Jakarta, Indonesia
e-mail: sula013@brin.go.id

I. I. Kusmini
Research Center for Applied Zoology, Research and Innovation Agency (BRIN), Jakarta, Indonesia

M. Mulyasari
Research Centre for Marine and Inland Waters Bioindustry, National Research and Innovation Agency (BRIN), Jakarta, Indonesia

E. Nugroho et al. (eds.), *Aquaculture Development in Archipelago*,
https://doi.org/10.1007/978-981-95-1084-9_12

2011). The degradation of its habitat is also caused by the palm oil industry. This issue has promoted our concern to develop kelabau fish domestication because it is one of the local fish that could potentially be a high economic value commodity. This chapter describes the domestication of kelabau fish, with emphasis on taxonomy and morphology, genetic diversity, and reproduction.

12.2 Taxonomy and Morphology

The taxonomy of kelabau (*O. melanopleurus*) from Indonesia was classified based on Kottelat et al. (1993).

Phylum:	Chordata
Subphylum:	Vertebrata
Subclass:	Neopterygii
Division:	Teleostei
Order:	Cypriniformes
Family:	Cyprinidae
Genus:	*Osteochilus*
Species:	*Osteochilus melanopleurus* (**Bleeker 1852**) (Fig. 12.1)

Kelabau has a similar appearance to Nilem (*Osteochilus hasselti*), but kelabau has a bigger size and can reach up to 1–3 kg. This fish is easy to identify by its mouth, elongated dorsal fin, and green and gray body color with silver and black spotted horizontally in the anterior part. This fish has a rounded and slim shape body. The head part is slightly flat, little mouth, and has an unbroken linea lateralis, a green dorsal part, and a white color in the stomach part. It has two antennas and has a maximum standard length (SL) of 30–35 cm.

Morphological determination was carried out on kelabau fish from waters in Riau Province, namely, the Kampar River, Siak River, and Rokan River. Asiah et al. (2019)) reported that kelabau fish from the Kampar River is morphometrically relatively different from fish from the Siak Rokan and Kampar Rivers. While based on meristic analyses the kelabau fish from the three locations are relatively the same.

The measurement of the morphological character of kelabau fish from Kampar River, Riau Province, has been carried out by Asiah et al. (2019) using truss

Fig. 12.1 Kelabau fish (*Osteochilus melanopleurus*). (Source: Kusmini et al. (2011))

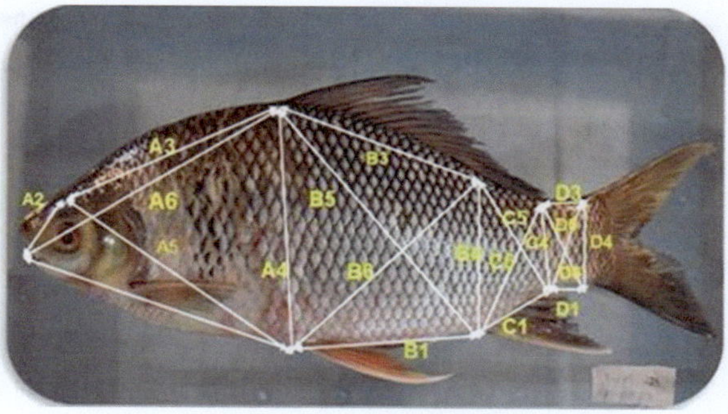

Fig. 12.2 Twenty-one (21) characteristics of morphometric and meristic of kelabau (Asiah et al., 2018, 2019).

morphometric and meristic methods. The measurement of truss morphometric is carried out based on the use of 10 truss points, which are then connected using lines into 21 morphometric characters. These points are (1) the beginning of the ventral fin, (2) the tip of the mouth, (3) the top of the eye, (4) the beginning of the hard dorsal fin, (5) the beginning of the anal fin, (6) the beginning of the soft dorsal fin, (7) the end of the soft dorsal fin, (8) the end of the anal fin, (9) the beginning of the upper caudal fin, and (10) the beginning of the lower caudal fin. Based on these lines, the body of the fish is then divided into four large parts and given codes, namely, A, B, C, and D. Each produces six characters so that from 10 truss points 21 characters are obtained (Fig. 12.2).

Meristic measurements were carried out by counting the number of soft fingers of the dorsal fin (D), the soft fingers of the pectoral fin (P), the soft fingers of the ventral fin (V), fingers of the anal fin (A), vertebrae segments, gills filter bones, number of scales on the sideline, number of scales above the sideline, number of scales below the sideline, number of scales around the body, number of scales of the tail shaft, and number of dorsal scales.

Research by Asiah et al. (2019) showed that the average morphometric character of kelabau fish from the Kampar River ranges from 1.37 to 2.06, with the lowest average value of morphometric character being character C1 while the highest character is character A6 (Table 12.1). Meanwhile, the variation coefficient of morphometric characters of kelabau fish ranges from 1.54 to 14.98 with the lowest variation coefficient in character A6 and the highest in character B3 (Table 12.2).

The meristic character of the kelabau fish from Kampar River, Riau Province, is presented in Table 12.2. Most of the meristic characters showed similarities between the sexes. A slight difference between male and female fish was observed especially in the number of gills, scales on the sideline and scales above the sideline, scales on the tail shaft and dorsal scales (D).

Research on the genetic diversity of kelabau fish was reported by Mulyasari et al. (2010) and Kusmini et al. (2011). Determination of genetic variation in these fish was conducted using the mitochondrial RFLP method (Mulyasari et al., 2010) and the RAPD method (Kusmini et al., 2011). The kelabau fish used in both studies came from West Kalimantan waters precisely in rivers in the Kapuas Hulu, Sekadau, Pontianak, and Sintang regions. The results of RFLP analysis show that kelabau fish

Table 12.1 The average and variation coefficient of morphometric characters of kelabau fish from Kampar River, Riau Province

Morphometric characters	Average[a]	Coefficient of variation
A1	2.03 ± 0.03	1.56
A2	1.47 ± 0.04	3.02
A3	1.95 ± 0.03	1.63
A4	1.94 ± 0.04	1.62
A5	1.97 ± 0.03	1.61
A6	2.06 ± 0.03	1.54
B1	1.83 ± 0.05	2.39
B3	0.91 ± 0.15	14.98
B4	2.03 ± 0.03	1.57
B5	2.05 ± 0.03	1.55
B6	1.93 ± 0.03	1.63
C1	1.27 ± 0.04	3.94
C3	1.91 ± 0.03	1.66
C4	1.66 ± 0.05	3.83
C5	2.05 ± 0.04	2.76
C6	1.77 ± 0.06	3.51
D1	1.47 ± 0.06	3.73
D3	1.63 ± 0.04	2.74
D4	1.53 ± 0.05	2.94
D5	1.78 ± 0.11	4.65
D6	1.68 ± 0.04	2.68

Source: Asiah et al. (2019)

[a]Values transformed according to Palma and Andrade (2002)

Table 12.2 Meristic characters of kelabau fish from Kampar River, Riau Province

Meristic characters	Male	Female
The soft fingers of the dorsal fin (D)	D. II.17–19	D.II.17–19
The soft fingers of the pectoral fin (P)	P. I.15–17	P. I.15–17
The soft fingers of the ventral fin (V)	V. I. 8	V. I.8
The soft fingers of the anal fin (A)	A. II.6.	II.6
Tail fin fingers	VI. 6.11–12	VI. 6.11–12
Vertebrae segments	30	30
Gill filter bones	36–37	36–38
Scales on the sideline	46–50	42–50
Scales above the sideline	9.5–11.5	10.5–11.5
Scales below the sideline	7.5–9.5	7.5–9.5
Scales around the body	40–46	40–46
Tail shaft scales	20–24	22
Dorsal scales (D)	19–21	19–20

Source: Asiah et al. (2019)

in West Kalimantan have an Mt DNA sequence length of 650–700 bp and have low haplotype or genetic variation. The highest variation of haplotype is produced by kelabau fish from Kapuas Hulu. The population of kelabau fish from Sintang is genetically different from the other three locations.

Like the results of RLFP analysis, the genetic variation of kelabau fish analyzed with the RAPD technique is also low. However, the highest genetic variation based on heterozygosity and polymorphic values is the kelabau fish from Sintang. This population is genetically different from kelabau fish from Pontianak and Kapuas Hulu. The latest information related to the genetics of kelabau is about the genetic variation of kelabau fish originating from waters in the Riau Province area. The study was conducted using DNA barcoding techniques using COI primers (Asiah et al., 2020). The study showed that kelabau fish in the waters of Riau Province have low nucleotide diversity, and the diversity of nucleotides in kelabau fish from the Rokan and Siak Rivers is higher than in the Kampar River. The low genetic diversity in kelabau fish both in Riau and West Kalimantan provinces is due to the limited migration of kelabau fish from one location to another, so the chance of gene exchange with other populations is very small.

12.3 Reproduction

Kelabau fish have sexual dimorphism. Male kelabau are slimmer than females. First maturity in female kelabau occurred in 2–8–30 cm total length, with 0.6–2.5 kg of weight, while in male kelabau, it occurred more than 26 cm of total length and 0.5–1.0 kg of weight. Generally, kelabau has a breeding season in the rainy season (December–April), and the peak of reproduction is held in February. The gonadal maturity index of male kelabau in the reared ponds did not significantly occur between March and June, but it rose significantly in October. Female kelabau has 3 steps of maturity. The first step is between March and June, then July–September, and the last step is in October, which is the peak season of breeding both males and females. Each year, there were spawning events outside of these months because there were mature fishes each month (Nasution et al., 2006).

12.4 Domestication

There was kelabau domestication in the Institute for Freshwater Aquaculture (BBAT Mandiangin) and the Research Installation of Freshwater Fish Germ Plasms, Cijeruk, Bogor. Domestication of kelabau fish in Cijeruk started by rearing 11 broodstocks of 529 g body weight, consisting of nine females and two males in a pond of $2 \times 4 \times 1$ m^3. The broodstocks were reared in a flowing water pond. They fed about 3% of total body mass with 30–40% protein (Kristanto et al., 2010).

Mature males were identified by ooze milt freely with the slightest pressure on their belly and females have reddish genital pores and mushy stomachs. The

Fig. 12.3 *Osteochillus kelabau* female. (Source: Juliansyah & dan Idrus, 2014)

breeding of kelabau can be bred by induced breeding using gonadotropins and anti-dopamine such as Ovaprim with 0.5–0.6 ml/kg doses to stimulate ovulation (Fig. 12.3).

The female was injected twice with 8–10 h intervals as amount 1:3 between two injections. The male was injected with 0.5 ml/kg of Ovaprim. Kelabau has a 2:1 male and female mating system. The spawning female will release the eggs at 5–10 h after the second injection. Kelabau spawns with semi-artificial or artificial breeding. In semi-artificial breeding, the spawning pond is equipped with flowing water, while in artificial breeding, eggs and sperm are mixed in one basket and stirred (Fig. 12.4).

The domestication of kelabau in BBAT Mandiangin produced 15,350 seeds, from 46,120 to 92,240 fecundity, and degree of fertilization about 84.23% to 97.92%; hatching rate 35.08–78.94%, survival rate in aquarium 86.30–91.94% (Juliansyah & dan Idrus, 2014). Another hormone used in the induced spawning of kelabau was HCG. Rachimi et al. (2015) stated that kelabau spawning with 900 IU/kg HCG and Ovaprim (1.2 ml/kg) could increase time ovulation and result in 74.53% of fertilization rate, 72.22% of hatching rate and about 36.27% survival rate. Redha et al. (2014) reported that embryo development in *cleavage*, morula, blastula, gastrula, and organogenesis stages were fastest at 32 °C temperature. The highest hatching rate was 88.33% with a 30 °C incubation temperature while hatching time at 32 °C was the fastest. Hatching larvae were reared in an aquarium and fed with *Artemia nauplii* after 4–5 days.

Fig. 12.4 Artificial spawning. (Source: Juliansyah and dan Idrus (2014))

References

Asiah N, Junianto J, Yustiati A, & Sukendi S. (2018). Morfometrik dan meristik ikan kelabau (*Osteochilus melanopleurus*) dari Sungai Kampar, Provinsi Riau.[Morfometric and meristic of Kelabau fish (*Osteochilus melanopleurus*) form Kampar river, Riau Province]. Jurnal Perikanan dan Kelautan, 23(1): 47–56.

Asiah, N., Sukendi, S., Junianto, J., Yustiati, A., & Windarti, W. (2019). *Truss* morfometrik dan karakter meristik ikan kelabau (*Osteochilus melanopleurus* Bleeker, 1852) dari tiga populasi di Sungai Kampar, Sungai Siak, dan Sungai Rokan, Provinsi Riau [Truss morphometric and meristic characters of kelabau fish (*Osteochilus melanopleurus* Bleeker, 1852) from three populations in Kampar, Siak, and Rokan Rivers, Riau Province]. *Jurnal Iktiologi Indonesia, 19*(2), 283–295. https://jurnal-iktiologi.org/index.php/jii/article/view/462

Asiah, N., Junianto, J., Yustiati, A., Sukendi S, Fahmi MR, Muchlisin ZA, Kadapi M, Windarti W (2020). Biometric and genetic differences in kelabau (*Osteochilus* spp.) as revealed using cytochrome c oxidase subunit 1 [version 3; peer review: 3 approved]. *F1000Research, 8*, 177. https://doi.org/10.12688/f1000research.17319.3

Juliansyah, Noor, M., and Idrus, M. I. (2014). Aspek bioekologi Reproduksi Ikan Kelabau (*Osteochilus melanopleurus* Bleeker) Sebagai Potensi Akuakultur untuk Mendukung Peningkatan Produksi Perikanan Budidaya. [Bioecology reproduction aspects of Kelabau fish (*Osteochilus melanopleurus,* Bleeker) for aquaculture potential to developing aquaculture production]. *Jurnal Balai Budidaya Air Tawar Mandiangin*, 2. https://bbatmandiangin.wordpress.com/2014/10/13/jurnal-aspek-bioeko-reproduksi-ikan-kelabau-osteochilus-melanopleurus-bleeker-sebagai-potensi-akuakultur-untuk-mendukung-peningkatan-produksi-perikanan-budidaya/

Kottelat, M., Whitten, A. J., Kartikasari, S. N., & Wirjoatmodjo, S. (1993). *Fresh water fish of western Indonesia and Sulawesi* (p. 293). Periplus Edition (HK) Ltd.

Kristanto, A. H., Asih, S., & Rasidi. (2010). Domestication of Kalabau fish (*Osteochilus melanopleura* Blkr) support for increasing of freshwater fish farming. In *Proceeding Forum Aquaculture Innovation Technology (FITA)* (p. 18). Fisheries Research Center.

Kusmini, I. I., Gustiano, R., & Mulyasari. (2011). Karakterisasi genetik ikan kelabau (*Osteochilus kelabau*) dari berbagai lokasi di Kalimantan Barat menggunakan metode RAPD (Random Amplified Polymorphism DNA). [Genetic characterization of Kelabau Fish (*Osteochilus kelabau*) from various locations in West Borneo using RAPD (*Random Amplified Polymorphism* DNA) method]. *Jurnal Berita Biologi, 10*(4), 449–454. https://biologyjournal.brin.go.id/index.php/berita_biologi/article/view/762/534

Merrifield, D., & Ringo, E. (2014). *Aquaculture nutrition: Gut health, probiotic and prebiotic* (p. 465). Wiley-Blackwell, John Wiley & Sons Ltd..

Mulyasari, I., Kristanto, A. H., & Huwoyon, G. H. (2010). Analisis variasi genotipe ikan Kalabau (*Osteochilus kalabau*) dengan metode Mitokondria-Restraction fragment length polymorphism (RFLP). [Genotype variation analysis of Kelabau fish (*Osteochilus kelabau*) using mitochondrial restriction fragment length polymorphism (RFLP) method]. *Jurnal Riset Akuakultur, 5*(1), 43–51. ISSN 1907-6754/e-ISSN 2502-6534. http://ejournal-balitbang.kkp.go.id/index.php/jra/article/view/2207

Nasution, S., & Nuraini. (2004). *Ekologi dan Kebiasaan Makan Ikan Kelabau dari Sungai Kampar*. [The ecology and feeding habits of Kelabau Fish from Kampar River]. *Research reports*. Universitas Riau.

Nasution, S., Nuraini, & Hasibuan, N. (2006). Potensl akuakultur ikan Kelabau (*Osteochilus Kelabau*) dari perairan Kabupaten Pelalawan, Propinsi Riau: Siklus Reproduksi. [The aquaculture potential of Kelabau fish (*Osteochilus kelabau*) from Pelalawan Waters, Riau Province: Reproduction cycle]. In *Fish National Seminar Proceeding IV* (pp. 301–308).

Nayak, S. K. (2010). Probiotics and immunity: A fish perspective: A review. *Fish & Shellfish Immunology, 29*, 2–14.

Palma, J., & Andrade, J. P. (2002). Morphological study of *Diplodus sargus*, *Diplodus puntazzo*, and *Lithognathus mormyrus* (Sparidae) in the Eastern Atlantic and Mediterranean Sea. *Fisheries Research, 57*(1), 1–8.

Rachimi, Raharjo, E. I., & Sudarsono, A. (2015). Pengaruh konsentrasi penyuntikan HCG dan ovaprim terhadap daya tetas telur dan sintasan larva ikan kelabau (*osteochilus melanopleura* blkr.). [Effect of concentration HCG hormone and Ovaprim on hatching hate and survival rate of Kelabau (*Osteochilus Melanopleura* Blkr.)]. *Jurnal Ruaya, 5*, 11–17. FPIK-Universitas Muhammadiyah Pontianak. ISSN 2541 – 3155.

Redha, A. R., Raharjo, E. I., & Hasan, H. (2014). Pengaruh suhu yang berbeda terhadap perkembangan embrio dan daya tetas telur ikan kelabau (*Osteochilus melanopleura*). [The effect of different temperature on embryonic development and hatching rate of Kelabau fish (*Osteochilus melanopleura*)]. *Jurnal Ruaya, 4*. Fakultas Perikanan dan Ilmu Kelautan, Universitas Muhammadiyah Pontianak. ISSN 2541 – 315. https://openjurnal.unmuhpnk.ac.id/JR/article/view/481

Roberts, T. R. (1989). The freshwater fishes of Western Borneo (Kalimantan Barat, Indonesia). *Memoirs of the California Academy of Sciences, 14*, 1–210. https://www.biodiversitylibrary.org/part/51675

Saanin, H. (1984). *The taxonomy and fish identification* (p. 508). Bina Cipta.

The Latest Development of Carp Farming in Indonesia

13

Didik Ariyanto, Erma Primanita Hayuningtyas, Suharyanto, and Listio Dharmawantho

13.1 Introduction

Common carp (*Cyprinus carpio*) culture in Indonesia is one of the oldest freshwater fish farming activities. Carp cultivation began in the mid-nineteenth century, since the introduction of this fish from Europe and China. The ease of rearing methods makes this commodity quickly popular and spread to almost all parts of Indonesia. Areas known as common carp farming centers include North Sumatra, West Sumatra, West Java, North Sulawesi, South Sulawesi, and West Nusa Tenggara. Common carp farming technology has also developed from the traditional system in the yard to super-intensive technology in running-water ponds and floating nets.

As a popular commodity, common carp farming has experienced a setback in terms of production due to outbreaks of diseases, namely *Aeromonas hydrophila* and *Koi herpesvirus* (KHV). Another obstacle in the development of carp production is the low quality of the seeds. This is due to the ease with which common carp breed so that the production of fry and brood stock becomes uncontrolled. Uncontrolled conditions for a long time resulted in low-quality seeds. However, several attempts have been made to overcome these problems, including through improvements in cultivation technology and improving the genetic quality (breeding) of carp. Until now, common carp is one of the most important commodities in Indonesia's freshwater fish farming industry.

D. Ariyanto (✉) · E. P. Hayuningtyas · Suharyanto · L. Dharmawantho
Research Center for Fishery, Research Organization for Earth Sciences and Maritime, National Research and Innovation Agency (BRIN), Jakarta, Indonesia
e-mail: didi031@brin.go.id

E. Nugroho et al. (eds.), *Aquaculture Development in Archipelago*,
https://doi.org/10.1007/978-981-95-1084-9_13

13.2 Taxonomy and Morphology

Based on Sa'anin (1968) in his book "Keys for Fish Identification," common carp that developed in Indonesia are classified as *Cyprinus carpio*, with the following morphological characteristics:

Phylum: Chordata
Class: Actinopterygii
Ordo: Cypriniformes
Family: Cyprinidae
Genus: *Cyprinus*
Species: *Cyprinus carpio* Linnaeus, 1758

Morphologically, common carp have a body shape that is slightly elongated and slightly compressed to the side (compressed) and has a gill cover or operculum (Fig. 13.1). Most of the carp's body is covered with scales. The scales on common carp are relatively large and are classified as cycloid or circular scales that are arranged in an orderly manner. The color of the scales on common carp varies according to race. Some are yellowish green, blackish gray, red, orange, and golden yellow, or a combination of these colors. Common carp have a complete lateral line or line that is in the middle of the body. The carp's body is bilaterally symmetrical, meaning that the body of this fish is the same size on the right and left. This can be proven when we split the carp's body into two parts. Thus, it will be seen that the right and left sides of the body look symmetrical or the same size. Common carp can grow up to 120 cm long and weigh up to 60 kg.

The carp's muzzle is located at the middle end (terminal) and can be popped up (protractile). We can see this kind of movement clearly when the fish tries to reach for the food in front of it. On the anterior, there are two pairs of antennae (Misae) and not serrated. Inside the mouth, there are pharyngeal teeth, which are shaped like the esophagus. This dental composition consists of three rows of molars. In addition to the mouth, common carp also have a nose as a differentiating identity from other types of fish. The common carp's nose is right at the top of the head.

The common carp has many fins that grow on several parts of its body, one of which is on the back which is elongated and is located opposite the surface of the pelvic fins. At the front of the fin, there is a kind of hard bone or thorn. Common

Fig. 13.1 The common carp (personal documentation)

carp also have a pair of symmetrical pectoral and pelvic fins. The common carp's anal fin is relatively the same as its dorsal fin, which has hard fingers with serrated edges. In addition, there is also a caudal fin, which is of the homocercal type because it consists of two branches of the same length.

13.3 Reproduction

Common carp are grouped into two types, namely males and females. The physical characteristics of male common carp are having a body that looks slimmer, moves faster, and, when stripped, releases white liquid (sperm). The reproductive organs in fish are often called gonads. The male fish gonads consist of a pair of testicles that are generally white or yellowish. The location of the testicles hanging on the wall in the middle of the abdominal cavity (abdomen). The function of the testes is to produce spermatozoa. The characteristics of a female carp are that the body on the abdomen looks distended or enlarged, movements tend to be slower, and when stripping is done, it will emit a yellow liquid. The reproductive organs or gonads of female fish consist of clear, reddish, jellylike ovaries. The ovaries are located near the intestines, filling two-thirds of the abdominal cavity or almost covering the organs of the body. The ovaries function to produce eggs or ova.

The reproductive system of carp is oviparous, namely sexual reproduction which is characterized by the release of male spermatozoa and female egg cells, and fertilization occurs outside the body. Naturally, common carp can reproduce by external fertilization. The fertilized eggs will attach to the aquatic substrate or aquatic plants in the waters. In cultivation activities, common carp hatchery is carried out naturally, semi-artificially, or fully artificially with the help of spawning stimulating hormones. Natural spawning is carried out by pairing mature male and female fish in spawning tanks with a male and female ratio between 2:1 and 3:1. Semi-artificial spawning was carried out by injecting male and female fish using the spawning stimulating hormone at a dose of 0.15 mg/kg male and 0.3 mg/kg female. Furthermore, male and female fish are paired in spawning tanks with the ratio of males and females equal to natural spawning. The third method is through fully artificial spawning, namely by stripping male and female fish after being injected with spawning stimulating hormone. Stripping males to produce sperm and females to produce eggs is carried out 6–12 h after injection.

The eggs are then fertilized with sperm that has been diluted with NaCl solution. Fertilization is done by mixing the sperm into a container containing the egg and stirring it with a soft object such as a cleaned chicken feather. Furthermore, the fertilized egg is attached to the egg attachment media. This media can be in the form of dry grasses, roots of aquatic plants or artificial media commonly called "kakaban," which is made from coconut tree fibers held together with bamboo clamps. The egg attachment media is placed in the egg-hatching tub until it hatches into fish larvae.

Common carp eggs are round and transparent, measuring 1.5–1.8 mm in diameter and weighing 0.17–0.20 mg. Egg size varies according to the parent's age, size, or weight. The eggs will hatch and develop into larvae within 2–3 days of fertilization. Common carp larvae have rather big yolk sacs, which act as food stores for them. The yolk sac will deplete within 2–4 days. Common carp larvae are connected and migrate vertically. Larval size ranges from 0.5 to 0.6 mm and weighs between 18 and 20 mg. Larvae develop into "kebul" (end-stage larvae) in 4–5 days. Common carp in the "kebul" stage requires an external source of food to survive. Rotifers, Moina, and Daphnia are the most common zooplankton that provide "kebul" natural food. The natural feed requirement for "kebul" in 1 day is approximately 60–70% of its weight.

After 2–3 weeks, the whitefly develops into a fry that measures 1–3 cm and weighs 0.1–0.5 g. After 2–3 weeks, the fry develops into seeds, measuring 3–5 cm long and weighing 0.5–2.5 g. The fish will grow to around 100 g in 2 months. After 6 months, the male fish can weigh up to 500 g. Meanwhile, the female parent can weigh 1.5 kg after 15 months. At that age, male common carp are ready to breed both naturally and artificially.

13.4 Domestication and Distribution

Common carp are not endemic to Indonesia, but are introduced from other countries for cultivation purposes. Common carp were first introduced in Europe and China in the 1800s. The purpose of this introduction is to meet the needs of European and Chinese nations for their protein needs. In 1860, evidence was found that common carp hatchery had been carried out in Galuh Village, Ciamis, with kakaban as a medium for attaching eggs. Carp culture in the area had also been carried out several years before.

In 1892, the first common carp were brought from Java to Sumatra and then developed in Bukittinggi (West Sumatra) and then developed in North Sumatra, especially in Lake Toba. In 1895, carp were brought and developed in Lake Tondano, Minahasa region, in North Sulawesi. The next development was in the South Bali area, precisely in Tabanan regency in 1903, Ende (Flores, NTT) in 1932, and in 1935 also developed in South Sulawesi.

Several years earlier, in 1927, the Dutch East Indies Government again imported a type of carp from the Netherlands, namely the Galician type (elephant carp). This Galician carp has a better taste, denser meat, fewer spines, and faster growth compared to carps that have developed in Indonesia before. In 1930 another carp was also imported, namely the Frankisia carp (glass carp).

During the independence era, namely in 1974, Indonesia imported fancy carp from Taiwan, Germany, and Japan. During its development, it became clear that maintaining the purity of the carp was challenging because it mingled with previous carp races in Indonesia, leading to crosses and the formation of new races. In 1977, Indonesia imported Yamato and Koi carp from Japan. The primary goal of introducing this variety of common carp is as a decorative fish in both aquariums and ponds.

Until now, the common carp is almost available in all regions of Indonesia. Several regions have become production centers of common carp, such as West

Java, West Sumatra, North Sumatra, Riau, South Sulawesi, North Sulawesi, and West Nusa Tenggara. These areas have developed into production centers because they are supported by suitable natural conditions, availability of feed ingredients, community preferences, and regional policies, which play an important role in assisting fish farmers in carrying out their business.

13.5 Development of Common Carp Culture

13.5.1 Culture Technology

Carp culture was originally carried out in the yard of the house. Usually, common carp farming is used as a part-time business for the local people to meet their daily side dishes. If there is excess, the proceeds are sold to supplement the family income. To increase yields, farmers usually increase the number of ponds used, depending on the availability of land they have. The cultivation technology used is relatively low, with relatively few inputs, such as feed, medicines, and others. This cultivation model is known as extensive cultivation.

Subsequent developments show that the harvest of carp is starting to be accepted in the market. This makes common carp farming grow. However, due to limited land, farmers are increasing their cultivation business by improving their cultivation technology. The first thing that was increased was the amount of seed stocking density in the rearing pond, which was 2–3 individuals/m to 10–12 individuals/m. Increasing the number of seeds used in aquaculture ponds requires a greater supply of fish feed. Therefore, farmers try to provide additional feed, ranging from leftover household food to self-made feed with locally available ingredients such as soybean meal, bran, leaves, and others.

The popularity of common carp as a food fish is getting higher. This is evidenced by the level of national carp production reaching the highest percentage in national freshwater fish farming in 1996–2000. The high demand for carp meat, especially in areas that are common carp centers, such as in West Java, West Sumatra, and North Sumatra, demands the use of high technology for the production process in intensive system carp culture. Fish culture with an intensive system with high stocking density and feeding rate affects several factors such as fish size, feed, and water quality (Serlina & Dahlia, 2022). The application of new technologies in the field of construction and fish culture models, feed, use of seeds, medicines, and others need to be included as new aquaculture inputs so that the yields obtained are higher. New fish culture models and constructions, such as the use of streams in running-water ponds and floating net cages in reservoirs, lakes, and other public waters, are alternative common carp farming systems. High stocking densities in intensive cultivation systems (100–200 fish/m^3) require the use of good artificial feed, high-quality seeds, as well as adequate medicines and technology, such as the addition of waterwheels and blowers. With the addition of quite a lot of aquaculture inputs, effective aquaculture management needs to be applied to the system. Mismanagement can lead to the failure of common carp harvests with intensive systems.

In addition to the above fish farming technologies, several technological innovations have also developed, such as bio floc technology, RAS (recirculating aquaculture system) technology, and aquaponics on limited lands. Bio floc or microbial floc are heterotrophic bacteria that are rich in protein and can be used as feed substitutes for shrimp and fish (Gunarto et al., 2009). Common carp farming technology through the application of bio floc with *Bacillus* sp. is used to suppress the incidence of KHV (Koi Herpes Virus) in common carp. At a bacterial floc density of 106–107 cfu/mL, the bio floc treatment could control the incidence of post-KHV disease by 0% or in the absence of KHV cases, compared to control common carp without bio floc treatment with an incidence of 83.3%. Maintenance with the application of bio floc also had a higher survival of 54.48% compared to the recirculation treatment (36.08%) and control (36.96%) (Hayuningtyas et al., 2014). This indicates that the application of bio floc, besides being able to increase immunity to disease, can also increase the survival of carp rearing.

Water quality management is a determinant of success in fish rearing with several parameters contained in it in the form of DO, PH, temperature, brightness, and other parameters (Indriati & Hafiludin, 2022). Fish culture technology that can be used to control the fish-rearing environment is carried out with the RAS system using a filter. The use of natural filters such as zeolite, coconut coir, and charcoal in neutralizing pH and ammonia (Nasir & Khalil, 2016). Another purpose of using natural filters is to serve as a place for microorganisms such as bacteria, algae, fungi, and protozoa to attach (Nababan et al., 2020). The use of natural filters in common carp rearing using activated charcoal, zeolite, coconut fiber, and a combination of the three resulted in lower feed conversion than the control. While the highest survival is in the treatment of coconut coir filters, reaching 100%, the best growth is in filter conditions with a combination of the three (Nasir & Khalil, 2016). This proves that the use of natural filters can control water quality thereby affecting growth, survival, use of feed, and good environmental conditions for the fish life cycle.

In urban communities, aquaponics technology is a recirculation technique that can get around the use of increasingly narrow cultivation land, increase fish stocking, and overcome water scarcity (Darwis et al., 2019). Efforts to control water quality with the application of biofilters can be carried out using aquatic plants, namely kale (*Brassica oleracea*). Maintenance of carp seeds at various densities in rearing media using a watercress biofilter resulted in the best growth (length and weight) and high survival rates at the highest density treatment, namely 25 fish/30 L of water (Serlina & Dahlia, 2022). The survival rate of common carp reared in aquaponic biofilter media is higher than in conventional systems without aquaponics. This is because biofilters play a role in controlling water quality in fish-rearing media (Nugroho et al., 2012).

13.5.2 The Types of Cultured Common Carp

Common carp can be grouped into two categories, namely food fish and ornamental fish. Carp as a food is cultured to meet food needs, especially nutrition or protein derived from animals. As an ornamental fish, the common carp serves as a decoration to fulfill human inner needs. Here are some types of consumption carp and ornamental fish that are commonly found in Indonesia.

13.5.2.1 Majalaya Carp

As the name implies, this common carp comes from the Majalaya area, Bandung Regency in West Java. The body is rather small, and the back is higher and sharper than in other common carp races. The ratio of length to height is 3.2:1. The character of this common carp is relatively docile and can swim on the surface of the water. The scales are gray green with dark edges, with the exception of the yellowish undersides of the gills and caudal fins. The fish scales become darker as they move further back. Majalaya carp have several advantages, including very rapid growth, resistance to *Aeromonas hydrophila* bacterial infection, and a nice and excellent flavor. Majalaya carp have a high fecundity or egg production rate, with 84,000–110,000 eggs per kg of brood stock.

13.5.2.2 Punten Carp

This strain was initially discovered in 1933 in Punten Village, Malang, East Java. The body is very short, but the dorsal side is tall and wide. As a result, the carp's body is short and round, with a large belly. The ratio of total length to body height is 2.3–2.4:1. The scales are dark green, the eyes are striking, and the body moves slowly and calmly.

13.5.2.3 Sinyonya Carp

The exact origin of this fish's name is unknown. Nonetheless, it is assumed that the Sinyonya carp strain was discovered for the first time in the Tasikmalaya region of West Java (Khairuman & Amri, 2008). The body shape is extended (long bodied), and the back is shorter than in Majalaya and Punten carp. The ratio of length to height is approximately 3.66:1. The scales are pale yellow. Fish's eyes are large while they are young, but narrow as they age. The Sinyonya carp is calmer than the Punten carp. Sinyonya carp swim on the water's surface. The fertility or number of Sinyonya carp eggs ranges from 85,000 to 125,000, with a dimension of 0.3 to 1.5 mm. The male Sinyonya carp matures first at 8 months, followed by the female at 18 months. This common carp is relatively resistant to Myxospora illness.

13.5.2.4 Red Carp

This carp's distinctive golden red scales set it apart. The movement is dynamic, not passive, and enjoys stirring the bottom of the pond. The body form is relatively elongated. It had a sharper and lower back than the Sinyonya strain. Its eyes are very striking.

13.5.2.5 Kumpay Carp

The striking feature of the Kumpay (tufted: Sundanese) carp is that all its fins are long and tufted so that they look beautiful when they are moving. The color of the scales varies, there are white, yellow, dark green, and red. Its body shape is elongated like Sinyonya's common carp. Kumpay carp growth is relatively slow. Because of its unique shape, this common carp is included in the ornamental fish group.

13.5.2.6 Kancra Domas Carp

The body shape is elongated, and the movement is not tame but very active. The scales are small and irregular in shape. The color of the scales varies, some are blue, green, or brown. The dorsal scales are dark. The further toward the belly, the lighter the golden or silvery color. This carp is also included in the ornamental fish group.

13.5.2.7 Glass Carp

This fish is remarkable in that it does not have scales on all its body. The side that is not coated with scales appears clear at first glance, like glass. Shiny white scales can be found along the lateral line (linea lateralis) and around the base of each fin. The scales vary in size and are huge.

13.5.2.8 Koi Carp

Koi carp are imported from Japan. This fish was found in Indonesia around 1980. Their body shape is round and elongated. The color of the scales varies, there are black, white, red, yellow, glowing red, or a combination of several of these colors. Koi hobbyists usually like the hybrid type Koi fish because of their beautiful color and spot pattern. Hobbyists like koi fish because their movements are quite tame and slow. Koi fish have various names according to the color and color pattern of the body; for example, Shusui Nishikigoi, Platinum Nishikigoi, Kohaku Taisho, Nishikigoi, Sushi Nishikigoi, and Sanshoku Nishikigoi (Sugiharto & Efendi, 2017).

13.5.3 Types of Genetically Improved Farmed Fish

Problems in the growth of common carp aquaculture require considerable attention. The low genetic quality of seeds requires prompt genetic enhancement through breeding efforts. Several breeding methods can be carried out, ranging from selective breeding, hybridization, and chromosome engineering to nucleic acid-level DNA engineering. Common carp have been selected from a variety of fish populations, including Rajadanu, Majalaya, Cangkringan, and composite common carp. The selective breeding program has produced several high-yielding varieties such as the Mantap (MMAF, 2015) and Jayasakti (MMAF, 2016a) for the Majalaya population, Mustika (MMAF, 2016b) and Super Rd common carp (MMAF, 2016c) for the Rajadanu, Najawa (MMAF, 2014) for the Cangkringan population, and Marwana carp (MMAF, 2016d) for the composite population. Apart from selective breeding, hybridization activities have also produced several populations of superior carp candidates, as was done by Ath-thar et al. (2011) and Ariyanto et al. (2020). Carp breeding through chromosome and DNA engineering has been carried out, among others, by Mukti (2005, 2007) through carp polyploidy and Syahputra et al. (2016) through the establishment of transgenic common carp. However, until now there has not been a population of superior genetically engineered carp used in aquaculture activities. The many and complex requirements to produce genetically engineered products, especially transgenic products, are thought to result in a longer production time for these products. Below are presented several types of genetically modified common carp that have been cultivated by farmers.

13.5.3.1 Mantap Carp

The Freshwater Culture Development Center in Sukabumi selected this excellent carp from the Majalaya common carp. The main advantage of Mantap carp is resistance to disease infections. Apart from that, Mantap carp also have higher tillers, reaching 140,000 fry with a survival rate of 80–90%. The female brood stock begins laying eggs at the age of 15 months and can reproduce six times. After being unproductive, the brood stock, which weighed 6 kg, still had high economic value and was hunted by fishing entrepreneurs as baboons. Mantap carp's feed conversion ratio is 1:2 (Fig. 13.2) (MMAF, 2015; Ridwantara et al., 2019).

13.5.3.2 Super RD Carp

This exceptional carp is a common carp family selection with the primary advantage of rapid growth. The fish selected by the Freshwater Fish Research and Development Center in Bogor have an efficient feed conversion ratio of 1:1–1:2. Super RD carp may thrive in ponds with salinities ranging from 0 to 12 g/L, with 5 g/L being the most suited. Super RD carp thrive in ponds with a dissolved oxygen content of 2.5–7.0 mg/L and an acidity level of approximately 5–6. The optimal temperature range is 28–32 °C. Adaptive super RD carp are cultivated in many types of ponds, including earthen ponds, stagnant water ponds, running-water ponds, and floating cages or net cages (Fig. 13.3) (MMAF, 2016c).

13.5.3.3 Mustika Carp

This fish, selected using molecular markers, offers significant benefits in disease resistance and growth rate. Mustika carp have a survival value percentage (SR) of

Fig. 13.2 Male and female brood stock of Mantap carp. (Source: Ministerial Decree No.24/2015 (MMAF, 2015))

Fig. 13.3 The brood stock of super RD carp. (Source: Ministerial Decree No.23/2016 (MMAF, 2016c))

Fig. 13.4 Male and female brood stock of Mustika carp. (Source: Suharyanto et al., 2019)

98.89%, which is higher than other carp species. This fish grows quickly and can tolerate a variety of situations. The feed conversion ratio of 1.24 is adequate to reduce organic waste and residual feed in the water. Mustika carp can gain up to 3.62% of their body weight per day. This shortens the harvest time to roughly 2.5 months, particularly for raising in floating net cages. Mustika carp have a higher output at the raising stage than other carp fish, reaching 67%. Mustika carp promotes the production of high-yielding common carp seeds that are resistant to the Koi herpes virus. Since 2015, Mustika carp brood stock have been dispersed to several production locations, including Central Java, West Java, Banten, Yogyakarta, West Sumatra, Riau, Lampung, and West Nusa Tenggara (Fig. 13.4) (Ariyanto et al., 2019, 2020; Suharyanto et al., 2019).

13.5.3.4 Marwana Carp

The Wanayasa Freshwater Fish Culture Center used molecular marker selection to generate Marwana carp. Marwana carp is a synthetic population of many common carp strains developed by farmers. The primary advantage of the Marwana carp is its rapid development rate and illness resistance (Fig. 13.5) (Fajarwati & Andriani, 2022; MMAF, 2016d).

13.5.3.5 Jayasakti Carp

Jayasakti carp is a common carp selected by the Jambi Freshwater Culture Center for disease resistance coding markers. The primary benefits of this new species of fish are disease resistance and high output (Fig. 13.6) (MMAF, 2016a).

13.5.3.6 Najawa Carp

The Marine and Fisheries Technology Development Center in Yogyakarta created the molecular markers used to select this superior common carp. The Cangkringan

Fig. 13.5 Male and female brood stock of Mawarna carp. (Source: Ministerial Decree No.27/2016 (MMAF, 2016d))

Fig. 13.6 Male and female brood stock of Jayasakti carp. (Source: Ministerial Decree No.25/ 2016 (MMAF, 2016a))

red carp provided most of the genetic material used to create this breed. The primary advantages of Najawa carp are their rapid development, relative disease resistance, tolerance to a variety of habitats, and high economic potential. The high economic value of this carp is due to its red-light color. Some Chinese people believe it is a fish that brings luck or good fortune. The farmers cultivate Najawa carp to produce new breeders to meet market demand. The male brood-stock fish have 0.5 kg in weight and 30 cm in length, while the females are heavier, around 1 kg in weight. Najawa carp is an endemic fish to Cangkringan, Yogyakarta (Fig. 13.7) (Iswarni, 2018).

13.6 National Production of Common Carp

National carp production has experienced ups and downs during its development. The peak of production in the first period occurred in 2001 with a total production of 418,910 tons. The next period was the gloomy period that occurred between 2002 and 2010. Apart from being triggered by a decrease in seed quality, the decline in national carp production was also caused by an outbreak of a disease caused by the herpesvirus, also known as the Koi herpesvirus (KHV). However, carp production

Fig. 13.7 Male and female brood stock of Najawa carp. (Source: Ministerial Decree No.41 (MMAF, 2014))

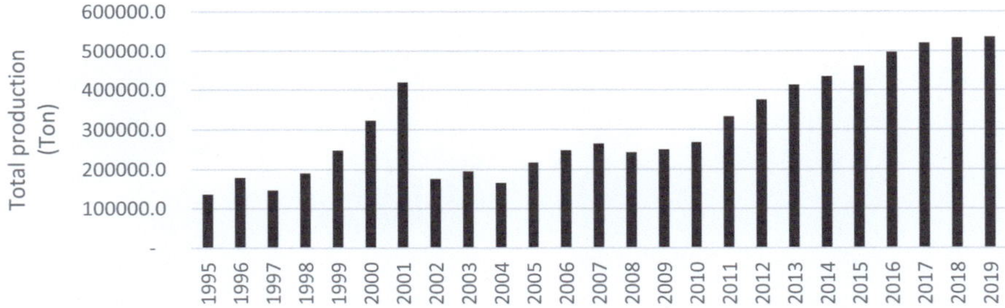

Fig. 13.8 National production of common carp from 1995 to 2019. (Source: Research personal data)

began to show its strength starting in 2011. Since then, national carp production has shown a consistent increase until now. This is strongly influenced by the use of carp fish obtained from breeding which grow fast, are disease-resistant, and have a relatively low food conversion ratio (FCR). In more detail, carp production from 1995 to 2019 is presented in Fig. 13.8.

References

Ariyanto, D., Himawan, Y., Syahputra, K., Palimirmo, F. S., & Suharyanto, S. (2019). Performa pertumbuhan dan produktivitas ikan mas strain mustika pada uji multi lokasi. [Growth and productivity performances of common carp, strain Mustika cultured in multi-locations conditions]. *Jurnal Riset Akuakultur, 14*(3), 139. https://doi.org/10.15578/jra.14.3.2019.139-144

Ariyanto, D., Syahputra, K., Himawan, Y., Sih Palimirmo, F., & Haryadi, J. (2020). Adaptabilitas dan stabilitas produksi ikan mas mustika di lingkungan budidaya berbeda. [Adaptability and production stability of common carp strain "Mustika" cultured in different farming systems and sites]. *Jurnal Riset Akuakultur, 15*(4), 221–227.

Ath-thar, M. H. F., Prakoso, V. A., Nugroho, E., & Gustiano, R. (2011). Heterosis, maternal, dan individual effect pada hibrida antara ikan mas Rajadanu, Majalaya, Subang, dan Kuningan. [Heterosis, maternal, and individual effect on hybridized carp of Rajadanu, Subang, Majalaya, and Kuningan]. *Jurnal Riset Akuakultur, 6*(3), 407. https://doi.org/10.15578/jra.6.3.2011.407-412

Darwis, D., Mudeng, J. D., & Londong, S. N. J. (2019). Budidaya ikan mas (*Cyprinus carpio*) sistem akuaponik dengan padat penebaran berbeda. [Aquaponic culture system of carp, *Cyprinus carpio*, at different stocking densities]. *E-Journal BUDIDAYA PERAIRAN, 7*(2), 15–21. https://doi.org/10.35800/bdp.7.2.2019.24148

Fajarwati, M., & Andriani, Y. (2022). Teknik pembenihan ikan mas (*cyprinus carpio*) di UPTD Balai Benih Ikan (BBI) Cimaja, Kabupaten Sukabumi, Jawa Barat. [Hatchery techniques of carp (*Cyprinus carpio*) in UPTD Balai Benih Ikan (BBI) Cimaja, Sukabumi District, West Java]. *Junal Media Akuakultur Indonesia, 2*(2), 86–98. https://journal.unram.ac.id/index.php/jmai/index

Gunarto, G., Mansyur, A., & Muliani, M. (2009). Aplikasi dosis fermentasi probiotik berbeda pada budidaya udang vaname (*Litopenaeus vannamei*) pola intensif. [Application of different doses of probiotic fermentation in intensive pattern vannamei shrimp (*Litopenaeus vannamei*) cultivation]. *Jurnal Riset Akuakultur, 4*(2), 241–255.

Hayuningtyas, E. P., Gunadi, B., & Ariyanto, D. (2014). Teknologi budidaya ikan mas (*Cyprinus carpio*) tahan KHV (Koi Herpes Virus) melalui aplikasi bioflok. [The cultivation technology of carp (*Cyprinus carpio*) is resistant to KHV (Koi Herpes Virus) through bio-floc application]. In *Prosiding Seminar Nasional Ikan Ke 8, Anonymous* (Vol. 2010, pp. 97–107).

Indriati, P. A., & Hafiludin, H. (2022). Manajemen kualitas air pada pembenihan ikan nila (*Oreochromis niloticus*) di Balai Benih Ikan Teja Timur Pamekasan. [Management of water quality for tilapia (*Oreochromis niloticus*) hatching fish at Balai Benih Ikan Teja Timur Pamekasan]. *Juvenil:Jurnal Ilmiah Kelautan Dan Perikanan, 3*(2), 27–31. https://doi.org/10.21107/juvenil.v3i2.15812

Iswarni, D. W. (2018). *Pertumbuhan ikan mas najawa (cyprinus carpio linn.) pada beberapa konsentrasi fermentasi kulit ari kedelai dalam pakan pelet komersil. [Growth of Najawa carp (Cyprinus carpio Linn.) in some soybean fermentation concentrations in commercial pellet feed]* (pp. 1–11). Fakultas Sains Dan Teknologi Terapa Universitas Ahmad Dahlan.

Khairuman, & Amri, K. (2008). *Buku Pintar Budi Daya 15 Ikan Konsumsi. [The smart book of cultivating 15 fish for consumption]* (1st ed.). Agro Media.

MMAF. (2014). *Keputusan Menteri Kelautan dan Perikanan Nomor 41/KEPMEN-KP/2014 tentang Pelepasan Ikan Mas Merah Najawa*. [Decree of The Minister of Marine Affairs and Fisheries of Republic Indonesia No. 41/Kepmen-KP/2014, The Release of The Red Carp of Najawa]. https://jdih.kkp.go.id/Homedev/DetailPeraturan/5594

MMAF. (2015). *Keputusan Menteri Kelautan dan Perikanan Nomor 24/KEPMEN-KP/2015 tentang Pelepasan Ikan Mas Mantap*. [Decree of The Minister of Marine Affairs and Fisheries of Republic Indonesia No. 24/KEPMEN-KP/2015, The Release of Mantap Carp]. https://jdih.kkp.go.id/Homedev/DetailPeraturan/5129

MMAF. (2016a). *Keputusan Menteri Kelautan dan Perikanan Nomor 25/KEPMEN-KP/2016 tentang Pelepasan Ikan Mas (Cyprinus carpio) Jayasakti* [Decree of The Minister of Marine Affairs and Fisheries of Republic Indonesia No. 25/Kepmen-Kp/2016, The Release of Jayasakti Carp]. https://jdih.kkp.go.id/Homedev/DetailPeraturan/4878

MMAF. (2016b). *Keputusan Menteri Kelautan dan Perikanan Nomor 24/KEPMEN-KP/2016 tentang Pelepasan Ikan Mas (Cyprinus carpio) Rajadanu Tahan Penyakit KHV*. [Decree of the Minister of Maritime Affairs and Fisheries Number 24/KEPMEN-KP/2016 concerning the Release of Rajadanu carp (*Cyprinus carpio*) Resistant to KHV Disease. https://jdih.kkp.go.id/Homedev/DetailPeraturan/4877

MMAF. (2016c). *Keputusan Menteri Kelautan dan Perikanan Nomor 23/KEPMEN-KP/2016 tentang Pelepasan Ikan Mas (Cyprinus carpio) Rajadanu Super RD*. [Decree of The Minister of

Marine Affairs and Fisheries of Republic Indonesia No. 23/Kepmen-KP/2016, The Release of Super RD Carp]. https://jdih.kkp.go.id/Homedev/DetailPeraturan/4876

MMAF. (2016d). *Keputusan Menteri Kelautan dan Perikanan Nomor 27/KEPMEN-KP/2016 tentang Pelepasan Ikan Mas (Cyprinus carpio) Marwana.* [Decree of The Minister of Marine Affairs and Fisheries of Republic Indonesia No. 27/KEPMEN-KP/2016, The Release of Marwana Carp. https://jdih.kkp.go.id/Homedev/DetailPeraturan/4880

Mukti, A. T. (2005). Differences in the success rate of polyploidization of carp (*Cyprinus carpio* Linn.) through heat shock. *Berkala Penelitian Hayati, 10*(2), 133–138. https://doi.org/10.23869/bphjbr.10.2.20058

Mukti, A. T. (2007). Comparison of growth and development of diploid and tetraploid carp (*Cyprinus carpio* Linn.) gonads. *Berkala Penelitian Hayati, 13*(1), 27–32. https://doi.org/10.23869/bphjbr.13.1.20074

Nababan, D., Sitorus, M. E. J., Brahmana, N. E. B., & Silitonga, E. M. (2020). Kemampuan biofilter anaerob berdasarkan jenis media dalam pengolahan air limbah domestik tahun 2016. [Anaerobic biofilter capacity based on media type in domestic wastewater treatment 2016]. *Jurnal Riset Hesti Medan Akper Kesdam I/BB Medan, 4*(2), 105. https://doi.org/10.34008/jurhesti.v4i2.143

Nasir, M., & Khalil, M. (2016). Pengaruh penggunaan beberapa jenis filter alami terhadap pertumbuhan, sintasan dan kualitas air dalam pemeliharaan ikan mas (*Cyprinus carpio*). [The effect of natural filters on the growth, survival, and water quality in ornamental goldfish (*Cyprinus carpio*) culture]. *Acta Aquatica: Aquatic Sciences Journal, 3*(1), 33. https://doi.org/10.29103/aa.v3i1.336

Nugroho, R. A., Pambudi, L. T., Chilmawati, D., & Haditomo, A. H. C. (2012). Aplikasi teknologi aquaponic pada budidaya ikan air tawar untuk optimalisasi kapasitas produksi. [Application of aquaponic technology to freshwater fish cultivation for production capacity optimization]. *SAINTEK PERIKANAN: Indonesian. Journal of Fisheries Science and Technology, 8*(1), 46–51. https://doi.org/10.14710/IJFST.8.1.46-51

Ridwantara, D., Buwono, I. D., Suryana, A. A., Lili, W., & Suryadi, I. B. (2019). Uji kelangsungan hidup dan pertumbuhan benih ikan mas mantap (*cyprinus carpio*) pada rentang suhu yang berbeda. [Survival and growth tests of Mantap carp (*Cyprinus carpio*) at different temperature ranges]. *Jurnal Perikanan Dan Kelautan, 10*(1), 46–54.

Sa'anin, H. (1968). Taksonomi dan Kuntji Identifikasi Ikan. [Keys for Fish Identification]. Binatjipta, *Jakarta*. 520 P.

Serlina, W., & Dahlia. (2022). Performa benih ikan mas (*Cyprinus carpio*) yang dipelihara pada tingkat kepadatan yang berbeda dengan media biofilter. [Performance of common carp (*Cyprinus carpio*) reared at different stocking densities integrated with biofilter media]. *Agrokompleks, 22*(2), 40–47. https://doi.org/10.51978/japp.v22i2.459

Sugiharto, U., & Efendi, M. (2017). In F. A. Nurrohmah & B. W. Prasetya (Eds.), *Koi fish* (1st ed.). Penebar Swadaya.

Suharyanto, A., Himawan, D. Y., & Palimirmo, F. S. (2019). *Petunjuk teknis budidaya ikan mas Mustika. [Technical Guide for Mustika Carp Culture]* (1st ed.). KKP.

Syahputra, K., Himawan, Y., Ariyanto, D., & Palim, S. F. (2016). Pewarisan marka Cyca-DAB1*05 dan keragaman genetik ikan mas (*Cyprinus carpio*) strain rajadanu tahan infeksi koi herpesvirus dan tumbuh cepat. [Inheritance of Cyca-DAB1*05 marker and genetic variation of Rajadanu strain of common carp resistant to koi herpesvirus infection and fast growth]. *Jurnal Riset Akuakultur, 11*(2), 115. https://doi.org/10.15578/jra.11.2.2016.115-123

The Development of Giant Freshwater Prawns in Indonesia: Domestication and Cultivation

14

Ikhsan Khasani, Fajar Anggraeni, and Estu Nugroho

14.1 Introduction

The cultivation of giant freshwater prawns (*Macrobrachium rosenbergii*) presents significant potential due to their favorable biological, technical, and economic characteristics. The rapid growth rate, large size, adaptability to diverse cultivation environments, low maintenance requirements, and substantial market demand have positioned giant freshwater prawns as a prominent commodity in several Indo-Pacific nations, including China, India, Bangladesh, Thailand, Vietnam, Malaysia, and Indonesia (New & Nair, 2012; Nugroho et al., 2008; Erlangga., 2016; Pillai et al., 2021). The prawns have been cultivated in Indonesia for over four decades due to their numerous advantages, including rapid growth rates, large body size, high-quality meat, omnivorous feeding habits, and robust domestic and international market potential (Marques & Valenti, 2012; Banu & Christianus, 2016; Indarjo et al., 2022). However, giant prawn farming in Indonesia is still experiencing ups and downs, influenced by some factors. Diseases that affect seed availability, decreased quality of the aquaculture environment, and decreased genetic quality of giant freshwater prawns are challenges that must be solved in an integrated manner (Fig. 14.1).

I. Khasani (✉) · F. Anggraeni · E. Nugroho
Research Center for Fishery, Research Organization for Earth Sciences and Maritime, National Research and Innovation Agency, Cibinong, Indonesia
e-mail: ikhs007@brin.go.id

E. Nugroho et al. (eds.), *Aquaculture Development in Archipelago*,
https://doi.org/10.1007/978-981-95-1084-9_14

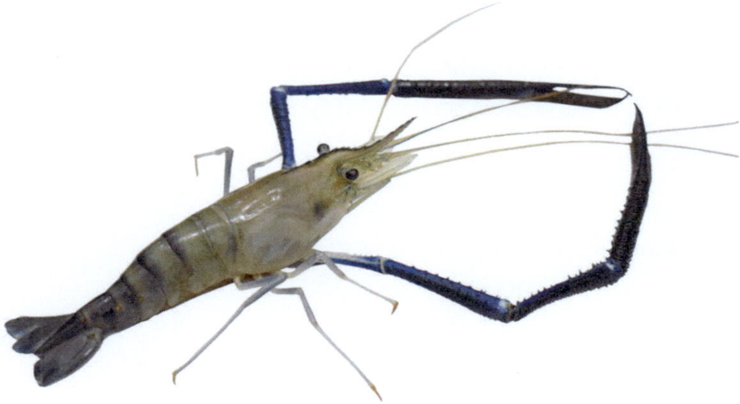

Fig. 14.1 Giant freshwater prawns (*Macrobrachium rosenbergii*). (Source: Personal Collection)

14.2 Taxonomy and Morphology

Most freshwater prawns developed in aquaculture belong to the *Macrobrachium* genus, the largest in the Palaemonidae family (Holthuis & Ng, 2010). Among the *Macrobrachium* genus, the giant freshwater prawn is the most developed species in the world, including Indonesia. The prawn with the main characteristic of a long second walking leg (reaching 1 1/2 body length) is the largest freshwater prawn. Male prawns can reach a size of 320 mm in length and a weight of 200 g (Ling, 1969), while the size of the second walking leg in female prawns is not too long (New, 2002). Giant freshwater prawns have morphological characteristics of internode bodies that each feature a pair of swimming legs, hard skin from chitin, and a second pleura covering the first and third pleura. The body consists of three parts, namely, the head and chest, that unite to form the chest head (cephalothorax), the body (abdomen), and the tail (uropod). Cephalothorax wrapped in carapace. Swordlike protrusions on the carapace are called rostrums, with the upper teeth being 11–15 pieces and the lower teeth being 8–14. Based on De Man (1879), taxonomically, giant freshwater prawn is classified as follows:

Phylum: Arthropoda
Class: Malacostraca
Ordo: Decapoda
Family: Palaemonidae
Genus: *Macrobrachium*
Species: *Macrobrachium rosenbergii* (De Man, 1879)

14.3 Distribution of Giant Freshwater Prawns

Indonesia is a country with high biodiversity (a huge biodiversity country), including the diversity of fresh prawns. Giant freshwater prawns (*M. rosenbergii*) are a species of freshwater prawn native to Indonesia that are evenly distributed in almost all regions of Indonesia, namely, the islands of Sumatra, Java, Kalimantan, Sulawesi, and Papua. The genetic resources of giant freshwater prawns are usually found in freshwater waters with calm to stagnant flows, such as rivers, lakes, and estuaries (Indarjo et al., 2022).

14.4 Reproduction

Reproduction is preserving a species' ability to defend itself in a natural ecosystem (Aprajita et al., 2014). The fertilization of giant freshwater prawns occurs externally. Compared to penaeid prawns, giant freshwater prawns have specific characteristics: a brood chamber (egg hardening site) formed on the abdomen's first, second, and third segments. This characteristic is not found in penaeid prawns because penaeid prawns do not carry eggs but instead lay eggs directly in the water. A few hours after spawning, the gonopore containing male prawn sperm is attached to the brood chamber of the female prawn, which is arranged by setae ovipositing. Setae ovipositing is a hair or bristle that is primarily permanent on the coxae of the last three pairs of pereiopods and pleopods and functions to regulate eggs during spawning. Setae ovigers exist only in the pre-spawning period and are used to tighten eggs to the pleopoda for incubation (New & Valenti, 2000).

The organs of the reproductive system of female giant freshwater prawns consist of paired ovaries, oviducts, gonophores, and sperm reception areas. The ovaries are identified at the maturation stage located on the cephalothorax in the upper part of the stomach and hepatopancreas. The oviduct comprises the lateral ovary next to the heart and opens through the gonophore in the abdominal cavity. Ovaries occupy the cephalothoracic area at the stage of gonad maturation. The color of the ovaries is translucent during the immature stage, then pale yellow to orange as the vitellogenesis progresses, and finally turns dark brown during the spawning stage (Nagaraju, 2010). During their five phases of development, giant freshwater prawns exhibit alterations in the size and color of their oocytes (Chang & Shih, 1995). A white gonad, the previtellogenic stage, characterizes the first level of gonadal maturity (TKG). The stage of vitellogenesis started in TGK II is marked with yellow on the gonads. The next stage of vitellogenesis, namely, TKG III and IV, is characterized by orange gonads with increasingly large oocyte diameters, and TKG V is characterized by reddish gonads and ready-to-lay eggs (Chang & Shih, 1995).

M. rosenbergii females can produce as many as 80,000–100,000 eggs per brood at each spawning with full egg conditions (not falling out), but they often reach at most 5000–20,000 eggs. Gonadal maturity in the ovaries can occur while the eggs are still infested in the brood chamber (New & Singholka, 1985; New & Valenti, 2000). Sureshkumar and Kurup's (1998) research showed that fecundity strongly correlates with total weight, total length, and carapace length. In the study, the average weight of females (n = 80) was 109.9 g, resulting in an average number of 95,687 eggs.

14.5 Domestication and Genetic Improvement of Giant Prawn in Indonesia

Indonesia is the center of origin for giant freshwater prawns because there are 19 species of the genus *Macrobrachium*. Many sources of genetic diversity offer the potential for forming new varieties of giant freshwater prawns with specific advantages (Hadie & Hadie, 2002). However, it still raises questions about its widespread and reasonable use in aquaculture in Indonesia. On the other hand, many genetic resources of freshwater giant prawns in natural waters have yet to be appropriately utilized. Efforts to produce superior seeds can be started by exploring various genetic resources of the GFP in Indonesia.

Using new genetic sources is expected to increase excellence in growth rate, disease resistance, and environmental tolerance (Liao & Huang, 2000). The initial stage for utilizing genetic sources in nature is domestication. Domestication is adapting wild species to a cultivation environment until they can be bred (Effendi, 2002).

To ensure a supply of high-quality giant freshwater prawn seed stock, domestication programs have been implemented utilizing various water bodies throughout Indonesia. Domestication efforts were initiated in 1998–2000, focusing on giant freshwater prawns from the Musi River in South Sumatra, the Kapuas River in West Kalimantan, and the Citanduy River in West Java (Hadie & Hadie et al., 2002). Subsequent domestication projects were undertaken between 2006 and 2008, targeting giant freshwater prawns from the Asahan River in North Sumatra, the Ogan River in South Sumatra, the Barito River in Central Kalimantan, the Ciasem River in West Java, and rivers in Papua (Imron et al., 2009). In 2013, domestication efforts were expanded to include giant freshwater prawns from Berau, East Kalimantan (Anggraeni et al., 2014). Domestication activities continued in 2014, focusing on giant freshwater prawns from the Mahakam River in East Kalimantan (Sesditjen, 2015). In 2018, domestication efforts were extended to the Bengawan Solo River (Nugroho et al., 2020) and in 2020, to water bodies in three districts of North Kalimantan: Bulungan, Nunukan, and Tana Tidung (Indarjo et al., 2022).

Realizing the strategic value of superior seeds in the development of aquaculture, including giant freshwater prawns, a national giant freshwater prawn network was formed in 2008, which aims to optimize the potential of genetic resources of the giant freshwater prawns in Indonesia and ensure the availability of superior prawn seeds and brood stock to support prawn farming. Through domestication and breeding, we produced several superior strains of the prawn, namely, GI Macro (genetic improvement of *Macrobrachium rosenbergii*), GI Macro II, Siratu, and Sijawa. The GI Macro strain resulted from a selection program of base populations that were formed through diallel crossing among three populations of the prawn from

Table 14.1 Superior strains of giant freshwater prawns in Indonesia

Strain	Superiority	Breeding methods	Year of release
GI Macro	Faster growth (32%) and a greater proportion of *edible portions* (20%) than its parents	Mass selection	2000
GI Macro II	67.61% faster weight growth than GI Macro strains	Mass selection	2014
Siratu	33.68% faster weight growth than its parents	Mass selection	2015
Sijawa	Seed production is 35.48% higher than the superior strain Siratu	Domestication	2022

Kalipucang (West Java), Cimanuk (West Java), and Musi rivers (South Sumatra) (Nugroho & Hadie, 2007) and was released in 2001. The GI Macro II (released in 2014) resulted from a mass selection program of basic populations formed from four populations: the GI Macro population, the prawn population from the Barito River, the prawn population from the Musi River, and the prawn population from the Ciasem River (Khasani et al., 2018). In 2015, the Siratu strain was released, which was formed through a selection program for the basic population formed from prawn populations from the Bone River (South Sulawesi), the Citanduy River (West Java), and the Mahakam River (East Kalimantan) (Sesditjen, 2015). In 2022, the SIJAWA strain was released, resulting from the domestication of three generations of giant freshwater prawns from the Bengawan Solo River in Central Java (Nugroho et al., 2020). The superior strains of giant freshwater prawns in Indonesia are presented in Table 14.1.

The domestication process of giant freshwater prawns in Indonesia faces several obstacles. Generally, the main constraints in domesticating aquaculture species often affect growth, reproduction, and disease immunity (Liao & Huang, 2000). The development of giant freshwater prawn larvae from the broodstock of natural populations is prolonged, cannibalism is high, and survival is low (Anggraeni & Krettiawan, 2015).

14.6 Development of Cultivation Technology

Giant freshwater prawns have been cultivated in Indonesia for more than four decades. In the early 1980s, prawn farming extensively used natural seeds (Djajadiredja et al., 1979). The successful development of an intensive giant freshwater prawn hatchery by Takuji Fujimura (New, 2010) has encouraged many countries, including Indonesia, to build several giant freshwater prawn seed production facilities (hatcheries). In the 1990s, the Indonesian government built several giant freshwater prawn hatcheries in several regions of Indonesia, such as West Java (Pangandaran, Pamarican), Jakarta (Pasar Minggu), Central Java (Cilacap), East Java (Prigi), Yogyakarta (Bantul), and Bali (Klungkung). Good seed availability encourages the extensibility of giant freshwater prawn farming in several regions by implementing semi-intensive cultivation systems, both monoculture and polyculture.

Giant freshwater prawn farming in Indonesia is generally developed in three segments. The first segment is the production of giant freshwater prawn seeds (fry) measuring 1 cm, carried out intensively and controlled in a closed (indoor) manner in the hatchery. In this segment, the average survival rate of fry is 30–50% after 45–50 days of rearing. Due to the high production costs and the use of Artemia (imported products) as natural feed, the fry production segment is underdeveloped and is only carried out by government-owned hatcheries. The second segment is the nursery to produce prawn fry with a size of 3–5 cm (maintenance for a month) and seeds measuring 5–7 cm (maintenance for 2 months), which is carried out intensively with an initial stocking density of 100–200 postlarvae/m^2 and a survival rate of 60–80%. The third segment is the grow-out phase, carried out for 3–4 months to produce consuming-size prawns (weighing 25–40 g per prawn).

For two decades, from 1990 to 2010, the grow-out of giant freshwater prawns has developed in most regions of Indonesia, including West Java, Central Java, East Java, Yogyakarta, Bali, Sumatra, and Kalimantan. The fish farmers generally carry out giant freshwater prawn farming in Indonesia by implementing a monoculture system, both in freshwater ponds and brackish water ponds. Such cultivation is carried out by applying semi-intensive technology, with an initial stocking density of seeds of 5–10 juveniles per square meter. In the system, farmers use seeds measuring 3–5 cm (age 2–3 months from hatching eggs) and are kept for 3 months to obtain consumption size prawn weighing 25–40 g. The average productivity in a semi-intensive giant freshwater prawn farming system is 1000–2000 kg/ha/crop (Khasani, 2008; Khasani et al., 2022). Besides extensive and semi-intensive systems, intensive giant freshwater prawn farming has also been developed in Bogor-West Java, with a stocking density of 30 juveniles per square meter (Ali, 2005). In the last system, several modifications were made to ensure the survival and growth of the prawn. One modification is using bamboo apartment shelters to increase the number of prawns. In addition, several paddle wheels are also placed to supply dissolved oxygen, which is the main limiting factor for giant freshwater prawns. However, the intensive giant prawn farming system did not develop because most farmers were constrained by capital and land area.

The polyculture system of giant freshwater prawn farming has also been developed in several regions of Indonesia. The prawn farmers in the southern regions of West Java (Garut, Tasikmalaya) and Yogyakarta raise giant freshwater prawns with gourami or carp. The average productivity in this polyculture was 600 kg of giant freshwater prawn and 2.0 tons of fish per ha/year. In addition, the giant freshwater prawn is also raised with rice plants and is called "Ugadi." In the early 1990s, the Ugadi system was implemented with a low stocking density, 1–2 prawns per square meter, and without artificial feed. In this system, the production of prawns was 200–300 kg/ha/crop (Tahir & Pasaribu, 2003). Nevertheless, the system did not develop due to a less significant increase in profitability. In early 2010, the Ugadi system was intensified by optimizing the land treatment system, increasing seed stocking density (5–10 juveniles/m^2), and using artificial feed. In the second system, the canal area (a deeper area, 50–80 cm, for prawn life) is expanded to 30% of the land area. Through the intensification of the system, significantly higher productivity of giant freshwater prawns was obtained, which is 1200–1700 kg/ha/season (Sarifin et al., 2013; Dewi et al., 2020). Although the intensive Ugadi system provides several advantages, including increased rice production and high selling values of the prawn, the farming system has only been developed in limited areas of Indonesia, such as Sleman Jogjakarta. The availability of good irrigation systems and the safety of prawns from predators and theft are obstacles to the development of Ugadi in Indonesia. Therefore, area-based giant freshwater prawn farming is one of the solutions for the development of Ugadi systems.

14.7 Giant Freshwater Prawn's Production Value in Indonesia

Indonesia has vast aquatic resources and the potential to develop giant freshwater prawns. The broad tolerance level of giant freshwater prawns to salinity (euryhaline) allows the commodity to be developed in many aquaculture environments with a salinity range of 0–15% (Hadie et al., 2004; Ali & Waluyo, 2015). Giant freshwater prawn farming in several countries (China, India, Bangladesh, Malaysia), including Indonesia, is generally carried out in freshwater ponds, low salinity ponds, rice cultivation land, and reservoirs, and in the floodland using a pen culture system (Nair et al., 2013; Dewi et al., 2020). Nevertheless, Aralar et al. (2007) reported that giant freshwater prawns can also be cultivated in floating nets.

Although giant freshwater prawn farming in Indonesia has been developing for over 40 years, and Indonesia is listed among the world's top 10 giant freshwater prawn-producing countries, its contribution to national fishery production is relatively small. National giant freshwater prawn production in 2018 was 3869 tons (FAO, 2020), a tiny proportion (0.015%) compared to national aquaculture production (26,046,394 tons). Farmers still face several obstacles in developing giant freshwater prawns in Indonesia. A mass outbreak of white-tailed disease caused by the MrNV (*Macrobrachium rosenbergi noda virus*) in 2010–2014 caused several giant freshwater prawn hatcheries to stop operation. The absence of seed supply causes national giant prawn production to drop drastically. White-tailed disease in several giant freshwater prawn-producing countries (Pillai & Bonami, 2012) has also led to a worldwide decline in giant freshwater prawn production (Fig. 14.2). The whitetail disease has an impact on the conversion of a large number of prawn ponds to other commodities, such as tilapia. A further impact of the reduction in giant freshwater prawn farming activities is the scarcity of commercial feed for giant freshwater prawns in many giant freshwater prawn farming areas. This condition caused a decrease in giant freshwater prawn production in 2017–2018.

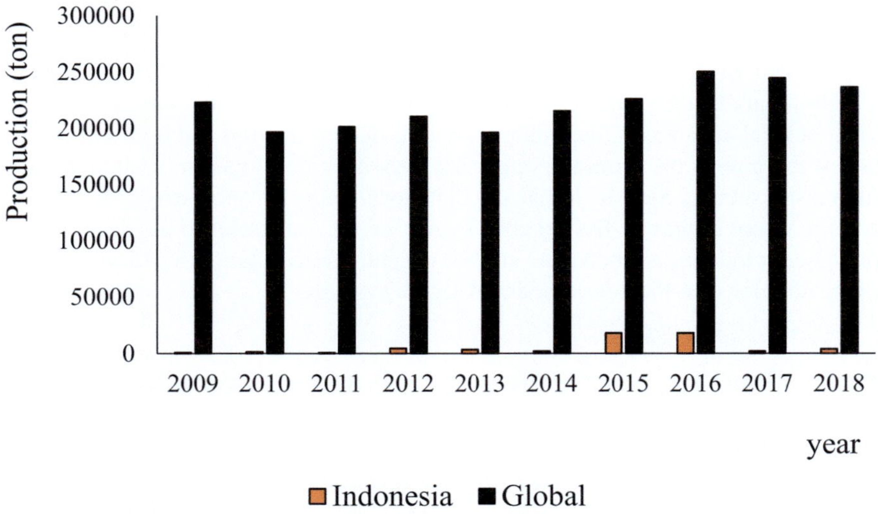

Fig. 14.2 National (Indonesia) and global giant freshwater prawn production in 2009–2018 (FAO, 2020)

References

Ali, F. (2005). The relationship between the use of artificial shields and the survival of giant freshwater prawns (*Macrobrachium rosenbergii*). *Limnotek, 12*(2), 66–725. [in Indonesian].

Ali, F., & Waluyo, A. (2015). Survival and growth rate of Indonesian giant freshwater prawns (*Macrobrachium rosenbergii* De Man) in saline media. *Limnotek, 22*(1), 42–51. [in Indonesian].

Anggraeni, F., & Krettiawan, H. (2015). *Evaluation of the development of giant prawn larvae (Macrobrachium rosenbergii) of Berau origin in the domestication process* (pp. 25–30). Proceedings of the Aquaculture Technology Innovation Forum. [in Indonesian].

Anggraeni, Krettiawanm, H., & Sopian, A. (2014). *Seed performance of F0 and F1 populations of giant freshwater prawns (Macrobrachium rosenbergii) of Berau strains from a mass selection* (pp. 709–719). Proceedings of the Aquaculture Technology Innovation Forum. [in Indonesian].

Aprajita, K., Pandey, A. K., Singh, R., & Chauhan, U. K. (2014). Reproductive endocrinal regulation in decapod crustaceans: A review. *Journal of Ecophysiology and Occupational Health, 14*(1 & 2), 55–65.

Aralar, M. A. L. A. C., Aralar, E., Laron, M., & Rosario, W. (2007). Culture of *Macrobrachium rosenbergii* (De Man 1879) in experimental cages in a freshwater eutrophic lake at different stocking densities. *Aquaculture Research, 38*, 288–294.

Banu, R., & Christianus, A. (2016). Giant freshwater prawn, *Macrobrachium rosenbergii*, farming: A review on its current status and prospective in Malaysia. *Journal of Aquaculture Research & Development, 7*, 423. https://doi.org/10.4172/2155-9546.1000423

Chang, C. F., & Shih, T. W. (1995). Reproductive cycle of ovarian development and vitellogenin profiles in the freshwater prawns, *Macrobrachium rosenbergii*. *Invertebrate Reproduction and Development, 27*(1), 11–20. https://doi.org/10.1080/07924259.1995.9672429

De Man, J. G. (1879). On some species of the genus Palaemon Fabr. with descriptions of two new forms. *Notes Leden Museum, 1*, 165–184.

Dewi, R. R. S. P., Krettiawan, H., Anggraeni, F., Kusnendar, E., & Kusdiarti. (2020). Productivity and profitability of rice- giant freshwater prawn culture in different stocking densities of prawn. *IOP Conference Series: Earth and Environmental Science, 521*, 1–8. https://doi.org/10.1088/1755-1315/521/1/012021

Djajadiredja, R., Suharso, H. H., & Ismail, A. (1979). A new technique of giant freshwater prawn hatchery, an appropriate pattern for small and mass businesses. *LPPD Reporter, 2*(1), 6–22. [in Indonesian].

Effendi, M. I. (2002). *Fisheries biology*. Pustaka Nusantara Foundation. [in Indonesian].

Erlangga. (2016). Distribution of giant freshwater prawn (*Macrobrachium rosenbergii*) in the east coast of Aceh. *Berkala Perikanan Terubuk, 44*(1), 56–68. [in Indonesian].

FAO (2020) Fishery and Aquaculture Statistics, Aquaculture Production. *Macrobrachium rosenbergii*: A review of global status. *Reviews in Aquaculture, 14*, 1–13. https://www.fao.org/fishery/static/Yearbook/YB2018_USBcard/index.htm

Hadie, L. E., & Hadie, W. (2002). *GI Macro giant freshwater prawn farming in irrigation ponds, rice fields, and ponds*. Self-help spreader, Jakarta, 88 pp. [in Indonesian].

Hadie, W., Sumantadinata, K., Noor, R. R. S., Carman, O., & Hadie, L. E. (2004). Phenotypic flexibility of giant freshwater prawns (*Macrobrachium rosenbergii*) strains of Musi, Barito, GIMacro, and their crosses in the maternity environment. *Journal Riset Akuakultur, 10*(2), 33–46. [in Indonesian].

Holthuis, B. L., & Ng, P. K. L. (2010). Nomenclature and taxonomy. In M. B. New, W. C. Valenti, J. H. Tidwell, L. R. D'Abramo, & M. N. Kutty (Eds.), *Freshwater prawns biology and farming* (pp. 12–16). Blackwell Publishing Ltd.

Imron, I., Iswanto, B., & Dewi, R. R. S. P. S. (2009). Morphological variability of several Indonesian populations of giant freshwater prawn, *Macrobrachium rosenbergii. Indonesian Aquaculture Journal, 3*(1), 1–12.

Indarjo, A., Salim, G., Nugraeni, C. D., Anggoro, S., Maryanto, T. I., Prakoso, L. Y., Pramono, B., Azis, M., Rансangan, J., & Khasani, I. (2022). Preliminary study of the domestication of giant freshwater prawn, *Macrobrachium rosenbergii* (De Man, 1879) from North Kalimantan, Indonesia. *AACL Bioflux, 15*(6), 3121–3130.

Khasani, I. (2008). Efforts to increase productivity in the enlargement of Galah Prawn (*Macrobrachium rosenbergii* de Man). *Media Akuakultur, 3*(1), 25–30. [In Indonesian].

Khasani, I., Krettiawan, H., Sopian, A., & Anggraeni, F. (2018). Selection response and heritability of growth traits of giant freshwater prawn (*Macrobrachium rosenbergii*) in Indonesia. *AACL Bioflux, 11*(6), 1588–1695.

Khasani, I., Sopian, A., Anggraini, F., & Pamungkas, W. (2022). Growth, survival rate, and reproductive performance of two superior strains of the giant freshwater prawn (*Macrobrachium rosenbergii*). *Omni-Akuatika* Special Issue The 4th KRIPIK-SciFiMaS, 18: 74–80.

Liao, I. C., & Huang, Y. S. (2000). *Methodological approach used for the domestication of potential candidates for aquaculture. Recent advances in Mediterranean aquaculture finfish species diversification* (pp. 97–107). CIHEAM.

Ling, S. W. (1969). The general biology and development of *Macrobrachium rosenbergii. FAO, Fisheries Report, 57*, 589–606.

Marques, H. L. D. A., & Valenti, P. M. C. M. (2012). REVIEW ARTICLE: Current status and prospects of farming the giant river prawn *(Macrobrachium rosenbergii (De Man 1879) and the Amazon river prawn Macrobrachium amazonicum (Heller 1862)) in Brazil. Aquaculture Research, 2012*(43), 984–992. https://doi.org/10.1111/j.1365-2109.2011.03032.x

Nagaraju, G. P. C. (2010). A review on reproductive regulators in decapods crustacean. *The Journal of Experimental Biology, 214*(1), 3–16.

Nair, C. M., Salin, K. R., Joseph, J., Aneesh, B., Geethalakshmi, V., & New, M. B. (2013). Organic rice–prawn farming yields 20% higher revenues. *Agronomy Sustainable Development*, 1–15. https://doi.org/10.1007/s13593-013-0188-z.A

New, M.B. & Singholka, S. (1985) Freshwater prawn farming. A manual for the culture of Macrobrachium rosenbergii. FAO Fisheries Technical Paper 225 (Rev 1). FAO, Rome.

New, M. B. (2002). *Farming freshwater prawns: A manual for the culture of the giant river prawn (Macrobrachium rosenbergii). FAO Fisheries Technical Paper. No 428* (212 p). FAO. https://doi.org/10.1136/bmj.1.6162.559-j

New, M. B. (2010). History and global status of freshwater prawn farming. In M. B. New, W. C. Valenti, J. H. Tidwell, L. R. D'Abramo, & M. N. Kutty (Eds.), *Freshwater prawns biology and farming* (pp. 1–9). Blackwell Publishing Ltd.

New, M. B., & Nair, C. M. (2012). Review article. The global scale of freshwater prawn farming. *Aquaculture Research, 43*, 960–969.

New, M. B., & Valenti, W. C. (2000). *Freshwater prawn culture: The farming of Macrobrachium rosenbergii*. Blackwell Science. https://doi.org/10.1002/9780470999554

Nugroho, E., & Hadie, L. E. (2007). Giant freshwater prawn culture in Indonesia. In I. M. R. R. Eguia & M. L. C. Aralar (Eds.), *Recent developments in the genetic improvement of the giant freshwater prawn (Macrobrachium sp.)* (pp. 4–7). Aquaculture Department, Southeast Asian Fisheries Development Center.

Nugroho, E., Mulyasari, Kristanto, A. H., Ali, F., & Gunawan. (2008). Evaluation of genetic variability of freshwater prawn collected from Makssar-Sulawesi, Pangkalan Bun-Kalimantan, Jambi-Sumatera, Sukabumi-Java, and GIMacro using MtDNA CO-1 markers. *Indonesian Aquaculture Journal, 3*(1), 23–28.

Nugroho, E., Astuti, A., Yulaeni, F., Praninda, P. Y., Putra, R. K., Suprayitno, S., Hariyanti, D. A., Sutandi, L., & Irvani, F. (2020). Co-1 MtDNA sequence, reproductive character, and tolerance to the environment of Bengawan Solo giant freshwater prawns. *Jurnal Riset Akuakultur, 15*(1), 11–18. [In Indonesian].

Pillai, D., & Bonami, J. R. (2012). Review article: A review on the diseases of freshwater prawns with a special focus on the white tail disease of *Macrobrachium rosenbergii. Aquaculture Research, 43*, 1029–1037.

Pillai, B. R., Ponzoni, R. W., Mahapatra, K. D., & Panda, D. (2021). Genetic improvement of giant freshwater prawn *Macrobrachium rosenbergii*: A review of global status. *Reviews in Aquaculture, 14*, 1–13. https://doi.org/10.1111/raq.12650

Sarifin, S. H., Rohmana, D., Wibowo, K. T., & Rosellia, S. (2013). *Cultivation of rice and giant freshwater prawns* (pp. 157–160). Proceedings of the Aquaculture Technology Innovation Forum. [In Indonesian].

Sesditjen. (2015). *Siratu giant freshwater prawn*. Directorate General of Aquaculture, www.djpb.kkp.go.id. Accessed on November 28, 2017. [in Indonesian].

Sureshkumar, S., & Kurup, B. M. (1998). Fecundity indices of giant freshwater prawn, *Macrobrachium rosenbergii* (De Man). *Journal of Aquaculture in the Tropics, 13*, 181–188.

Tahir, A. G., & Pasaribu, A. M. (2003). Study of the adaptation of giant freshwater prawns (*Macrobrachium rosenbergii*) and carp (*Cyprinus carpio*) with the jajar legowo rice mina system in irrigated rice fields. *Jurnal Penelitian dan Pengembangan Teknologi Pertanian, 6*(2), 167–175. [in Indonesian].

Development of Bonylip Barb (*Osteochilus vittatus*) in Indonesia

15

Wahyu Pamungkas and Dewi Puspaningsih

15.1 Introduction

Fifteen species of the genus *Osteochilus* are found in Indonesia's inland waters, including *O. borneensis, O. kelabau, O. vittatus, O. melanopleura, O. schlegli, O. kahajenensis, O. repang, O. triporus, O. brachynotopterus, O. hasselti, O. kappeni, O. waandersi, O. brevicauda, O. spilurus, and O. harrisoni*. Among these, *O. vittatus* inhabits Lake Maninjau, Lake Singkarak, the Koto Panjang Reservoir, and the Kampar Kanan River (Kurniawan et al., 2020, 2021). *Osteochilus vittatus*, commonly known as the bonylip barb, is an endemic species to Indonesia, found in freshwater environments such as rivers and swamps in Java and Sumatra (Imron et al., 2024; Rochmatin et al., 2014), also in Madura Island (Hasan et al., 2019) and Bali (Parawangsa et al., 2022). This species belongs to the Cyprinidae family. Its natural habitat includes water bodies where the periphyton groups Cyanophyceae and Chlorophyceae are abundant, providing a source of natural food. The diet of *O. vittatus* consists of detritus and periphyton-associated organisms, including aquatic algae, cyanobacteria, and heterotrophic microorganisms found beneath the water's surface (Siahaan et al., 2024; Aryani et al., 2017). The bonylip barb is also referred to by various local names, including lehat, magut, regis, milem, muntu, trough, pawas, puyau, asang, penopa, and carp.

Bonylip barb fish farming is economically profitable, environmentally sustainable, and highly productive, making it a promising candidate for superior aquaculture products (Jusmaldi et al., 2020; Fira et al., 2021). Due to its numerous advantages, the bonylip barb has the potential to become a key freshwater aquaculture commodity, helping to boost domestic fish production. Therefore, it is essential to implement strategies to enhance bonylip barb production, including promoting local aquaculture practices and improving the genetic quality of the fish (Andriani et al., 2020).

W. Pamungkas (✉) · D. Puspaningsih
Research Center for Fishery, Research Organization for Earth Sciences and Maritime, National Research and Innovation Agency (BRIN), Jakarta, Indonesia
e-mail: wahy064@brin.go.id

E. Nugroho et al. (eds.), *Aquaculture Development in Archipelago*,
https://doi.org/10.1007/978-981-95-1084-9_15

Once bonylip barb fish are processed into products such as baby fish, dendeng, pindang, smoked fish, and canned fish, their market value increases (Rahardjo & Marliani, 2007). The fish eggs are well-regarded for their excellent flavor and have potential as an export product (Winarlin et al., 2006; Subagja et al., 2006a, b). As a biocleaning agent, bonylip barb fish are effective in cleaning floating net cages by consuming periphyton and debris from the environment. In terms of aquaculture, bonylip barb fish are easy to maintain in various water conditions and exhibit high survival and reproduction rates (Cholik et al., 2005). Additionally, they are resistant to numerous diseases (Subagja et al., 2006a, b).

15.2 Taxonomy and Morphology

15.2.1 Classification

Kingdom: Animalia
Phylum: Chordata
Class: Actinopterygii
Order: Cypriniformes
Family: Cyprinidae
Genus: *Osteochilus*
Species: O. *vittatus*

The bonylip barb is a medium-sized fish, reaching a total length of up to 260 mm. The body height at the beginning of the dorsal fin is 3–3.7 times the standard length (excluding the caudal fin). The head length is 4.1–4.5 times the standard length. It has a blunt, rounded mouth with puckered lips that can protrude. The maxillary barbels are about as long as the eye diameter, while the rostral barbels are shorter. The dorsal fin begins around the 8th or 9th lateral line scale and is separated from the caudal fin by 10–12 scales. The dorsal fin has three hard spines and 10–13 soft rays, with the front rays being as long as or longer than the head length. The pectoral fin formula is I.13–16, the pelvic fin is I.8, and the anal fin is III.5. The lateral line has 33–34 scales (Weber & de Beaufort, 1916) (Fig. 15.1).

Fig. 15.1 Bonylip barb (*Osteochilus vittatus*). (Source: Personal collection)

15.3 Distribution

The bonylip barb is distributed across Malaysia, Thailand, Vietnam, Cambodia, and Indonesia, including regions such as Tonkin, Siam, the Malayan Peninsula, Borneo, Sumatra, Sulawesi, and Java (Soulivongsa et al., 2020; Serdiati et al., 2024). It is primarily farmed for consumption in Java. Recently, bonylip barb has also been introduced to several lakes in Sulawesi (Hasan et al., 2019).

15.4 Reproduction

The bonylip barb is a freshwater species known for its year-round spawning and high reproductive potential. Male brood stock can begin spawning at 1 year old, with a length of 20 cm and a weight of 100–150 g. Female brood stock can produce approximately 80,000–110,000 eggs per kilogram of body weight.

The bonylip barb exhibits a total spawning pattern, releasing all its eggs at once during spawning, and can do so year-round (Putri et al., 2015). The eggs are transparent and tend to sink to the bottom of the water. They have a diameter ranging from 0.8 to 1.2 mm (Fig. 15.2).

15.5 Domestication and Breeding Program

Fish aquaculture can begin with the domestication process (Azrita et al., 2015). A comprehensive database for the domestication of *O. vittatus* is available, covering aspects such as morphological characterization (Syandri et al., 2014a), genetic variation (Azrita et al., 2014), growth models (Syandri et al., 2014b), feeding habits (Syandri et al., 2014c), and reproductive aspects (Syandri et al., 2015a). Domestication can start at various stages, including larval, fingerling, and brood stock phases. For larval rearing, live diets play a crucial role in influencing growth and survival rates (Sarkar et al., 2006; Le et al., 2011).

The genetic quality of bonylip barb fish has been improved through a breeding program aimed at producing superior brood stock. This program involves selection, hybridization, and genetic manipulation. Selection and hybridization between

Fig. 15.2 Bonylip barb (*Osteochilus vittatus*) brood stock. (Source: Personal collection)

different bonylip barb strains are conducted to obtain superior brood stock with faster growth, disease resistance, and high reproductive potential. Additionally, genetic manipulation is used to create functional males that produce monosex females, as female bonylip barbs exhibit faster growth compared to males due to sexual dimorphism.

The adaptation phase of the bonylip barb domestication process significantly affects survival rate and growth. A study using a recirculation system for 30 days found that a water flow rate of 0.11 l per minute resulted in an 88.33% survival rate, with an average length increase of 0.31 cm and weight gain of 0.93 g (Kurniawan et al., 2019).

15.6 Aquaculture Technology

The fisheries industry has identified significant potential in bonylip barb fish culture, particularly given the high market demand for this species. Despite this demand, there are relatively few bonylip barb fish farmers. However, this does not suggest that cultivating bonylip barb is difficult; in fact, it is an accessible endeavor that beginners can easily undertake. Bonylip barb fish are primarily raised for human consumption, especially in Java. Moreover, they are often viewed as a by-product in ponds where tilapia, carp, and other cyprinids are cultivated. Nonetheless, this does not diminish their value as a primary aquaculture product, as the market demand remains consistently strong while the number of suppliers is still quite limited. The aquaculture technology used in bonylip barb fish farming has not received much attention in scientific journals. Several scientific documents mentioned about dietary regimen (Syandri et al., 2015b), grown using biofloc technology (Mumpuni et al., 2022), grown in a circular water-flowing container (Yustiati et al., 2022), vitamin E's impact on the quality of eggs (Tarigan et al., 2020).

Bonylip barb fish are often harvested from the farms at 3 months of age, which is a rather short culture period. Bonylip barb fish are raised from brood stock. To produce eggs, superior male and female breeders are spawned. The eggs are then grown until they are 3 months old, at which point they are harvested and put on the market (Fira et al., 2021). Anyone can grow bonylip barb fish since it is simple to maintain, does not require a lot of capital, produces fish promptly, and is quite simple to market (Jusmaldi et al., 2020).

References

Andriani, Y., Buwono, I. D., Subhan, U., & Mandhasia, A. (2020). Genetic kinship of nilem sih strains (*Osteochilus hasselti*). *Asian Journal of Biochemistry, Genetic and Molecular Biology, 3*(1), 28–37. https://doi.org/10.9734/AJBGMB/2020/v3i130077

Aryani, N., Mardiah, A., Azrita, & Syandri, H. (2017). Influence of different stocking densities on growth, feed efficiency, and carcass composition of bonylip barb (*Osteochilus vittatus*, Cyprinidae) fingerlings. *Pakistan Journal of Biological Sciences, 20*(10), 489–497. https://doi.org/10.3923/pjbs.2017.489.497

Azrita, Syandri, H., & Junaidi. (2014). Genetic variation among asang fish (*Osteochilus vittatus*) populations using random amplified polymorphic DNA (RAPD) markers. *International Journal of Fisheries and Aquatic Studies, 1*, 213–217.

Azrita, Basri, Y., & Syandri, H. (2015). Preliminary study on domestication of bluespotted snakehead (*Channa lucius*, Channidae) in concrete tank. *Journal of Aquaculture Research & Development, 6*, 1–5.

Cholik, F., Jagatraya, A. G, Poernomo, R. P., & Jauzi, A. (2005). *Akuakultur: Tumpuan Harapan Masa Depan Bangsa. Masyarakat Perikanan Nusantara dan Taman Akuarium Air Tawar-TMII* [*Aquaculture: The pillar of hope for the nation's future*. Nusantara Fisheries Society and Freshwater Aquarium Park-TMII], Jakarta, 415 hlm.

Fira, D., Wiradana, P. A., Ansori, A. N., Susilo, R. J., & Sabdoningrum, E. K. (2021). Ectoparasite inventorisation of nilem fish (*Osteochilus hasselti*) fingerlings cultured on ponds in Sukabumi, West Java, Indonesia. *Iraqi Journal of Veterinary Sciences, 35*(3), 605–609. https://doi.org/10.33899/ijvs.2020.127031.1440

Hasan, V., Soemarno, Widodo, M. S., & Wiadnya, D. G. R. (2019). First record of *Osteochilus vittatus* (Cypriniformes: Cyprinidae) in Madura Island, Indonesia. *AACL Bioflux, 12*(1), 338–342.

Imron, Anggraeni, F., Pamungkas, W., Marnis, H., Himawan, Y., Astuti, D. N., Palimirmo, F. S., Arifin, O. Z., Subagja, J., Mokodongan, D. F., & Hidayat, R. (2024). Intraspecific diversity and phylogeography of bonylip barb (*Osteochilus vittatus*) in Sundaland, as revealed by mitochondrial cytochrome oxidase I (mtCOI). *Fisheries and Aquatic Sciences, 27*(3), 145–158. https://doi.org/10.47853/FAS.2024.e15

Jusmaldi, Hariani, N., Hendra, M., Wulansari, N. A., & Sarah. (2020). Some reproductive biology aspects of bonylip barb (*Osteochilus vittatus* Valenciennes, 1842) in the waters of Benanga Reservoir, East Kalimantan. *Indonesian Journal of Ichthyology, 20*(3), 217–233.

Kurniawan, A., Azhari, M., & Prasetiyono, E. (2019). Domestication of *Osteochilus spirulus*: Survival and growth in recirculated water. *IOP Conference Series: Earth and Environmental Science, 230*, 012116. https://doi.org/10.1088/1755-1315/230/1/012116

Kurniawan, A., Hariati, A. M., Kurniawan, A., Haryono, & Wiadnya, D. G. R. (2020). Morphological comparison of *Osteochilus spilurus* (Cyprinidae) from three Sundaland Islands in Indonesia using geometric morphometrics. *Egyptian Journal of Aquatic Biology & Fisheries, 24*(7), 835–845.

Kurniawan, A., Hariati, A. M., Kurniawan, A., & Wiadnya, D. G. R. (2021). First genetic record and the phylogenetic relationship of *Osteochilus spilurus* (Cyprinidae: Labeoninae) originating from Bangka and Belitung Islands, Indonesia. *Biodiversitas, 22*(2), 794–802. https://doi.org/10.13057/biodiv/d220233

Le, Y., Yun, Y. S., Ming, Z. X., Liu, M., Yi, L. J., & Chang, W. K. (2011). Effect of temperature on survival, development, growth, and feeding of larvae of yellowtail clownfish *Amphiprion clarkia* (Pisces: Perciformes). *Acta Ecologica Sinica, 31*, 241–245.

Mumpuni, F. S., Muarif, Y., Yulianti, N., & Hilmy, A. M. (2022). The growth, feed efficiency, and survival rate of bonylip barb (*Osteochilus hasselti*) in biofloc media C/N ratio 10 with different stocking densities. *Journal of Aquaculture and Fish Health, 11*(2), 227–237. https://doi.org/10.20473/jafh.v11i2.30814

Parawangsa, I. N. Y., Artha, G. A., & Tampubolon, P. A. (2022). Morphoregression and reproduction aspects of bonylip barb (*Osteochilus vittatus* Valenciennes, 1842) in Tamblingan Lake, Bali Island. *Scientific Journal of Fisheries and Marine Sciences, 14*(2), 272–284. https://doi.org/10.20473/jipk.v14i2.34629

Putri, M. R. A., Sugianti, Y., & Krismono. (2015). Some biological aspects of bonylip barb (*Osteochilus vittatus*) in Lake Talaga, Central Sulawesi. *BAWAL, 7*(2), 111–120.

Rahardjo, A. A. & Marliani, L. (2007). Bonylip barb: Diolah Naik Derajat [Bonylip barb: Processed and elevated in value]. Trubus. http://www.trubus.com [21 Maret 2008]. *Jurnal Riset Akuakultur, 5*(2), 175–182. Tahun 2010.

Rochmatin, S. Y., Solichin, A., & Saputra, S. W. (2014). The growth aspects and reproduction of nilem (*Osteochilus hasselti*) in Rawa Pening, District of Tuntang, Semarang Regency. *Diponegoro Journal of Maquares, 3*(3), 153–159. http://ejournal-s1.undip.ac.id/index.php/maquares.

Sarkar, U. K., Lakra, W. S., Deepak, P. K., Negi, R. S., Paul, S. K., & Srivastava, A. (2006). Performance of different types of diets on experimental larval rearing of endangered *Chitala chitala* (Hamilton) in recirculatory system. *Aquaculture, 261*, 141–150.

Serdiati, N., Ndobe, S., Rosyida, E., Gani, A., Hermawan, R., Nurdin, M. S., Sari, D. E., Herlina, S., Pawaro, M. F. D., & Danty, A. R. (2024). Length-weight relationship, condition factor, and DNA barcoding of bonylip barb, *Osteochilus vittatus* (Valenciennes, 1842) in Dampelas Lake, Central Sulawesi, Indonesia. *Omni Akuatika, 20*(1), 23–37. https://doi.org/10.20884/1.oa.2024.20.1.1111

Siahaan, D. P., Subandiyono, Amalia, R., & Hastuti, S. (2024). Impact of dietary protein on total feed consumption, utilization, protein retention, and growth performance of Nilem carp (*Osteochilus vittatus*) fingerlings. *Journal of Advances in Food Science & Technology, 11*(4), 12–18. https://doi.org/10.56557/jafsat/2024/v11i48858

Soulivongsa, L., Tengjaroenkul, B., & Neeratanaphan, L. (2020). Effects of contamination by heavy metals and metalloids on chromosomes, serum biochemistry and histopathology of the Bonylip Barb fish near Sepon Gold-Copper Mine, Lao PDR. *International Journal of Environmental Research and Public Health, 17*, 9492. https://doi.org/10.3390/ijerph17249492

Subagja, J., Gustiano, R., & Djajasewaka, H. (2006a). Penentuan dosis hormon steroid dan teknik pemberian untuk feminisasi ikan Bonylip barb (*Osteochilus hasselti*). [Determination of steroid hormone dosage and administration techniques for feminization of Bonylip barb *(Osteochilus hasselti)*]. *Laporan Hasil Riset Balai Riset Perikanan Budidaya Air Tawar Tahun Anggaran 2006.* Balai Riset Perikanan Budidaya Air Tawar Bogor, Badan Riset Kelautan dan Perikanan, hlm, pp. 300–312.

Subagja, J., Gustiano, R., & Winarlin, L. (2006b). Pelestarian ikan Bonylip barb (*Osteochilus hasselti* CV) melalui teknologi perbenihannya. [Conservation of Bonylip barb (Osteochilus hasselti CV) through seed technology]. *Prosiding Lokakarya Nasional Pengelolaan dan Perlindungan Sumberdaya Genetik di Indonesia,* Bogor, 20 Desember 2006, hlm, pp. 279–286.

Syandri, H., Azrita, & Junaidi. (2014a). Morphological characterization of asang fish (Osteochilus vittatus, Cyprinidae) in Singkarak Lake, Antokan River, and Koto Panjang Reservoir, West Sumatra Province, Indonesia. *Journal of Fisheries and Aquaculture, 1*, 158–162.

Syandri, H., Azrita, & Junaidi. (2014b). Biological parameters of Bonylip barb (Osteochilus vittatus, Cyprinidae). In *Proceedings of the national seminar on incentive research SINas*. Ministry for Research and Technology, Republic of Indonesia.

Syandri, H., Azrita, & Junaidi. (2014c). Food habits of Bonylip barb (Osteochilus vittatus, Cyprinidae). In *Proceedings of the annual national seminar XI fisheries and marine research.* Department of Fisheries, Faculty of Agriculture, University of Gajahmada.

Syandri, H., Azrita, & Junaidi. (2015a). Fecundity of Bonylip barb (Osteochilus vittatus, Cyprinidae) in different water habitats. *International Journal of Fisheries and Aquatic Studies, 2*, 157–163.

Syandri, H., Azrita, N., & Junaidi. (2015b). Preliminary study on the feeding schedule of laboratory-reared bonylip barb larvae, Osteochilus vittatus (Cyprinidae). *Journal of Aquaculture Research & Development, 6*(10), 1000368. https://doi.org/10.4172/2155-9546.1000368

Tarigan, N., Affandi, R., & Meiyasa, F. (2020). Effect of vitamin E on the egg quality of Bonylip barb fish (*Osteochilus vittatus*, Valenciennes, 1842). *Aceh Journal of Animal Science, 5*(2), 112–116. www.jurnal.unsyiah.ac.id/AJAS.

Weber, M., & de Beaufort, L. F. (1916). *The fishes of the Indo-Australian Archipelago III* (pp. 131–133, as Osteochilus vittatus). E.J. Brill.

Winarlin, L., Setiadi, E., Widiyati, A. & Djajasewaka, H. (2006). Pengaruh tingkat kedalaman air terhadap perkembangan pakan alami untuk pertumbuhan benih ikan Bonylip barb (*Osteochilus hasselti* CV). [The effect of water depth on the development of natural feed for the growth of Bonylip Barb (*Osteochilus hasselti* CV) seedlings]. *Laporan Hasil Riset Balai Riset Perikanan Budidaya Air Tawar Tahun Anggaran 2006*. Balai Riset Perikanan Budidaya Air Tawar Bogor, Badan Riset Kelautan dan Perikanan, hlm, pp. 313–332.

Yustiati, A., Rojai, M. T. S., Suryadi, I. B. B., & Iskandar. (2022). Growth and survival of bonylip barb (*Osteochilus vittatus*) in round water flowing container. *World News of Natural Sciences, 42*, 44–55.

The Status of Domestication and Aquaculture Systems for Tilapia (*Oreochromis niloticus*) in Indonesia

16

Bambang Gunadi, Adam Robisalmi, and Estu Nugroho

16.1 Introduction

Nile tilapia (*Oreochromis niloticus*), part of the Cichlidae family, is a freshwater fish introduced to various tropical, subtropical, and temperate regions worldwide since the late twentieth century. It can be farmed using a broad range of methods, from small-scale systems that use organic waste as feed to highly intensive production techniques. According to El-Sayed (2006), Nile tilapia is widely farmed in developing countries due to its fast growth, adaptability to different environmental conditions, resilience to stress and disease, ease of breeding, cost-effectiveness, and status as a valuable protein source, often nicknamed "aquatic chicken."

Tilapia is widely recognized for its resilience, ease of breeding, and adaptability to a wide range of water conditions and temperatures. In aquaculture, Nile tilapia primarily feeds on plant-based diets and shows a high level of dietary flexibility, consuming a variety of food sources, including natural organisms, garden waste, greens, and formulated feeds. Globally, Nile tilapia and its hybrids are among the most cultivated freshwater fish species after carps. Specifically, Nile tilapia is the most important species, accounting for over 90% of total tilapia production (Fitzsimmons, 2016).

In Indonesia, tilapia is the leading freshwater aquaculture commodity, ranking highest in terms of production. However, the industry faces several challenges, including deteriorating water quality that leads to frequent disease outbreaks, making cultivation increasingly difficult and causing financial losses. Additionally, there is a lack of high-quality brood stock and superior seeds, which hinders productivity.

B. Gunadi (✉) · A. Robisalmi · E. Nugroho
Research Center for Fishery, Research Organization for Earth Sciences and Maritime, National Research and Innovation Agency, Cibinong, Indonesia
e-mail: bambang.gunadi@brin.go.id

E. Nugroho et al. (eds.), *Aquaculture Development in Archipelago*,
https://doi.org/10.1007/978-981-95-1084-9_16

As a result, farmers often neglect the importance of using quality seeds, leading to reduced yields in the field.

16.2 Taxonomy and Morphology

The name 'tilapia' originates from an African Bushman word meaning "fish" (Trewavas, 1982). Tilapias, part of the Cichlidae family, comprise a diverse group of freshwater fish species. Within the suborder Labroidei, which includes the families Cichlidae, Embiotocidae, Pomacentridae, and Labridae, Tilapia was first described by Smith in 1840. It was later divided into two subgenera based on breeding and feeding behaviors: *Tilapia* (substrate spawners) and *Sarotherodon* (mouthbrooders). Mouthbrooders incubate fertilized eggs and newly hatched fry within the mouth of one or both parents. Eventually, *Sarotherodon* was elevated to genus status and split into two genera: *Oreochromis*, where females are the primary mouthbrooders, and *Sarotherodon*, where males or both sexes participate in mouthbrooding.

Tilapia has a streamlined body structure with a compressed and deep shape. Their bodies are covered with large, firmly attached cycloidal scales. The dorsal and anal fins are equipped with strong spines and flexible rays, while the pectoral and pelvic fins are well-developed and located toward the front of the body, aiding in precise swimming and maneuvering. These fins, supported by red muscles, facilitate sustained but relatively slow movement (Ross, 2000). Tilapia typically exhibit muted body colors with vertical bars that blend into their overall hue, allowing them to change color in response to stress and environmental conditions by regulating the activity of chromatophores in their skin. They also possess well-developed sensory organs, including a prominent snout, a clearly visible lateral line, and relatively large eyes that provide excellent visual acuity.

Tilapia are often herbivorous or omnivorous, with physical characteristics such as jaw structure, mouth size and shape, dentition, and the number of gill rakers adapting to their varied diet. Depending on species, fish size, and food availability, *O. niloticus*, *O. aureus*, and *O. mossambicus* primarily consume phytoplankton, periphyton, zooplankton, insects, small shrimp, aquatic plants, detritus, benthic fauna, and even fish eggs. Their digestive systems are relatively unspecialized, consisting of a simple, long, coiled intestine that can be 7–13 times the length of the fish, connected to a small, saclike stomach via a short esophagus. Tilapia have two types of teeth: jaw teeth and pharyngeal teeth, which classify them as filter feeders.

Based on Saanin (1968) in his book *Taxonomy and Identification Key for Fish*, tilapia that develop specific characteristics are classified under the genus *Oreochromis*, with the following taxonomic classification (Fig. 16.1):

Kingdom:	Metazoa
Phylum:	Chordata
Subphylum:	Vertebrata
Class:	Actinopterygii
Order:	Perciformes

Fig. 16.1 The Nile tilapia. (Source: Personal documentation)

Family:	Cichlidae
Genus:	*Oreochromis*
Species:	*Oreochromis mossambicus, Oreochromis niloticus*

16.3 Reproduction

Most domesticated tilapia species exhibit maternal mouthbrooding behavior, a fascinating reproductive strategy. Breeding begins with males building and guarding a spawning area, followed by a brief courtship with a female. The female then releases a string of 20–50 eggs, and the male fertilizes them by releasing a cloud of milky milt over the eggs after pressing his genital papilla on the nest or tank bottom. The female immediately collects the fertilized eggs in her mouth (Rothbard, 1979; Rana, 1990). In Nile tilapia, females release the eggs into the water and collect them into their mouths before fertilization. The male then approaches, releasing milt near the female's mouth, which she sucks in. This process lasts 4–6 min before the female leaves the spawning area (Mansour, 2001).

Nile tilapia's natural spawning behavior is similar to that of lekking animals, where groups of males defend individual "nests" for mating and oviposition. Females enter these areas when ready to mate and can mate with one or more males. Unlike group spawning species, tilapia typically exhibit a one-to-one mating pattern between male and female.

Tilapia is known for their relatively low fecundity and large egg size. Nile tilapia, for instance, produces fewer than 100 to over 3,000 eggs per spawn, with egg sizes ranging from under 2 to 7.9 mm (De Graaf et al., 1999). Substrate spawners, on the other hand, lay larger quantities of smaller eggs. In intensive hatcheries, fry production for Nile tilapia involves placing males and females in tanks or hapas suspended in fertilized ponds. The fertilized eggs are collected from the females' mouths and artificially incubated in tank-based hatcheries. Stocking densities for optimal fry production depend on factors like species, broodstock size, culture systems, and water quality.

Little (1992) found that a stocking density of 6 fish per square meter in hapas suspended in fertilized ponds was ideal for Nile tilapia. Ridha and Cruz (1999) observed that a lower density of 4 fish per square meter led to better seed production and more synchronized spawning compared to higher densities. Successful breeding depends on the sex ratio, which varies based on species and rearing systems, ranging from 1 male to 5 females in ponds to 1 male to 10 females in tanks or hapas (Little & Hulata, 2000). A common ratio is 1 male to fewer than 3 females when synchronized spawning is desired. Studies indicate that Nile tilapia females have an inter-spawning interval of 18–27 days, with younger and smaller females spawning more frequently (Guerrero & Guerrero, 1985; Ponzoni et al., 2007). In commercial seed production, group mating is common, with sex ratios of 1 male to 2 females (Barman & Little, 2006) or 1 male to 3–4 females (Mires, 1982). Many small-scale production systems adopt a 1:2 male-to-female ratio (Barman & Little, 2006; Bhujel, 2000).

Edwards (2009) found a tilapia seed production system in Sukabumi, West Java, consisting of four phases managed by a different group of farmers. The first group of farmers managed broodstock spawning in small shallow ponds, with male and female brood stocks stocked at a 1:4 ratio, and fry harvested every 15 days. The other three groups of farmers managed three stages of fry nursing. In the first stage, tilapia fry was stocked at 200–250 fry/m^2 to grow to a total length (TL) of 2–3 cm. In the second stage, fry was stocked at 150–200 fry/m^2 to reach a TL of 3–5 cm. The third stage, fry was reared at 75–100 fry/m^2 stocking density until they reached a TL of 5–8 cm. Fingerlings were then sold for further rearing, primarily in reservoir pens or floating cages.

16.4 Domestication and Distribution

Nile tilapia is naturally found in lake and river ecosystems across Asia and Africa, including Egypt, Palestine, Israel, and Syria. Due to its favorable traits such as fast growth, disease resistance, ease of reproduction, and tolerance to environmental stress, it has been intentionally introduced to tropical, subtropical, and even temperate regions worldwide, establishing itself successfully in many global locations. In Indonesia, before the introduction of Nile tilapia, the primary cultured fish species were Mozambique tilapia (*O. mossambicus*) and common carp (*Cyprinus carpio*). Nile tilapia was first introduced to Indonesia from Taiwan in 1969, followed by red Nile tilapia in 1981 and Chitralada black Nile tilapia in 1984. To further improve productivity, the Indonesian government, through the Research Institute for Freshwater Fisheries (Balitkanwar), imported the third generation (G3) of GIFT (Genetic Improvement of Farmed Tilapia) in 1994, followed by the sixth generation (G6) in 1995 and 1997 (Widiyati & Sudarto, 1997). In 2002, the West Java Provincial Government, in collaboration with the Bureau of Fisheries and Aquatic Resources (BFAR) of the Philippines, imported genetically enhanced tilapia (GET) brood stocks (Arifin & Kurniasih, 2007).

According to Phillips et al. (2015), Indonesia has approximately 927,755 household-scale fish farmers involved in freshwater aquaculture. Among these farmers, 65.8% have land areas of less than 0.1 ha, and about 34% cultivate Nile tilapia. This makes Nile tilapia a significant contributor to Indonesia's aquaculture sector, with expectations for increased employment opportunities in the future. Various Nile tilapia strains are currently being developed in Indonesia for cultivation in freshwater, brackish water, and marine environments.

16.5 Development of Tilapia Culture

16.5.1 Culture Technology

The impact of culture systems and environmental conditions on tilapia's response to water temperature is significant. El-Sayed et al. (2006) observed that the growth and survival of Nile tilapia, reared in earthen ponds, were strongly influenced by pond depth and water temperature. Their study revealed that fish growth was poorest and mortality rates highest at a pond depth of 50 cm. Conversely, the best growth performance and survival rates occurred at depths of 100–200 cm. At the shallow depth of 50 cm, about 85% of the fish died, while increasing the depth to 200–300 cm provided a refuge from extreme temperatures, greatly enhancing both growth and survival.

Tilapia can produce high yields in monoculture systems due to their ability to occupy various feeding niches, unlike selective feeders such as carp. Polyculture systems, where tilapia are cultured alongside crustaceans, are widely practiced worldwide and are believed to yield higher net production than monoculture systems by utilizing multiple ecological niches. Understanding the interactions between fish species and their environment is crucial for optimizing tilapia production in polyculture systems. This knowledge allows fish farmers and farm managers to make informed decisions regarding species selection, fish size, stocking density, fertilization regimes, nutrient inputs, and other key management practices.

16.5.2 Cage Culture

Tilapia cage culture, both experimental and commercial, has been practiced since the early 1970s, as noted by Bardach et al. (1972). One of the key advantages of tilapia cage culture is its versatility, allowing it to be adapted to a wide range of water bodies, including rivers, lakes, reservoirs, swamps, ponds, thermal effluents, wastewater, and even seawater. Fixed cages made from synthetic netting are commonly used in shallow water bodies. These cages, which vary in size from 5 m × 5 m × 3 m to 15 m × 10 m × 3 m, are anchored to bamboo poles driven into the bottom and positioned to ensure optimal water circulation. Floating cages are preferred in deeper water bodies such as lakes, reservoirs, lagoons, and bays. Constructed with bamboo or steel frames and synthetic netting, floating cages typically range in size from 5 m × 5 m × 2.5 m to 10 m × 10 m × 4 m. When stocking these cages, tilapia fingerlings weighing between 0.5 and 2 g are introduced at densities of 15–45 fish per cubic meter in fixed cages. In contrast, floating cages can support higher densities, ranging from 55 to 80 fish per cubic meter. These fish have supplemental or commercial feeds to meet their nutritional needs.

Tilapia cage culture is widely practiced in Indonesia in various water bodies, including rivers, irrigation canals, lakes, and reservoirs. For instance, floating cages measuring 7 m × 7 m × 2 m are used in the Jatiluhur Reservoir, stocked with 100–150 kg of Nile or red tilapia. After a cultivation period of 60–110 days, these cages yield a production range of 626–1200 kg (Hardjamulia & Rukyani, 2000). Similarly, Eguia and Romana-Eguia (2004) found that large net cages, measuring 10 m × 20 m × 2.5 m, can be stocked with fingerlings weighing 10 g at densities of 10–15 fish per cubic meter. In areas rich in plankton, the fingerlings can rely on natural food sources, although pelleted feeds (with 25–30% crude protein) become essential when stocking densities increase to 50–200 fish per cubic meter.

16.6 Genetic Improvement Programs

Genetic improvement programs have significantly boosted the production of Nile cichlid (Komen & Trong, 2014). Over the years, Nile cichlids have undergone numerous group and family-based breeding programs in aquaculture, leading to remarkable genetic advancements (Gjedrem et al., 2012; Komen & Trong, 2014). These breeding programs play a crucial role in producing superior strains, characterized by improved growth, adaptability to environmental fluctuations, and resistance to diseases. The process of producing these superior seeds involves several stages, starting with the collection, characterization, and evaluation of breeding stock, followed by both conventional and unconventional selection programs. Utilizing superior seeds in cultivation can significantly boost production or reduce costs, leading to higher profit margins (Nugroho et al., 2013).

In Indonesia, the rapid expansion of tilapia farming has been accompanied by uncontrolled breeding practices, resulting in inbreeding and a decline in genetic quality. This decline is evident in slower growth, higher mortality rates, and early maturation. The decreased quality of seeds and brood stock can be attributed to genetic-environmental interactions and the segregation of hybrid alleles. Efforts to improve tilapia varieties in Indonesia include collecting and characterizing various strains, as reported by Nugroho and Maskur (2002) and Arifin et al. (2007). Since 2010, the Indonesian government has programmatically released new superior tilapia varieties to address this issue (Nugroho, 2018).

In 2010, Nugroho introduced a superior fish variety, tested on a field scale to increase productivity in floating net cages (Nugroho et al., 2013). Further advancements were achieved by enhancing genetic diversity and heritability through selection methods (Nugroho et al., 2014, 2016, 2017). Quality improvements in tilapia have also been achieved through hybridization, as measured by heterosis value (Robisalmi et al., 2020; Gunadi et al., 2021). Optimization of tilapia farming has been further refined by applying biofloc technology (Nugroho et al., 2020), adding supplements, and using fasting periods (Robisalmi et al., 2021). Combining these research outcomes with practical aquaculture techniques (such as biofloc and superior seeds) ensures sustainable tilapia harvesting, making it a promising business.

To further increase the productivity of Nile tilapia farming, several superior varieties have been developed and officially released for cultivation through hybridization and selection methods, as outlined below:

(a) *JICA tilapia*

JICA tilapia is a genetically engineered fish developed through a collaboration between the Freshwater Aquaculture Center (BBAT) in Jambi and the Japan International Cooperation Agency (JICA), a Japanese donor agency. Compared to GIFT (Genetically Improved Farmed Tilapia), JICA tilapia exhibits 20% faster growth and greater feed efficiency, which can reduce feed costs by up to 25%. This strain of tilapia was officially released through the Ministry of Marine Affairs and Fisheries of the Republic of Indonesia under Decree No. 52 of 2004 (MMAF, 2004) (Fig. 16.2).

(b) *GESIT*

GESIT, short for Genetically Supermale Indonesian Tilapia, is the result of a collaboration between researchers from the Agency for the Assessment and Application of Technology (AAAT), the Research Institute for Freshwater Aquaculture (RIFF), and Bogor Agricultural University (IPB University). When bred with the same GESIT variety, these tilapia produce predominantly male offspring, achieving a male-to-female ratio of 98%. They also exhibit rapid growth, reaching up to 600 g within 6 months. GESIT tilapia was officially released under Decree No. 44 of 2006 (MMAF, 2006a) (Fig. 16.3).

(c) *Nirwana tilapia*

Nirwana tilapia results from selective breeding conducted at the Freshwater Fisheries Development Office in Wanayasa, under the Provincial Department of Marine Affairs and Fisheries of West Java Province. This variety is known for

Fig. 16.2 The JICA tilapia. (Source: https://www.sibina.pasolusi.com)

Fig. 16.3 The GESIT Tilapia. (Source: https://www.sibina.pasolusi.com)

Fig. 16.4 The Nirwana tilapia. (Source: MMAF, 2006b)

Fig. 16.5 The BEST tilapia. (Source: MMAF, 2009a)

its fast growth, reaching a size of 650 g within 7 months. Nirwana tilapia was officially released Decree No. 45 of 2006 (MMAF, 2006b) (Fig. 16.4).

(d) *BEST tilapia*

BEST, which stands for Bogor Enhanced Strain Tilapia, is the product of crossbreeding four strains of tilapia. It belongs to the GIFT (Genetically Improved Farmed Tilapia) group and shares the same coloration as GIFT tilapia. BEST tilapia offers several advantages, including resistance to extreme

Fig. 16.6 The Larasti tilapia. (Source: MMAF, 2009b)

Fig. 16.7 The Srikandi tilapia. (Source: MMAF, 2012a)

environmental conditions, faster growth, higher egg production (3–5 times more than other tilapia), relatively larger larvae, disease resistance, and a survival rate exceeding 90%. This strain was officially released under Decree No. 77 of 2009 (MMAF, 2009a) (Fig. 16.5).

(e) *Larasati tilapia*

Larasati tilapia results from crossbreeding between black tilapia and red tilapia, specifically a male Pandu tilapia and a female Kunti tilapia. This variety was developed at the Freshwater Fish Nursery and Aquaculture Center in Klaten, under the Provincial Department of Marine Affairs and Fisheries of Central Java Province. Larasati tilapia is known for its high survival rate of up to 90%, faster growth than other tilapia, and a lower feed conversion ratio (FCR) than local tilapia varieties. MMAF (2009b) officially released the Larasati tilapia under Decree No. 79 of 2009 (Fig. 16.6).

(f) *Srikandi tilapia*

Srikandi tilapia is a hybrid fish produced by crossbreeding female Nirwana tilapia (*O. niloticus*) with male blue tilapia (*O. aureus*). This variety is known for its ability to tolerate salinity levels up to 30 ppt and its faster growth than

Fig. 16.8 The Sultana tilapia. (Source: MMAF, 2012b)

Fig. 16.9 The Anjani tilapia. (Source: MMAF, 2012c)

other tilapia species. Additionally, Srikandi tilapia shows moderate resistance to *Streptococcus agalactiae* bacteria and has a low feed conversion ratio (FCR) (Setyawan et al., 2022). MMAF (2012a) officially released the Srikandi tilapia under Decree No. 09 of 2012 (Fig. 16.7).

(g) *Sultana tilapia*

This Sultana tilapia has several advantages over other tilapia, including a strong immune system, higher number of eggs, and relatively fast growth. This tilapia was officially released through Decree No. 28 of 2012 (MMAF, 2012b) (Fig. 16.8).

(h) *Anjani tilapia*

Anjani tilapia is the result of individual selection by the Freshwater Fish Nursery and Aquaculture Office Aik Mel, under the Provincial Department of Marine Affairs and Fisheries of West Nusa Tenggara Province. This tilapia variety stands out for its ability to reach a harvest weight of approximately 128 g in just 100 days and for its thicker meat texture. The Anjani tilapia was officially released through Decree No. 46 of 2012 (MMAF, 2012c) (Fig. 16.9).

Fig. 16.10 The Nilasa tilapia. (Source: MMAF, 2012d)

Fig. 16.11 The Salina tilapia. (Source: MMAF, 2014)

(i) *Nilasa tilapia*

The name "Nilasa" tilapia is derived from the Javanese term "Nila Satria," meaning "hero tilapia." This variety was developed at the Institute for Marine and Fisheries Technology Development, under the Provincial Department of Marine Affairs and Fisheries of Yogyakarta Province. Nilasa tilapia is the result of crossbreeding Citralada tilapia, Philippine tilapia, Singaporean tilapia, and Red NIFI tilapia. It is known for its rapid growth, high adaptability to new environments, and high fecundity.

Nilasa tilapia has demonstrated significant genetic improvement, with a weight value of 0.63 (male) and 0.65 (female), resulting in an average genetic gain of 18.72% for males and 2.01% for females. The selection process yielded a total response of 95.22% in males and 55.94% in females, improving growth from 205.6 g to 401.4 g in males and from 240.6 g to 386.4 g in females over four generations (Nugroho et al., 2016). MMAF (2012d) officially released the Nilasa tilapia under Decree No. 47 of 2012 (Fig. 16.10).

(j) *Salina tilapia*

Salina tilapia is a hybrid variety known for its tolerance to high salinity, which was developed through genetic improvement at the Agency for the Assessment and Application of Technology. This tilapia can thrive in salinity

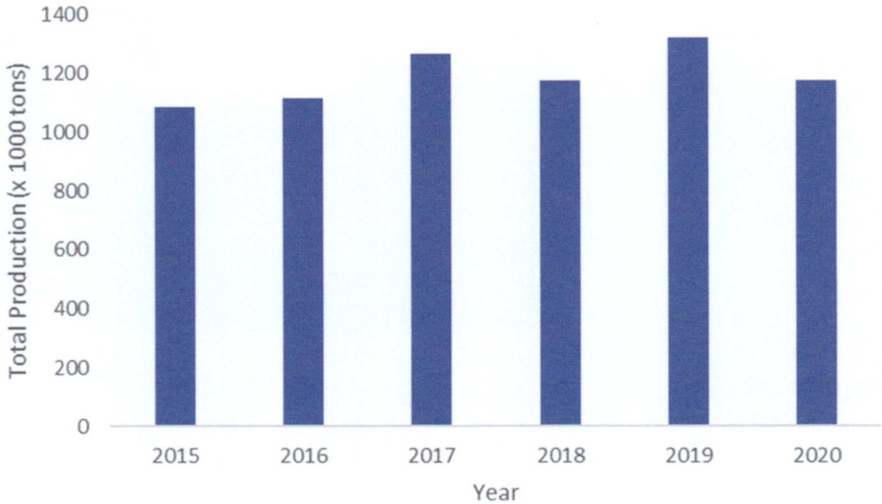

Fig. 16.12 National production of tilapia from 2015 to 2020. (Source: FAO, 2020)

levels of 20–25 ppt and features a dominant red color. It is also resistant to *Streptococcus* sp. bacteria and has a low feed conversion ratio (FCR), making it efficient for aquaculture. MMAF (2014) officially released the Salina tilapia under Decree No. 22 of 2014 (Fig. 16.11).

16.7 National Tilapia Production

Indonesia is now recognized as the second-largest producer of tilapia in the world, following China. Tilapia farming also rapidly expands in other Asian countries, such as Bangladesh, Philippines, Thailand, and Vietnam. These six countries contributed 95.3% of Asia's tilapia production in 2017. In Indonesia, tilapia aquaculture experienced relatively slow growth from 1970 to 2002. However, production increased significantly from 139,651 tons in 2004 to 1,084,281 tons in 2015, with an annual growth rate of 20%, reaching 1,172,633 tons in 2020, reflecting an annual average growth of 2.08% between 2015 and 2020 (FAO, 2020; MMAF, 2022) (Fig. 16.12). Due to its adaptability to a wide range of salinity levels, tilapia is cultivated in freshwater, brackish, and even saltwater environments, though freshwater farming remains the most prevalent.

The top three fish species produced in global aquaculture are grass carp (*Ctenopharyngodon idellus*) at 10.5%, silver carp (*Hypophthalmichthys molitrix*) at 8.8%, and Nile tilapia (*O. niloticus*) at 8.3% (FAO, 2020). The tilapia culture has expanded significantly and is now practiced in 125 countries, with global tilapia production reaching 5,880,586 tons in 2017. The leading producers include China (26.9%), Indonesia (22.4%), Egypt (16.5%), Bangladesh (5.7%), and Brazil (4.9%). According to FAO statistics, 16 tilapia/cichlid groups, along with some unidentified cichlids, have been used in aquaculture production, with the three most commonly farmed species being *O. niloticus* (70.24%), *O. aureus x O. niloticus* hybrid (6.74%), and *O. mossambicus* (1.05%) (El-Sayed, 2020a, b; FAO, 2019a; b).

In recent years, tilapia has become a key commodity in aquaculture development in Indonesia, contributing to the nation's fisheries industrialization program. The top five tilapia-producing provinces in Indonesia in 2017 were West Java (344,583.06 tons), South Sumatra (160,594.19 tons), West Sumatra (114,391.16 tons), North Sulawesi (91,571.39 tons), and North Sumatra (51,228.37 tons) (MMAF, 2018). West Java leads in tilapia production, largely due to the utilization of diverse ecosystems, such as the reservoirs at Cirata, Saguling, and Jatiluhur; running-water ponds in Subang; and brackish water ponds in the northern coastal region, where a polyculture system with shrimp is practiced. Tilapia farming in these reservoirs typically uses floating net cages (Hadie et al., 2018).

References

Arifin, O. Z., & Kurniasih, T. (2007). Variasi genetik tiga populasi ikan nila (*Oreochromis niloticus*) berdasarkan polimorfisme Mt-DNA [Genetic variation of three Tilapia (*Oreochromis niloticus*) populations based on Mt-DNA polymorphism]. *Jurnal Riset Akuakultur, 2*(1), 67–75. https://doi.org/10.15578/jra.2.1.2007.67-75

Arifin, O. Z., Nugroho, E., & Gustiano, R. (2007). Keragaman genetik populasi ikan nila (*Oreochromis niloticus*) dalam program seleksi berdasarkan RAPD [Genetic diversity of the tilapia (*Oreochromis niloticus*) population in a selection program based on RAPD]. *Berita Biologi, 8*(6), 465–471.

Bardach, J. E., Ryther, J. H., & McLarney, W. O. (1972). *Aquaculture: The farming and husbandry of freshwater and marine organisms* (868 pp). Wiley Interscience.

Barman, B. K., & Little, D. C. (2006). Nile tilapia (*Oreochromis niloticus*) seed production in irrigated rice-fields in Northwest Bangladesh an approach appropriate for poorer farmers? *Aquaculture, 261*(*1*), 72–79. https://doi.org/10.1016/j.aquaculture.2006.06.018

Bhujel, R. C. (2000). A review of strategies for the management of Nile tilapia (*Oreochromis niloticus*) broodfish in seed production systems, especially hapa-based systems. *Aquaculture, 181*, 37–59. https://doi.org/10.1016/S0044-8486(99)00217-3

De Graaf, G. J., Galemoni, F., & Huisman, E. A. (1999). Reproductive biology of pond reared Nile tilapia, *Oreochromis niloticus* L. *Aquaculture Research, 30*(1), 25–33. https://doi.org/10.1046/j.1365-2109.1999.00295.x

Edwards, P. (2009). Changes in traditional inland aquaculture in West Java, Indonesia. *Aquaculture Asia, 14*, 3–8.

Eguia, R. V., & Romana-Eguia, M. R. (2004). *Tilapia farming in cages and ponds* (40 pp). SEAFDEC Aquaculture Department.

El-Sayed, A. F. M. (2006). *Tilapia culture*. CABI Publishing, CAB International. Wallingford, Oxfordshire, UK. 277 p. https://doi.org/10.1079/9780851990149.0000

El-Sayed, A. F. M. (2020a). Environmental requirements. *Tilapia Culture*, 47–67. https://doi.org/10.1016/b978-0-12-816509-6.00004-5

El-Sayed, A. F. M. (2020b). The role of tilapia culture in rural development. *Tilapia Culture*, 275–295. https://doi.org/10.1016/b978-0-12-816509-6.00012-4

[FAO] Food and Agriculture Organization of the United Nations. (2019a). *Fishery and Aquaculture Statistics 2019/FAO annuaire. Statistiques des pêches et de l'aquaculture 2019/FAO anuario. Estadísticas de pesca y acuicultura 2019*. In *FAO Yearbook. Fishery and Aquaculture*

Statistics 2019/FAO annuaire. Statistiques des pêches et de l'aquaculture 2019/FAO anuario. Estadísticas de pesca y acuicultura 2019 (110 p). FAO. https://doi.org/10.4060/cb7874t

[FAO] Food and Agriculture Organization of the United Nations. (2019b). *Global aquaculture production 1950–2016.* http://www.fao.org/fishery/statistics/global-aquacultureproduction/query/en

[FAO] Food and Agriculture Organization of the United Nations. (2020). *The state of world fisheries and aquaculture*. https://doi.org/10.4060/ca9229en

Fitzsimmons, K. M. (2016). *Tilapia aquaculture 2016 and where will we be in 2026.* 11th International Symposium for Tilapia Aquaculture (ISTA), World Aquaculture Society-Asian Pacific Conference, 2016, Surabaya, Indonesia, April 26–29, 2016.

Gjedrem, T., Robinson, N., & Rye, M. (2012). The importance of selective breeding in aquaculture to meet future demands for animal protein: A review. *Aquaculture, 350–353*, 117–129. https://doi.org/10.1016/j.aquaculture.2012.04.008

Guerrero, R. D. I., & Guerrero, L. A. (1985). Effect of breeder size on fry production of nile tilapia in concrete pools. *Transactions of the National Academy of Science and Technology (Philippines), 7*, 63–66.

Gunadi, B., Setyawan, P., & Robisalmi, A. (2021). Pertumbuhan, hubungan panjang-bobot, dan faktor kondisi ikan nila NIFI (*Oreochromis* sp.) dan Srikandi (*Oreochromis aureus x niloticus*) pada pembesaran di tambak bersalinitas tinggi. [Growth, length-weight relationship, and condition factor of NIFI tilapia (*Oreochromis* sp.) and Srikandi tilapia (*Oreochromis aureus x niloticus*) in high-salinity pond culture]. *Jurnal Iktiologi Indonesia, 21*(2), 117–130.

Hadie, L. E., Kusnendar, E., Priono, B., Sinarni Dewi, R. R. S. P., & Hadie, W. (2018). Strategi dan kebijakan produksi pada budidaya ikan nila berdaya saing [Production strategies and policies in competitive tilapia aquaculture]. *Jurnal Kebijakan Perikanan Indonesia, 10*(2), 75. https://doi.org/10.15578/jkpi.10.2.2018.75-85

Hardjamulia, I. A., & Rukyani, A. (2000). Improved Nile tilapia (GIFT strain) cultured in Indonesia. *Naga: The ICLARM Quarterly, 23*, 38–39.

Komen, J., & Trong, T. Q. (2014). *Nile tilapia genetic improvement: Achievements and future directions*. Proceedings of the 10th International Symposium on Tilapia in Aquaculture (ISTA10), Jerusalem, Israel, 6–10 October 2014, pp. 1–9.

Little, D. C. (1992). *The development of fry production and nursing strategies for the Nile tilapia Oreochromis niloticus* (24 pp). Final report. Asian Institute of Technology.

Little, D. C., & Hulata, G. (2000). Strategies for tilapia seed production. In M. C. M. Beveridge & B. J. McAndrew (Eds.), *Tilapias: Biology and exploitation* (pp. 267–326). Kluwer Academic Publishers.

Mansour, C. R. (2001). *Nutritional requirements of Nile tilapia broodstock reared at different water salinities*. [PhD thesis, Alexandria University, Alexandria, Egypt].

Mires, D. (1982). A study of the problems of the mass production of hybrid tilapia fry, pp. 317–329. In R. S. V. Pullin & R. H. Lowe-McConnell (Eds.), *ICLARM Conference* (p. 432). International Center for Living Aquatic Resources Management.

[MMAF] Minister of Marine Affairs and Fisheries of the Republic of Indonesia. (2004). *Keputusan Menteri Kelautan dan Perikanan Nomor: KEP. 52/MEN/2004 tentang Pelepasan varietas ikan Nila JICA sebagai varietas baru [Decree Number KEP. 52/MEN/2004 on the release of JICA tilapia as a new variety*].

[MMAF] Minister of Marine Affairs and Fisheries of the Republic of Indonesia. (2006a). *Keputusan Menteri Kelautan dan Perikanan Nomor: KEP.44/MEN/2006 tentang Pelepasan Varietas Ikan Nila Gesit sebagai Galur Unggul Induk Jantan [Decree Number KEP.44/MEN/2006 on the release of Gesit tilapia as a superior male broodstock strain*].

[MMAF] Minister of Marine Affairs and Fisheries of the Republic of Indonesia. (2006b). *Keputusan Menteri Kelautan dan Perikanan Nomor KEP.45/MEN/2006 Tentang Pelepasan Varietas Ikan Nila Nirwana sebagai Varietas Unggul Induk Penjenis [Decree Number KEP.45/MEN/2006 on the release of Nirwana tilapia as superior Great Parent Stock variety*].

[MMAF] Minister of Marine Affairs and Fisheries of the Republic of Indonesia. (2009a). *Keputusan Menteri Kelautan dan Perikanan Nomor KEP.77/MEN/2009 tentang Pelepasan Varietas Ikan*

Nila BEST Sebagai Baku Unggul Ikan Nila [Decree Number KEP.77/MEN/2009 on the release of BEST tilapia as superior variety].

[MMAF] Minister of Marine Affairs and Fisheries of the Republic of Indonesia. (2009b). *Keputusan Menteri Kelautan dan Perikanan Nomor KEP.79/MEN/2009 tentang Pelepasan Varietas Ikan Nila Larasati Sebagai Benih Bermutu [Decree Number KEP.79/MEN/2009 on the release of LARASATI tilapia as superior seed*].

[MMAF] Minister of Marine Affairs and Fisheries of the Republic of Indonesia. (2012a). *Keputusan Menteri Kelautan dan Perikanan Nomor KEP.09/MEN/2012 tentang Pelepasan Ikan Nila Srikandi [Decree Number KEP.09/MEN/2012 on the release of Srikandi tilapia*].

[MMAF] Minister of Marine Affairs and Fisheries of the Republic of Indonesia. (2012b). *Keputusan Menteri Kelautan dan Perikanan Nomor KEP.28/MEN/2012 tentang Pelepasan Ikan Nila Sultana [Decree Number KEP.28/MEN/2012 on the release of Sultana tilapia*].

[MMAF] Minister of Marine Affairs and Fisheries of the Republic of Indonesia. (2012c). *Keputusan Menteri Kelautan dan Perikanan Nomor KEP.46/MEN/2012 tentang Pelepasan Ikan Nila Anjani [Decree Number* KEP.*46/MEN/2012 on the release of Anjani tilapia*].

[MMAF] Minister of Marine Affairs and Fisheries of the Republic of Indonesia. (2012d). *Keputusan Menteri Kelautan dan Perikanan Nomor KEP.47/MEN/2012 tentang Pelepasan Ikan Nila Merah Nilasa* [*Decree Number KEP.47/MEN/2012 on the release of Nilasa red tilapia*].

[MMAF] Minister of Marine Affairs and Fisheries of the Republic of Indonesia. (2014). *Keputusan Menteri Kelautan dan Perikanan Nomor 22/KEPMEN-KP/2014 tentang Pelepasan Ikan Nila Salina [Decree No. 22/KEPMEN-KP/2014 on the release of Salina tilapia*].

[MMAF] Ministry ofMarine Affairs and Fisheries of the Republic of Indonesia. (2018). *Kelautan dan Perikanan dalam Angka Tahun 2018* [Marine and Fisheries in Figures 2018]. The Center for Data, Statistics and Information. Ministry of Marine Affairs andFisheries of the Republic of Indonesia. 356 p.

[MMAF] Minister of Marine Affairs and Fisheries of the Republic of Indonesia. (2022). *Marine and fisheries in figures 2022* (348 p). The Center for Data, Statistics and Information. Ministry of Marine Affairs and Fisheries of the Republic of Indonesia.

Nugroho, E. (2018). *Aplikasi Ilmu Genetika dalam Program Pemuliaan di Perikanan Air Tawar [The application of genetic science in breeding programs for freshwater fisheries]* (40 p). AMAFRAD Press.

Nugroho, E., & Maskur. (2002). Benarkah ikan nila merah adalah hasil hybrid? Melacak asal usul nila merah dengan menggunakan molecular genetic marker [Is ed tilapia a result of hybridization? Tracing the origins of red tilapia using molecular genetic markers]. *Warta Penelitian Perikanan Indonesia, 8*(1), 8–11.

Nugroho, E., Saepudin, & Bajar, M. (2013). Kajian lapang penggunaan benih nila (*O. niloticus*) hasil pemuliaan di keramba jaring apung Jatiluhur [Field study on the use of cultivated nile tilapia (*O. niloticus*) seed in floating net cages in Jatiluhur]. *Jurnal Riset Akuakultur, 8*(1), 43–49.

Nugroho, E., Rustadi, Priyanto, D., Sulistyo, H., Susila, Sunaryo, & Wasito, B. (2014). Penurunan Keragaman genetik pada F-4 ikan nila merah "cangkringan" hasil pemuliaan dideteksi dengan marker genetik [Reduction of genetic diversity in F-4 of "Cangkringan" red tilapia from selective breeding detected using genetic markers]. *Jurnal Riset Akuakultur, 9*(1), 25–30.

Nugroho, E., Kusdiarti, Rustandi, Priyanto, D., & Sulstyo, H. (2016). Nilai heritabilitas dan genetic gain pada karakter bobot ikan nila hasil pemuliaan di Jogjakarta [Heritability and genetic gain values for weight traits on nile tilapia resulted from breeding program in Yogyakarta]. *Prosiding Inovasi Teknologi Akuakultur*, Pusat Riset Perikanan Budidaya, Surabaya, pp. 519–524.

Nugroho, E., Mayadi, L., & Budileksono, S. (2017). Heritabilitas dan perolehan genetik pada bobot ikan nila hasil seleksi [Heritability and genetic gain in weight of selectively bred nile tilapia]. *Berita Biologi, 16*(2), 129–135.

Nugroho, E., Khakim, A., & Dewi, R. S. S. P. S. (2020). The performance of tilapia culture in biofloc technology (BFT) and floating net cage (KJA) system as a candidate for the next tilapia culture in Indonesia. *Proceeding of IOP Conference Series: Earth and Environmental Science. Ecological intensification: A new paragon for sustainable aquaculture, 521*, 0120202.1–0120202.8. https://doi.org/10.1088/1755-1315/521/1/012020. IOP Publishing.

Phillips, M., Henriksson, P., Tran, N., Chan, C., Mohan, C., Rodriguez, U.-P., Suri, S., Hall, S., & Koeshendrajana, S. (2015). *Exploring Indonesian aquaculture futures* (p. 39). Program report.

Ponzoni, R. W., Nguyen, N. H., & Khaw, H. L. (2007). Investment appraisal of genetic improvement programs in Nile tilapia (*Oreochromis niloticus*). *Aquaculture, 269*(1–4), 187–199. https://doi.org/10.1016/j.aquaculture.2007.04.054

Rana, K. J. (1990). Reproductive biology and the hatchery rearing of tilapia eggs and fry. In J. F. Muir & R. J. Roberts (Eds.), *Recent advances in aquaculture* (Vol. 3, pp. 343–406). Croom Helm.

Ridha, M. T., & Cruz, E. M. (1999). Effect of different broodstock densities on the reproductive performance of Nile tilapia, *Oreochromis niloticus* (L.), in a recycling system. *Aquaculture Research, 30*(3), 203–210. https://doi.org/10.1046/j.1365-2109.1999.00311.x

Robisalmi, A., Gunadi, B., & Setyawan, P. (2020). Evaluasi performa pertumbuhan dan heterosis persilangan antara ikan nila Nirwana (*Oreochromis niloticus*) betina dengan ikan nila biru (*Oreochromis aureus*) jantan F2 pada kondisi tambak hipersalinitas [Evaluation of growth performance and heterosis in crossbreeding Nirwana tilapia (*Oreochromis niloticus*) females with F2 blue tilapia (*Oreochromis aureus*) males under hypersaline pond conditions]. *Berita Biologi, 19*(1), 1–11.

Robisalmi, A., Alipin, K., & Gunadi, B. (2021). Growth performance and microstructure in the muscle fiber of red tilapia (*Oreochromis* spp.) under different cycles of fasting and refeeding. *Aquaculture, Aquarium, Conservation & Legislation – International Journal of the Bioflux Society, 14*(3), 1498–1512.

Ross, L. G. (2000). Environmental physiology and energetics. In M. C. M. Beveridge & B. J. McAndrew (Eds.), *Tilapias: Biology and exploitation* (pp. 89–128). Kluwer Academic Publishers.

Rothbard, S. M. (1979). Observation on the reproductive behaviour of Tilapia zillii and several *Sarotherodon* spp. under aquarium conditions. *Bamidgeh, 31*, 35–43.

Saanin, H. (1968).Taksonomi dan Kuntji identifikasi ikan [Taxonomy and Identification Key for Fish]. *Binatjipta*, Bandung. 262 p

Setyawan, P., Gunadi, B., Listiyowati, N., & Robisalmi, A. (2022). *Petunjuk Teknis Budidaya Ikan Nila Srikandi [Technical Guidelines for Cultivating Srikandi Tilapia]* (1st ed., 21 p). AMAFRAD Press.

Trewavas, E. (1982). Tilapias: Taxonomy and speciation. In R. S. V. Pullin & R. H. Lowe-McConnell (Eds.), *The biology and culture of Tilapias, ICLARM conference proceedings 7* (pp. 3–15). International Center for Living Aquatic Resources Management.

Widiyati, A., & Sudarto. (1997). Evaluasi pertumbuhan beberapa strain ikan nila (Nila 69, Nila GIFT, dan Nila Chitralada) [Growth evaluation of several strains of nile tilapia (69 tilapia, GIFT tilapia, and Chitralada nile tilapia)]. *Prosiding Hasil Penelitian Perikanan Air Tawar* 1995/1996. Balitkanwar. Sukamandi, pp. 44–49.

The Status of Domestication and Aquaculture System on the Giant Gourami *Osphronemus goramy* in Indonesia

17

Sularto, Nunuk Listiyowati, and Nurdiansyah

17.1 Introduction

The giant gourami *Osphronemus goramy* Lacépède (1801) is a freshwater fish native to Indonesia. Giant gourami has high economic value with a price on the market of around Rp.30,000–40,000/kg. The giant gourami has been reared for decades in Indonesia especially Java, Sumatra, and Borneo islands as main gourami producer. Five countries in Southeast Asia that are also being gourami producers are Thailand, Myanmar, Malaysia, the Philippines, and Singapore (FAO, 2023). This fish has become one of the main species being farmed and is in great demand in the food aquaculture industry (Amornsakun et al., 2014).

However, gourami is known as a slow growth fish, so it becomes an obstacle in cultivation of gourami development culture. The Indonesian government through the Ministry of Marine Affairs and Fisheries (MMAF) has released some strains of gourami with a fast growth character to support the increase of the gourami production. The newest released strain was the BIMA strain, which was a hybrid gourami of Majalengka (female) and Jambi (male) populations (Fig. 17.1). This BIMA strain was officially released by the Minister of Marine Affairs and Fisheries under Decree Number 56 of 2021. So far, with an estimated production in excess of 145,000 tons in 2017, Indonesia is the most important producer of giant gourami and its production represents more than 98% of the total world production.

Sularto (✉) · N. Listiyowati · Nurdiansyah
Research Center for Fishery, Research Organization for Earth Sciences and Maritime, National Research and Innovation Agency, Jakarta, Indonesia
e-mail: sula015@brin.go.id

E. Nugroho et al. (eds.), *Aquaculture Development in Archipelago*,
https://doi.org/10.1007/978-981-95-1084-9_17

Fig. 17.1 Giant gourami BIMA *Osphronemus goramy*, Lacépède 1801, a hybrid gourami strain of Majalengka and Jambi population. (Source: personal documentation)

17.2 Taxonomy and Morphology

Giant gourami fish has scientific name: *Osphronemus goramy* Lacépède, 1801; Synonym name: *Trichopus goramy*, *Osphronemus gouramy*, *Osphronemus gorami*, *Osphronemus olfax*, *Osphronemus notatus*; English common name: Giant gourami; Indonesian common name: Ikan gurami, gurame. Based on Integrated Taxonomic Information System Report, *Osphronemus goramy* Lacepède, 1801 TSN 638762, Taxonomic Hierarchy was as follows:

Phylum: Chordata
Class: Teleostei
Order: Perciformes
Family: Osphronemidae Bleeker, 1859
Genus: *Osphronemus* Lacepède, 1801
Species: *Osphronemus goramy* Lacepède, 1801

The morphometric and meristic characters of Indonesian gourami populations have been studied extensively. According to earlier reports by Setijaningsih et al. (2007), Azrita and Syandri (2015), Suharyanto et al. (2016), Arifin et al. (2017), and Radona et al. (2017), the morphometric and meristic characteristics of Indonesian gourami populations generally match the values of morphological characteristics of the key identification of the gourami species *Osphronemus goramy* published on the Fishbase website (2015). These include the dorsal spines (total): 12–14; dorsal soft rays (total): 10–13; anal spines: 9–13; anal soft rays: 18–21; and vertebrae: 30–31. These values serve as critical reference points for identifying this species based on morphometric characteristics. The specific studies mentioned above likely explored these parameters in greater depth but focused on broader aspects such as distribution patterns and genetic variations among Indonesian gourami populations. The juvenile body has 8–10 complete black vertical stripes, while in the adult phase the vertical black stripes are absent (disappear) or sexual dichromatism, both sexes drab; transverse scale rows usually 6.1.12; dorsal fin spines usually 12–13 (rarely 11 or 14); soft-rayed portion of anal fin greatly enlarged, its distal margin parallel to distal margin of caudal fin; caudal fin rounded or obtusely rounded, not truncate or emarginate. Pelvic fins with first soft ray prolonged into a threadlike tentacle reaching posteriorly to or beyond hind margin of caudal fin. Gourami have the ability to breathe air through a specific organ called a labyrinth that allows fish to survive in deoxygenated areas. Rather than the air bubbles that distinguish smaller labyrinth fishes, this species reproduces by building a nest out of twigs and plant remnants collected from the aquatic vegetation (FAO, 2023).

17.3 Distribution

The giant gourami is an important freshwater commercial fish native to Indonesia. This species belongs to the family Osphronemidae and the suborder Anabantoidei, and it has been introduced and disseminated throughout Southeast Asia and China (Welcomme, 1985; Setijaningsih et al., 2007; Nugroho & Kusmini, 2007; Low & Lim, 2012). Froese and Pauly (2007), in Nuryanto et al. (2018), retrieved that in Indonesia, the geographical distribution of gourami starts from Java, Sumatra, and the Kalimantan islands. The populations of wild giant gourami are now very rare, but some residual populations are known in the Batanghari River in Sumatra and in the rivers of Kalimantan. Giant gourami prefers calm waters, especially slow-moving areas such as lakes and large rivers. Distinguish gourami found in West Java Province into six strains based on egg production, growth performance, and maximum size or weight of adult gourami. The six strains are soang, japanese, blue saphire, paris, bastar, and porcelain. In some provinces of Sumatra Island, gourami can be found with different morphological characteristics. At the Batanghari River of the Jambi Province, a strain of gourami was found with the characteristics of thick meat with a wide posture and can grow better in swampy waters, which was later named Batanghari gourami.

The distribution of the giant gourami in Southeast Asia in the year 2000 based on four environmental variables, that is, NDVI (Normalized Difference Vegetation Index), precipitation, surface temperature, and river flow accumulation and human population density showed that giant gourami are distributed in a wide range of habitat types, including relative warmer surface temperatures, a range of flows, and in areas with range of precipitation. The distribution of the giant gourami is not significantly influenced by human population density. Furthermore, it is anticipated that the distribution of this species will remain relatively stable in the face of projected climate warming, although this assertion remains untested (Stephani, 2016).

17.4 Reproduction

Natural spawning of gourami is commonly carried out by farmers. Artificial spawning in gourami with ovaprim injection has been initiated, but the success rate was still very low (Arfah et al., 2006). Giant gourami reaches full sexual maturity when the body weight exceeds 2.5 kg or after 2–3 years old. Some recommendations can be taken into consideration to maximize sexing success: Sexually mature individuals should look for the hump on the upper part of the head, the thickening of the lower jaw, and the pigmentation on the pectoral fin to identify the black phenotype. The pigmentation on the pectoral fin is not a relevant criterion for identifying the white phenotype. The observation of other morphological criteria (such as body shape, caudal fin edging, or fish behavior) generally used by fish farmers does not improve sexing success. Therefore, particular attention should be given to the examination of the criteria listed above. Sexing should be confirmed by urogenital cannulation and intraovarian biopsies, when the necessary skills and materials are available. Whenever possible, and under good fish welfare, the cannulation method should always be preferred for sexing gourami brood stock. Preselection of future brood stock from sexually immature individuals should be based on the reliable morphological criteria identified in the present study, emphasizing color at the base of the pectoral fin in individuals with the black phenotype. Sexing should be confirmed when the sexual maturity is reached and before using these individuals for reproduction. Slembrouck et al. (2019) showed that sexing errors can occur when selecting giant gourami brood stock for reproduction based on observations of morphological criteria.

Gourami fish can spawn all year round, with no seasonal effect. The most facilities used for the reproduction stage are earthen or concrete ponds. Spawning ponds are partitioned with nets or bamboo fences into several compartments generally with a size of 20 m^2 for 4–5 broods, or 1 brood stock requires an area of 4 m^2. The typical sex ratio in giant gourami farming is one male to three females (1:3); however, it can also range from 1:2 to 4 depending on the farmer's experience and the availability of brood stock.

To improve reproductive efficiency fish farmers always select larger males than females. In spawning ponds, males and females are introduced at the same time in order to avoid the occurrence of strong male aggressiveness. Once the breeding pair is matched (from 1 day to 1 month or more), both male and female start to build the nest. In Java, fish farmers often use small bamboo fiber baskets (called "sosog") as nest supports, and vegetable fibers, generally of palm trees (Arenga species), are provided for the brood stock to make their nests. Fish can also nest in holes dug by fish farmers along the pond banks. Relative fecundity may vary according to the strain ranging from 1300 to 2500 eggs per kg of female; egg diameter is around 2–3 mm; eggs are buoyant due to the presence of yolk and a very large oil globule. After spawning, brood stock close the nest with palm fibers to protect the eggs, and then the male guards it by showing territorial behavior. Fish farmers monitor the construction of the nest daily and they harvest the eggs immediately after spawning or at a maximum of 2 days later. Eggs are incubated in static water in covered plastic basins, which are simply placed on the pond bank, under a shed, or in a small hatchery. However, some artisanal hatcheries use more sophisticated structures that are equipped with aquaria, wooden or fiberglass tanks, and aerated water. Even recirculating aquaculture system (RAS) may also be used. The egg density during incubation is highly variable, but 4–5 eggs per square centimeter is recommended by the Indonesian National Standard (SNI).

Giant gourami eggs are valuable commodities. For example, in Central Java, a cluster of fish farmers specialized in egg production can send millions of eggs per week across the country. Hatching occurs around 35–40 h after fertilization, and larvae can swim rapidly, but yolk resorption continues until 13–15 days after hatching (dah). Larvae can be fed with zooplankton as early as 4–5 dah. But generally, they are fed from 8 to 12 dah. At this age, larvae can be sold or transferred to nurseries. The male and female gourami are distinguished by their dorsal fins and body color; the male's dorsal fin ends in a point, and the body is darker, turning nearly black during spawning. The eggs will be laid within the next 24 h after the nest is built, and the water in the tank should be lowered to about 20 cm (8 in) deep. The giant gourami is an egg layer that constructs a nest from plant fibers (Arifin et al., 2020). The female should be moved to a different tank after spawning because the male will jealously protect the eggs in a captivity setting and may even become hostile toward the female. The eggs need to be kept in a dark aquarium/round plastic tub once they hatch, which takes 24 h (FAO, 2023).

17.5 Domestication and Breeding Program

Giant gourami, in most cases probably originating from Indonesia, have been introduced widely into other tropical and temperate countries, including tropical islands. There have been 24 documented translocations worldwide, with the first being in the nineteenth century and going to Algeria, France, India, and Madagascar. According to FishBase, giant gourami are present, either native or introduced, in 20 countries.

Gourami is a slow-growing fish, so breeding efforts are needed to produce new strains of gourami having superior growth character. Efforts to improve the growth performance of gourami at the Sukamandi Research Institute for Fish Breeding (RIFB), MMAF had been carried out through a breeding program entitled "Formation of Superior Varieties of Fast-Growing Gourami Fish," which began in 2014. The breeding program began with collecting gourami from several different locations, namely, Majalengka and Tasikmalaya in West Java Province, Jambi in Jambi Province, and Banjar in South Kalimantan Province. The results of morphological and genotypic characterization showed that the Majalengka and Jambi populations have the most distant kinship, so they have the potential to be used in hybridization programs. The cross between the male broods of the Jambi gourami and the female broods of the Majalengka gourami had a high heterosis value, measuring 32.68% at 14 months, according to the results of the evaluation of the growth performance of the two-way crosses (diallel crossing) of the four populations. This shows that the hybrid gourami resulting from the crossing is prospectively developed as a high-growing variety of superior gourami. In 2018, the hybrid gourami were evaluated through multisite tests, disease resistance tests, and tolerance tests to environmental conditions. Multilocation tests in Tulungagung (East Java Province) and Cilacap (Central Java Province) showed that the hybrid gourami had a growth rate of 26.97% and 21.72% higher than that of local gourami, and in Sukamandi, the hybrid gourami also had a growth of 17.18% higher than the Galunggung Super gourami. In Bogor, hybrid gourami was more resistant to disease infection, with survival rate reaching 96% at a time when local gourami had experienced high mortality (survival rate of 3.81%). The challenge test with bacterial infections *Aeromonas hydrophila* at LD50 6.05 × 10^6 CFU/mL and *Mycobacterium fortuitum* at LD50 8.86 × 10^8 CFU/mL showed that the hybrid gourami fry had a higher resistance (44.39% survival after 1 day of *A. hydrophila* injection and 73.33% post-4-week injection of *M. fortuitum*) than the Galunggung Super gourami fry released by Institute for Gurame and Nilem Fish Farming Development Center Singaparna. The results of the environmental tolerance test to exposure to temperatures of 40 °C, salinity of 12 and 26 g/L, and an alkaline pH of 10.25 showed that the hybrid gourami fry had a relatively higher tolerance resistance than Galunggung strain. In addition to having faster growth, the overall test showed that hybrid gourami between Majalengka female and Jambi males, also have higher tolerance to environmental stress and better resistance to disease. In 2021, the hybrid gourami was officially released by MMAF under Decree number 56 of 2021, which named the gourami as “BIMA.”

17.6 Aquaculture System

The significant expansion of the fisheries and aquaculture production came with many transformations, including changes in the source of production being increasingly dependent on aquaculture. In the last three decades, aquaculture has been the main driver of the increase in fisheries and aquaculture production, with an average growth of 5.1% per year in the period 2000–2020, reaching a record of 87.5 million ton in 2020. Aquaculture accounted for 49% of the global total fish production. Yet, its share in total production varied significantly across continents, going from over 62% in Asia in 2020 to 20% in the Americas, 19% in Europe and Africa, and 13% in Oceania (FAO, 2022). Gourami farming technology is carried out ranging from traditional, semi-intensive, and intensive techniques. In Indonesia, there is a national standard for each production phase of the rearing of giant gourami and Indonesian good aquaculture practice (GAP) is available. Indonesian authorities promote the application of GAP and certify compliance. So far, genetic studies have shown that local strains presenting slight phenotypical differences are genetically undifferentiated and all strains are the same species (*O. goramy*). Partial sequencing of the cytochrome b gene has shown that all these strains have 99% similarity with *O. goramy*. The production cycle is the same, whatever strains are reared (FAO, 2023).

Typically, gourami grow-out production in West Java occurs in 4–5 static water small earthen ponds (300–1000 per square meter) with a depth of 0.5–1.5 m. The most important criterion of farming is the “strain” of giant gourami, but other criteria are also employed (e.g., price, size, and fish condition). Fish are usually stocked at 5–10/m^2, during grow-out, fish may be graded and their density is reduced if necessary. Growth rate is quite slow and giant gourami may reach commercial size

(500–750 g) within 9–14 months. To achieve larger sizes (1–2 kg), an extended period of grow-out rearing is required, generally comprising 25–35 months in total. Although strong variations exist according to strain, rearing conditions and water temperature, the time required to reach commercial sizes mainly depends on the feeding strategy. Giant gourami are fed with commercial floating pellets (extruded) containing ~30–33% of crude protein. The feeding rate is generally 2–3% of biomass. Commercial fish food and raw plant material are the rule for feeding giant gourami during grow-out. Giant taro (*Alocasia machrorrhizos*), taro (*Colocasia esculenta*), and cassava (*Manihot esculenta*) are the plants mostly used, but other species of plants may also be used. Plants may be distributed 2–7 days per week; sometimes they are cut, but usually they are distributed entirely on the water surface. The ratio between pellets and vegetable material varies according to fish size and feed cost and may range between 5% and 100%. Generally, fish farmers tend to use much more raw vegetables when fish are nearly commercial size. Feed is generally distributed twice a day, 7 days a week, but in some cases, the distribution of feed, particularly pellets, may be completely ceased, due to economic reasons (FAO, 2023).

17.7 National Production of Giant Gourami

Indonesia and five other countries including Thailand, Myanmar, Malaysia, the Philippines, and Singapore are reported as gourami producers according to FAO; however, other Southeast Asian countries such as Vietnam and the Lao People's Democratic Republic are known to produce unreported amounts of gourami (Azrita, 2020). Based on data of fishery and aquaculture statistics global production in FishStatJ (FAO, 2023), the global farmed production of giant gourami reached 241.326 tons in 2017, and Indonesia accounts for more than 98.4% of total production, having expanded nearly ninefold since its production of 16.438 tons in 2002. Based on Statistik KKP, in the 2019–2023 time frame, the national production volume of giant gourami farming in Indonesia ranged from 149,170 to 190.289 tons, with the highest production in 2019 (Fig. 17.2). In Indonesia the production of giant gourami is spread throughout all the major islands of the country; however, Java and Sumatra together produce more than 94% of the total national production (60% and 34%, respectively) (BPS, 2013).

Market segmentation is a characteristic of gourami production in Indonesia. Giant gourami may be sold at different sizes (from eggs to brood stock) in which 11 commercial sizes are officially identified in the country. For each segment, the market price varies according to the season, size, and quality of fish. The market price increases with size and is enhanced when fish are sold live. The market size may be up to 500 g; however, there has been a trend toward the purchase of smaller fish (300 g), which are now available in some markets (Fig. 17.2).

The market size is still increasing and prices are stable. Nevertheless, it is possible that production expansion may be limited by the shortage of juveniles, both in quantity and quality. Diversified products (fillets) are present in urban markets. A

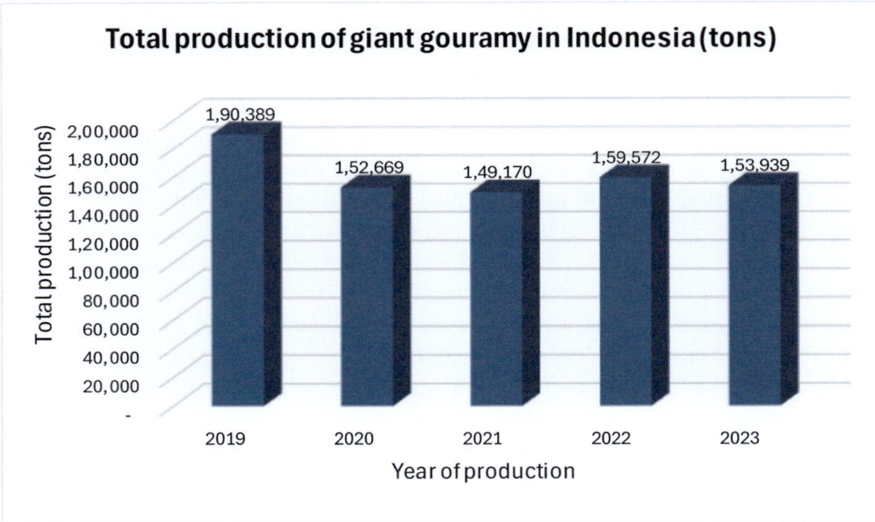

Fig. 17.2 National production volume of giant gourami farming in Indonesia (tons). (Source: Statistik KKP, 2019–2023)

sustainable approach toward small-scale farming is essential to maintain and increase gourami production. Providing technical assistance to the fish farmers to find specific solutions in response to the challenges generated by a changing and competitive environment is required.

References

Amornsakun, T., Kullai, S., & Hassan, A. (2014). Some aspects in early life stage of giant gourami, *Osphronemus goramy* (Lacepède) larvae. *Songklanakarin Journal of Science and Technology, 36*(5), 493–498.

Arfah, H., Maftucha, L., & Carman, O. (2006). Pemijahan Secara Buatan Pada Ikan Gurame Osphronemus gouramy Lac. Dengan Penyuntikan Ovaprim [Artificial spawning in gourami fish Osphronemus gouramy Lac. with ovaprim injection]. *Jurnal Akuakultur Indonesia, 5*(2), 103–112. http://journal.ipb.ac.id/index.php/jai. http://jurnalakuakulturindonesia.ipb.ac.id.

Arifin, O. Z., Imron, Muslim, N., Hendri, A., Aseppendi, & Yani, A. (2017). Karakteristik fenotipe dan genotipe ikan gurami, Osphronemus goramy, strain Galunggung Hitam, Galunggung Putih dan hibridanya [Phenotype and genotype characteristics of gourami fish, Osphronemus goramy, Black Galunggung, White Galunggung strains and their hybrids]. *Jurnal Riset Akuakultur, 12*(2), 99–110.

Arifin, O. Z., Slembrouck, J., Subagja, J., Pouil, S., Yani, A., Asependi, A., Kristanto, A. H., & Legendre, M. (2020). New insights into giant gourami (*Osphronemus goramy*) reproductive biology and egg production control. *Aquaculture, 519*, 734743. https://doi.org/10.1016/j.aquaculture.2019.734743. ISSN 0044-8486.

Azrita. (2020). *Gurami Sago (Osphronemus goramy Lac): Komiditi Unggulan Perikanan Air Tawar Untuk Pangan dan Ikan Hias [Superior commodity of freshwater fisheries for food and ornamental fish]*. LPPM Universitas Bung Hatta, Mei 2020. ISBN 978-623-93573-3-7. 163 hal.

Azrita, & Syandri, H. (2015). Morphological character among five strains of giant gourami, Osphronemus gouramy Lacépède, 1901 (Actinopterygii: Perciformes: *Osphronemidae*) using a truss morphometric system. *International Journal of Fisheries and Aquatic Studies, 2*(6), 344–350.

Badan Pusat Statistik. (2013). Jumlah rumah tangga usaha budidaya bukan ikan hias menurut wilayah dan jenis ikan utam [Number of non-ornamental fish farms by region and fish species]. https://st2013.bps.go.id/dev2/index.php/site/tabel?tid=57&wid=0. Accessed September 13, 2019.

Fishbase. (2015). *Osphronemus goramy* Lacépède, 1801. www.fishbase.org

Food and Agriculture Organization. (2022). *World food and agriculture – Statistical yearbook 2022*. https://doi.org/10.4060/cc2211en

Food and Agriculture Organization. (2023). *Osphronemus goramy*. Cultured Aquatic Species Information Programme. Text by Caruso, D., Arifin, Z.O., Subagja, J., Jacques Slembrouck, J. and New, M. Fisheries and Aquaculture Division [online]. Rome. [Cited Sunday, March 12th 2023]. https://www.fao.org/fishery/en/culturedspecies/osphronemus_goramy/en

Low, B. W., & Lim, K. K. P. (2012). Gouramies of the genus Trichopodus in Singapore. *Nature in Singapore, 5*, 83–93.

Nugroho, E., & Kusmini, I. I. (2007). Evaluasi variasi genetik tiga ras ikan gurame (*Osphorenemus gouramy*) dengan metodee isozyme. *Jurnal Riset Akuakultur, 2*, 51–57.

Nuryanto, A., Amalia, G., Khairani, D., Pramono, H., & Bhagawati, D. (2018). Molecular characterization of four giant gourami strains from Java and Sumatra. *Biodiversitas, 19*(2), 578–584. https://doi.org/10.13057/biodiv/d190228

Radona, D., Nafiqoh, N., & Arifin, O. Z. (2017). Studi biometri berdasarkan meristik dan morfometrik ikan gurami galur Bastar dan Bluesafir [Biometry study based on meristic and morphometric gourami fish strains Bastar and Bluesapphire]. *Berita Biologi, 16*(2), 121–127.

Setijaningsih, L., Arifin, O. Z., & Gustiano, R. (2007). Karakterisasi tiga strain ikan gurami (*Osphronemus gouramy* Lac.) berdasarkan metode truss morfometriks [Characterization of three strains of gourami fish (*Osphronemus gouramy* Lac.) based on the morphometric truss method]. *Jurnal Iktiologi Indonesia, 7*(2), 23–30.

Slembrouck, J., Arifin, O. Z., Pouil, S., Subagja, J., Yani, A., Kristanto, A. H., & Legendre, M. (2019). Gender identification in farmed giant gourami (*Osphronemus goramy*): A methodology for better broodstock management. *Aquaculture, 498*, 388–395.

Statistik KKP. Volume Produksi Perikanan Budidaya Pembesaran per Komoditas Utama (ton). [Production Volumeof Aquaculture Farming by Major Commodity (ton).https://prtaldata.kkp.go.id/portals/data-statistik/prod-ikan/tblstatis/d/53

Stephani, I. (2016). Exploring giant gourami distribution in Southeast Asia. GEO599/GEO584-Advanced Spatial Statistics and GIS, 2013–2016 (oregonstate.edu).

Suharyanto, Febrianti, R., & Sularto. (2016). Karakterisasi empat populasi ikan gurami (*Osphronemus goramy* Lac.) dan persilangannya berdasarkan metode truss morfometriks [Characterization of four gourami (*Osphronemus goramy* Lac.) populations and their crosses based on the truss morphometric method]. *Jurnal Riset Akuakultur, 11*(2), 125–135.

Welcomme, R. L. (1985). *River fishery* (p. 330). FAO Fish. Tech. Paper (262).

Review of *Clarias* Catfish Culture in Indonesia

18

Bambang Iswanto, Huria Marnis, Rommy Suprapto, and Imron

18.1 Introduction

Clarias catfish (ikan lele) is one of the primary commodities of freshwater aquaculture in Indonesia, which plays an essential role in supporting national food security. *Clarias* catfish has been cultivated and consumed in almost all Indonesian regions. Its production ranks second after its growth and resistance to being cultivated in the area with limited space and water availability (Gustiano et al., 2021).

The development of *Clarias* catfish farming in Indonesia has gone through a long process. *Clarias* catfish farming began in Java in the 1970s using a local species *Clarias batrachus* (Zonneveld et al., 1988). The farming activities overgrown with the introduction of a new African catfish *C. gariepinus* from Taiwan in 1985, which was named the Dumbo. The development of Dumbo African catfish farming encouraged the subsequent introductions of the African catfish from other countries (Iswanto, 2013).

The development of African catfish farming also encourages efforts to increase the efficiency and productivity in aquaculture. Efforts to increase the efficiency and productivity of African catfish farming are carried out through improvements in cultivation technology and genetic improvement (breeding). Improvements in aspects of cultivation technology are performed in the fields of feed (nutrition), pathology, and environment. As a result of these efforts, several African catfish farming systems have been established in Indonesia. On the other hand, the increasing productivity is achieved through genetic improvement applied by selective breeding, crossbreeding, and genetic-engineering methods. Currently, most of the

B. Iswanto (✉) · H. Marnis · R. Suprapto · Imron
Research Center for Fisheries, Research Organization for Earth Sciences and Maritime, National Research and Innovation Agency, Cibinong, Indonesia
e-mail: bamb068@brin.go.id; huria.marnis@brin.go.id

E. Nugroho et al. (eds.), *Aquaculture Development in Archipelago*,
https://doi.org/10.1007/978-981-95-1084-9_18

African catfish cultivated in Indonesia are superior varieties resulting from the genetic improvement program.

18.2 Taxonomy and Distribution of Cultured *Clarias* Catfish

Indonesia has many species of local *Clarias* catfish, but only the *C. batrachus* has been used in aquaculture since the 1970s (Zonneveld et al., 1988; Knud-Hansen et al., 1990). Later, *C. meladerma,* another local *Clarias* catfish, has also been domesticated since the 1990s (Sudarto, 1993). However, farming of these local species had not developed, because Asian *Clarias* catfish species generally have low disease resistance and growth (Lenormand et al., 1999). Mostly, *Clarias* catfish farmers in Indonesia preferred in cultivating the introduced African catfish, due to having superior growth, disease resistance, and environmental tolerance (Viveen et al., 1985; Na-Nakorn & Brummet, 2009). Thus, the *Clarias* catfish species that have ever been used in aquaculture in Indonesia are the local *Clarias* catfish *C. batrachus* and *C. meladerma*, and the introduced African catfish. Besides the African catfish, the local Thai *Clarias* catfish *C. macrocephalus* was also introduced and used in hybridization activities in Mojokerto Regency, East Java Province, by PT Matahari Sakti in 2010 (2011, "personal observation"). However, no further information exists related to the developmental status of the introduced *C. macrocephalus* in Indonesia.

In general, the taxonomy of Asian and African catfish species used in aquaculture in Indonesia are as follows (Ferraris, 2007):

Phylum:	Chordata
Class:	Teleostei
Order:	Siluriformes Cuvier, 1816
Family:	Clariidae Bonaparte, 1846
Genus:	*Clarias* Scopoli, 1777
Species:	*Clarias batrachus* (Linnaeus, 1758)
	Clarias meladerma Bleeker, 1846
	Clarias gariepinus (Burchell, 1822)

As with other fish species, the systematics of the *Clarias* catfish changed several times since it was first described. A systematic revision of the African catfish species based on the biometric-morphological characteristics has been conducted (Teugels, 1986). The African catfish whose potential for aquaculture species has been introduced and widely cultivated in various countries worldwide (Na-Nakorn & Brummet, 2009). The African catfish is naturally present in almost all regions of the African continent and Minor Asia (Teugels, 1986). Initially, the species of African catfish widely cultivated in Africa was known by three different names, namely *C. gariepinus*, *C. lazera*, and *C. mossambicus*. However, the valid scientific name is *C. gariepinus* (Teugels, 1984).

The African catfish was first introduced to Indonesia in 1985 from Taiwan and the Netherlands. However, only introduction from the Netherlands was recorded in the Database of Invasive Alien Species (DIAS) of Food and Agricultural Organization (FAO) (Na-Nakorn & Brummet, 2009). The African catfish from the Netherlands was mainly used for collaborative research between the Agricultural University of Wageningen and University of Brawijaya in Malang Regency. More recently, a genetic study implied that the Dutch African catfish from the Netherlands was likely the hybrid between African catfish *C. anguillaris* and *C. gariepinus* (Anene & Gao, 2007). On the other hand, the *Clarias* catfish introduced from Taiwan was developed for aquaculture use and became widely known as the Dumbo (Fig. 18.1). Unfortunately, the taxonomic status of the Dumbo *Clarias* catfish was not well documented. In the official governmental documents, their scientific names were written differently, either as the African catfish *C. gariepinus* species or a hybrid between the Taiwanese local *Clarias* catfish *C. fuscus* and the African catfish *C. gariepinus* (Iswanto, 2013). In addition, limited study has been done in Indonesia to ascertain the taxonomic status of the Dumbo *Clarias* catfish. Based on microsatellite DNA analysis, Sudarto (1999) concluded that the Dumbo *Clarias* catfish remained a distinct species of African catfish. However, prior to that another African catfish has been introduced from the Netherlands and Thailand (Iswanto et al., 2015a), so it was still possible to result in a sampling misidentification. Furthermore, biometric-morphological analysis revealed that the Dumbo *Clarias* catfish was found to be identical to the African catfish (Iswanto et al., 2017). In addition, hybridization experiments between the local *Clarias* catfish *C. fuscus* and African catfish in China suggested that the male hybrid was sterile (Wu et al., 1990). In contrast, male broodstocks of the Dumbo *Clarias* catfish are fertile and could be used to

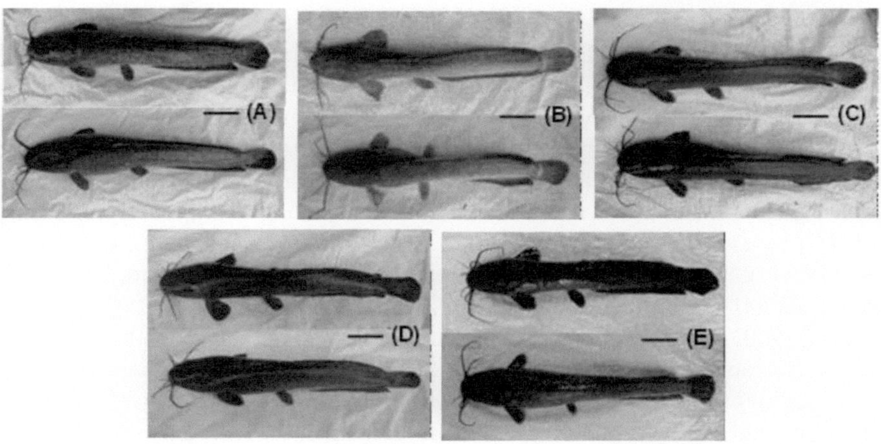

Fig. 18.1 African catfish *Clarias gariepinus* introduced to Indonesia. Note: Dumbo African catfish (**a**), Paiton African catfish (**b**), Egyptian African catfish (**c**), Masamo African catfish (**d**), and Kenyan African catfish (**e**) (upper = male, lower = female, bar scale = 10 cm). (Source: Personal documentation)

produce seeds. Those results suggested that the Dumbo *Clarias* catfish should belong to the species of African catfish.

Besides introduced from the Netherlands and Taiwan, the African catfish *C. gariepinus* was also introduced to Indonesia on multiple occasions from Thailand in 1998 and 2010, from Egypt in 2007 and from Kenya in 2010 (Iswanto et al., 2015a). These introductions of African catfish were not documented in DIAS records (Na-Nakorn & Brummet, 2009). These introduced strains were then popular among the fish farmers under the given names of Egyptian African catfish, Kenyan African catfish, Paiton African catfish, and Masamo African catfish (Fig. 18.1). However, the farmers usually use the outdated term Dumbo African catfish to refer to African catfish whose identity and ancestry are unknown. In addition, biometric-morphological characterization generally revealed no distinctions between the introduced African catfish strains (Iswanto et al., 2015a).

Currently, the African catfish used in aquaculture in Indonesia are mainly the varieties resulting from the breeding program. Four superior varieties of the African catfish have been officially released by the Ministry of Marine Affair and Fisheries (MMAF), namely Sangkuriang African catfish in 2004, Mandalika African catfish in 2013, Sangkuriang 2 African catfish in 2014 and Mutiara African catfish in 2015. Those African catfish varieties, mainly the Mutiara African catfish, have been distributed and cultivated throughout nearly all of Indonesia regions (Pangestika et al., 2019). Therefore, it has been one of the most important freshwater aquaculture commodities in Indonesia, which contributes significantly to the Indonesian aquaculture production and improves the welfare of Indonesian people. In addition to being normal colored, there are albino African catfish in Indonesia. The albino African catfish has been characterized in terms of biometric-morphology, growth, embryonic, and larval development (Iswanto et al., 2015b, 2016c, 2020a, 2021), among others that were studied by previous workers from other countries.

As with the African *Clarias* catfish species, a systematic revision of the Asian *Clarias* catfish species has also been carried out based on biometric-morphological and genetic characteristics (Sudarto, 2003; Sudarto & Pouyaud, 2005). In addition to redescribing the morphological characteristics (as the identification keys) of local Asian catfish species already recorded, the systematic revision led to the discovery of new local *Clarias* catfish species in Indonesia. Once upon a time there are 15 known local *Clarias* catfish species in Indonesia, including *C. batrachus* (Linnaeus, 1758), *C. nieuhofii* Valenciennes, 1840, *C. meladerma* Bleeker, 1846, *C. leiacanthus* Bleeker, 1851, *C. teijsmanni* Bleeker, 1857, *C. olivacea* Sudarto, Teugels & Pouyaud, 2003, *C. pseudoleiacanthus* Sudarto, Teugels & Pouyaud, 2003, *C. nigricans* Ng, 2003, and *C. pseudonieuhofii* Sudarto, Teugels & Pouyaud, 2004 (Ferraris, 2007). Additionally, subsequent morphological identification studies have led to the discovery of two new local Indonesian *Clarias* catfish species, namely *C. microspilus* Ng & Hadiaty, 2011 and *C. rennyae* Low, Ng & Tan, 2022. Thus, 17 species of local *Clarias* catfish have already been identified in Indonesia.

Among other local *Clarias* catfish species, *C. batrachus* is the most popular local species for aquaculture in Indonesia, due to its extensive natural distribution. Nonetheless, the taxonomic status of the *C. batrachus* is still problematic. It has been for a long time ago there were many *Clarias* catfish known under the name *C. batrachus* in countries ranging from India, Sri Lanka, Bangladesh, the countries in Indochina region, Malaysia, and Indonesia (Ferraris, 2007). Ng and Kottelat (2008) redescribed *C. batrachus* specimens from various locations and concluded that four species share the name *C. batrachus*. In a systematic revision study of Asian catfish species, Sudarto (2003) proposed *C. punctatus* Cuvier & Valenciennes, 1826 as the scientific name for the *Clarias* catfish *C. batrachus* in Southeast Asia, including Indonesia.

Clarias catfish *C. meladerma* is a native species with a fairly extensive range in Indonesia, specifically in the eastern parts of Sumatra from north to south and the southern parts of Kalimantan from west to center (Sudarto, 2003; Ferraris, 2007). The *C. meladerma* was considered potential to be developed as a new aquaculture species, thus the domestication has been initiated since the 1990s (Sudarto, 1993). However, its farming had not developed since the aquaculture performance was still inferior to the African catfish. Likewise, the *C. meladerma* has been used in hybridization studies with the African catfish, and it was demonstrated that the hybrid was superior to their parents in terms of growth performance (Lenormand et al., 1999).

18.3 Domestication and Breeding

Initially, the local *Clarias* catfish *C. batrachus* was cultivated in Indonesia using seeds obtained from natural spawning in ponds equipped with hollow (nest) and kakaban (Zonneveld et al., 1988; Knud-Hansen et al., 1990; Sudarto, 1993). In 1987, the artificial spawning of *C. batrachus* was begun using hormonal stimulation with carp pituitary extracts (Zonneveld et al., 1988). In the 2000s, hormonal stimulation in the artificial spawning of the *Clarias* catfish were conducted either using human chorionic gonadotropin (hCG) or luteinizing hormone-releasing hormone (LHRH) combined with anti-dopamine (Lenormand et al., 1999; Gustiano et al., 2021). In broad terms, the artificial spawning technology for the *C. batrachus* in Indonesia is similar to that of African catfish practiced in the past in Africa (Viveen et al., 1985) and nowadays practiced in Indonesia (Iswanto et al., 2021). Moreover, the larval rearing technology for the *C. batrachus* in hatcheries was developed in 1989 using the natural feed of *Artemia* sp. *nauplii* and cladoceran (Knud-Hansen et al., 1990).

The spawning and larval rearing technology used in subsequent African catfish farming in Indonesia was relatively similar to that of *C. batrachus*. African catfish farmers in Indonesia, on the other hand, generally use natural spawning methods, either with or without hormonal stimulation. The initial feed given to the larvae was usually *Tubifex* sp. (Suprapto et al., 2012), but some farmers use commercial feed with a crude protein of 60–65% (Iswanto et al., 2018). Furthermore, commercial feed with a crude protein of around 40% was used in the subsequent nursery phase. The larval rearing phase lasted for 2–3 weeks, and the nursery phase lasted for 1–1.5 months in tarpaulin-covered ponds, either with clear or green water (rich in phytoplankton) as high as 10–30 cm. Clear water systems employed high stocking densities of up to 10,000 individuals/m^2, whereas green water systems applied low stocking densities of 500–1000 individuals/m^2. When the seeds reached a size of about 5–7 cm or 7–9 cm of the standard length, they were ready to be reared further in the grow-out phase.

The grow-out phase of the African catfish for food fish in Indonesia is generally carried out for 2–3 months until it reaches a size of around 100–150 g. There are several grow-out systems for African catfish in Indonesia with various characteristics (Iswanto et al., 2015c; Gustiano et al., 2021). However, grow-out phase of the African catfish is generally carried out in earthen ponds of 50–200 m^2, with a stocking density of about 100–150 fish/m^2 (water depth of about 70–100 cm) and using commercial feed with a crude protein of around 30%, which is given in the morning and evening. Some African catfish are reared in small, round or square tarpaulin-covered ponds. The circular tarpaulin covered ponds are generally used for rearing African catfish using a biofloc system. Moreover, an aquaponics system, recirculation system and IMTA (integrated multi-trophic aquaculture) are other technologies in grow-out phase of the African catfish in Indonesia. However, these systems have not been widely adopted and practiced by fish farmers in Indonesia. Despite varying levels of fish productivity, the various African catfish grow-out technologies typically generate a relatively similar low proportion of economic benefits of about 5–20% due to highly commercial feed expenses (Iswanto et al., 2019).

As with other cultured species, one of the main efforts in *Clarias* catfish farming in Indonesia is to improve the efficiency and productivity. This was attempted by improving the aquaculture technology and genetics through breeding programs. Most improvements in aquaculture technology involve fish feed, environmental management, and disease control. Genetic improvement is conducted to improve the performance of traits related to aquaculture productivity. The well-management of the aquaculture environment has been the subject of extensive research to increase the productivity of catfish farming in Indonesia. Application of biofloc technology for rearing African catfish has been evaluated for broodstock (Ekasari et al., 2016) as well as the rearing of larvae (Nugraha et al., 2020) and fry at the nursery phase (Fauji et al., 2018; Ekasari et al., 2019) and grow-out phase (Fajrin et al., 2020), resulting in better performance than without a biofloc system.

In Indonesia, efforts to increase the productivity of African catfish farming through genetic improvement focus primarily on enhancing the growth performance. The genetic improvements are conducted using conventional breeding programs and genetic engineering. A breeding program to improve the growth performance of the African catfish in Indonesia has been conducted using individual selection method for three generations, resulting in a superior varieties namely Mutiara African catfish with significant advantages in terms of growth and other aquaculture important characters (Iswanto et al., 2012, 2014a, b, 2016a, b; Iswanto & Suprapto, 2015; Marnis et al., 2014a; Imron et al., 2020a). Crossing and backcrossing techniques have also been used to improve growth performance and other characteristics of the African catfish, resulting in the Sangkuriang African catfish, Sangkuriang 2 African catfish, and Mandalika African catfish. The Sangkuriang catfish was produced by backcrossing the male broodstocks of the sixth generation of

Dumbo African catfish with the female broodstocks of the second generation (Sunarma et al., 2005), while Mandalika African catfish resulted from a crossbreeding between female broodstocks of Sangkuriang African catfish and male broodstocks of Masamo African catfish (Nugroho et al., 2015). Then, Sangkuriang 2 African catfish was obtained from a crossbreeding between female broodstocks of Sangkuriang African catfish and male broodstocks of the Thai African catfish (Ade Sunarma, 2012, "personal communication").

Genetic improvement to increase the growth performance of the African catfish in Indonesia were also implemented using the transgenesis method by inserting a growth hormone gene from the tilapia *Oreochromis niloticus* (Gusrina et al., 2009), the striped catfish *Pangasianodon hypophthalmus* (Dewi et al., 2013; Marnis et al., 2012, 2013, 2014b, 2015, 2016a, b, c), and African catfish itself (Buwono et al., 2019). However, transgenic African catfish have not grown significantly faster than non-transgenic African catfish, therefore further studies are needed to find out the possible causes.

Along with improvements of the growth performance, the genetic improvement to increase the productivity of African catfish in Indonesia also focuses on the disease resistance characteristics, particularly resistance to *Aeromonas hydrophila* bacterial infection-related diseases. Utilizing major histocompatibility complex (MHC) gene markers, efforts were made to increase the disease resistance of African catfish (Azis et al., 2015a, b; Alimuddin et al., 2018; Suprapto et al., 2017, 2020; Imron et al., 2020b). This genetic improvement is still in the early stages of research, and the results have not been shared with the catfish farmers.

Since the farming of the African catfish during grow-out phase in Indonesia generally resulted in a low economic return due to the highly cost of the commercial feeds, a genetic improvement has also been initiated for low maintenance requirement in terms of dietary protein level using an individual selection method (Iswanto et al., 2019, 2020b). However, this genetic improvement still faced several problems, mainly related to the availability of commercial feeds with suitable levels of low dietary protein. Thus, further research is needed to refine the individual selection method.

18.4 Future Research Needs

Research in developing *Clarias* catfish culture should be continuously conducted. In particular, Indonesia has many local *Clarias* catfish species, which its potency as the new candidates for aquaculture species need to be explored. The local *Clarias* catfish species are still potential to be developed as the food fish commodities, since the African catfish dominantly cultured in Indonesia has a main disadvantage related to the meat quality, for which its taste is inferior to the local Asian *Clarias* catfish species (Csavas, 1994; Lenormand et al., 1999; Na-Nakorn & Brummet, 2009), thus it has lower price in the market. On the other hand, domestications of the local *Clarias* catfish species are also essential to be conducted for conservation purposes, since the availability in the wild tends to decrease either due to the highly captured activities or habitat destruction (Sudarto, 2003).

Since grow-out phase (the final phase in producing the food fish) of the African catfish in Indonesia generally resulted in low economic benefits, mainly due to the highly commercial feed expenses and low meat quality, research on genetic improvement and improvement of the aquaculture technology for the African catfish should focus to overcome the problems. A selective breeding program to establish a new strain of the African catfish capable of growing in minimum maintenance conditions such as low dietary protein intake (Iswanto et al., 2019, 2020b) should be prioritized in order to reduce the high cost of commercial feed used in the grow-out phase. In addition, research in feed nutrition should also focus on creating feed with lower cost production that supports the growth performance of the African catfish during grow-out phase.

Despite domesticating and developing the culture of local *Clarias* catfish species, increasing the meat quality could also be attempted via interspecific hybridization between local *Clarias* catfish species and the African catfish. Several such interspecific hybridizations resulted in superior hybrids. Interspecific hybridization between a local Asian *Clarias* catfish *C. macrocephalus* and African catfish has been carried out for a long time in Thailand, resulting in a hybrid with a better meat quality than that of the African catfish and better growth and disease resistance than *C. macrocephalus*, thus it has commercially farmed (Na-Nakorn, 1993, 1999; Bartley et al., 2001; Na-Nakorn & Brummet, 2009). Besides Thailand, such hybridization has also been carried out in Vietnam (Csavas, 1994), Malaysia (Kechik, 1995), Bangladesh, and Myanmar (Tripathi, 1996) as well as in the Philippines (Na-Nakorn, 1999). A superior *Clarias* catfish hybrid has also been resulted from an interspecific hybridization between a local Asian *Clarias* catfish *C. fuscus* and African catfish in China (Zheng et al., 1988; Wu et al., 1990; Csavas, 1994), Vietnam (Csavas, 1994), and Taiwan (Huang et al., 2005). In addition, a superior *Clarias* catfish hybrid has also been resulted from an interspecific hybridization between African catfish *C. gariepinus* and a local *Clarias* catfish *C. batrachus* in Bangladesh (Rahman et al., 1995; Khan et al., 2000) and India (Sahoo et al., 2003). In Indonesia, research in interspecific hybridization of the *Clarias* catfish is still scarce. The research on interspecific hybridization between African catfish and local *Clarias* catfish *C. batrachus* resulted in no viable offspring, while hybridization between African catfish *C. gariepinus* and a local *Clarias* catfish *C. meladerma* resulted in the hybrid with better growth performance than that of the local *Clarias* catfish *C. meladerma* (Lenormand et al., 1999). Research of the *Clarias* catfish interspecific hybridizations still needs to be carried out, since Indonesia has many local *Clarias* catfish species whose characteristics are still unknown.

References

Alimuddin, Putri, F. M., Wahjuningrum, D., Hardiantho, D., Sunarma, A., & Nuryati, S. (2018). Resistance against *Aeromonas hydrophila* infection and growth of second generation (F2) African catfish (*Clarias gariepinus*) using selected molecular markers. *Biotropia, 25*(2), 95–102.

Anene, N. S., & Gao, T. X. (2007). Is the Dutch domesticated strain of Clarias gariepinus (Burchell, 1822) a hybrid? *African Journal of Biotechnology, 6*(8), 1072–1076.

Azis, Alimuddin, Sukenda, & Junior, M. Z. (2015a). Identification of the MHC I marker candidate in catfish (*Clarias* sp.) resistant to *Aeromonas hydrophila* infection. *Jurnal Riset Akuakultur, 10*(2), 261–269. (in Indonesian with English abstract).

Bartley, D. M., Rana, K., & Immink, A. J. (2001). The use of inter-specific hybrids in aquaculture and fisheries. *Reviews in fish biology and fisheries, 10*(3), 325–337.

Azis, Alimuddin, Sukenda, & Junior, M. Z. (2015b). MHC I molecular marker inheritance and first generation catfish (*Clarias* sp.) resistance against *Aeromonas hydrophila* infection. *Pakistan Journal of Biotechnology, 12*(2), 131–137.

Buwono, I. D., Junianto, J., Iskandar, I., & Alimuddin. (2019). Growth and expression level of growth hormone in transgenic Mutiara catfish second generation. *Journal of Biotech Research, 10*, 102–109.

Csavas, I. (1994). *Status and perspectives of culturing catfishes in East and Southeast Asia* (pp. 2–10). FAO Aquaculture Newsletter December 1994 – Number 8. Food and Agriculture Organization.

Dewi, R. R. S. P. S., Marnis, H., Suprapto, R., & Syawalia, N. (2013). Production of F-0 generation of fast growing African catfish by transgenesis. *Jurnal Riset Akuakultur, 8*(2), 173–180. (in Indonesian with English abstract).

Ekasari, J., Suprayudi, M. A., Wiyoto, Hazanah, R. F., Lenggara, G. S., Sulistiani, R., Alkahfi, M., & Junior, M. Z. (2016). Biofloc technology application in African catfish fingerling production: The effects on the reproductive performance of broodstock and the quality of eggs and larvae. *Aquaculture, 464*, 349–356.

Ekasari, J., Setiawati, R., Ritonga, F. R., Setiawati, M., & Suprayudi, M. A. (2019). Growth and health performance of African catfish *Clarias gariepinus* (Burchell 1822) juvenile fed with graded levels of biofloc meal. *Aquaculture Research, 50*(2), 1802–1811.

Fajrin, M., Utomo, N. P., Ekasari, J., & Nuryati, S. (2020). Evaluation of health status and meat quality of dumbo catfish *Clarias gariepinus* maintained using the biofloc system. *Jurnal Akuakultur Indonesia, 19*(2), 190–198.

Fauji, H., Budiardi, T., & Ekasari, J. (2018). Growth performance and robustness of African Catfish *Clarias gariepinus* (Burchell) in biofloc-based nursery production with different stocking densities. *Aquaculture Research, 49*(2), 1339–1346.

Ferraris, C. J. (2007). Checklist of catfishes, recent and fossil (Osteichthyes: Siluriformes), and catalogue of siluriform primary types. *Zootaxa, 1418*, 1–628.

Gusrina, Alimuddin, Sumantadinata, K., & Widyastuti, U. (2009). Transfer of gene encoding tilapia growth hormone (mBP-tiGH) in catfish (*Clarias* sp.) by microinjection method. *Jurnal Riset Akuakultur, 4*(3), 333–340. (in Indonesian with English abstract).

Gustiano, R., Prakoso, V. A., Radona, D., Dewi, R. R. S. P. S., Saputra, A., & Nurhidayat. (2021). A sustainable aquaculture model in Indonesia: Multibiotechnical approach in *Clarias* farming. *IOP Conference Series: Earth and Environmental Science, 718*, 012039.

Huang, C. F., Lin, Y. H., & Chen, J. D. (2005). The use of RAPD markers to assess catfish hybridization. *Biodiversity and Conservation, 14*, 3003–3014.

Imron, Iswanto, B., Suprapto, R., & Marnis, H. (2020a). Development of genetically improved farmed African catfish, *Clarias gariepinus*; A review and lessons learned from Indonesian fish breeding program. *IOP Conference Series: Earth and Environmental Science, 593*, 012032.

Imron, Marnis, H., Iswanto, B., & Suprapto, R. (2020b). Development of a PCR marker for the identification of resistance to Motile Aeromonad Septicaemia disease in African catfish (*Clarias gariepinus*). *AACL Bioflux, 139*(3), 1255–1267.

Iswanto, B. (2013). Tracing the identity of the Dumbo *Clarias* catfish. *Media Akuakultur, 8*(1), 85–95. (in Indonesian).

Iswanto, B., & Suprapto, R. (2015). Morphological abnormality of the seeds of Mutiara strain of the African catfish (*Clarias gariepinus*). *Media Akuakultur, 10*(2), 51–57. (in Indonesian with English abstract).

Iswanto, B., Imron, Suprapto, R., & Marnis, H. (2012). *Establishment of a rapid growth strain of African catfish Clarias gariepinus (Burchell, 1822) by means of individual selection: Production the base population* (pp. 1197–1210). Prosiding Forum Inovasi Teknologi Akuakultur (FITA). (in Indonesian).

Iswanto, B., Imron, Suprapto, R., & Marnis, H. (2014a). Establishment of a rapid growth strain of African catfish *Clarias gariepinus* (Burchell, 1822) by means of individual selection: Production the first-generation population. *Jurnal Riset Akuakultur, 9*(3), 343–352. (in Indonesian with English abstract).

Iswanto, B., Imron, Suprapto, R., Marnis, H., & Ridzwan, N. S. (2014b). *Establishment of a rapid growth strain of African catfish Clarias gariepinus (Burchell, 1822) by means of individual selection: Production the second generation population* (pp. 837–847). Prosiding Forum Inovasi Teknologi Akuakultur. (in Indonesian).

Iswanto, B., Imron, Suprapto, R., & Marnis, H. (2015a). Morphological characterization of the African catfish (*Clarias gariepinus* Burchell, 1822) strains introduced to Indonesia. *Indonesian Aquaculture Journal, 10*(2), 91–99.

Iswanto, B., Imron, Suprapto, R., & Marnis, H. (2015b). Embryonic and larval development of a red strain of the Egyptian African catfish (*Clarias gariepinus* Burchell, 1822). *Indonesian Aquaculture Journal, 10*(1), 19–31.

Iswanto, B., Suprapto, R., Marnis, H., & Imron. (2015c). *Field test performance of the third generation of the fast growing African catfish (Clarias gariepinus)* (pp. 95–102). Prosiding Forum Inovasi Teknologi Akuakultur. (in Indonesian).

Iswanto, B., Imron, Marnis, H., & Suprapto, R. (2016a). Response to selection for body weight in the third generation of mass selection of the African catfish (*Clarias gariepinus*) at Research Institute for Fish Breeding Sukamandi. *Indonesian Aquaculture Journal, 11*(1), 15–21.

Iswanto, B., Suprapto, R., Marnis, H., & Imron. (2016b). Reproductive performances of Mutiara strain of the African catfish (*Clarias gariepinus*). *Media Akuakultur, 11*(1), 1–9. (in Indonesian with English abstract).

Iswanto, B., Suprapto, R., Marnis, H., & Imron. (2016c). Morphological characteristics of a red strain of the Egyptian African catfish (*Clarias gariepinus* Burchell 1822). *Indonesian Aquaculture Journal, 11*(2), 49–59.

Iswanto, B., Imron, Suprapto, R., & Marnis, H. (2017). Biometric characterization of the lele Dumbo compared to that of African catfish (*Clarias gariepinus* Burchell, 1822). *Berita Biologi, 18*(2), 223–232. (in Indonesian with English abstract).

Iswanto, B., Suprapto, R., & Suwargono, P. (2018). *Growth performance of the African catfish (Clarias gariepinus) offspring resulted from a bidirectional selection* (pp. 483–491). Prosiding Seminar Nasional Ikan Ke-10. (in Indonesian).

Iswanto, B., Suprapto, R., & Imron. (2019). Establishing a base population of individual selection for low maintenance requirement in term of dietary protein level in the African catfish (*Clarias gariepinus*). *Indonesian Aquaculture Journal, 14*(1), 23–29.

Iswanto, B., Suprapto, R., & Suwargono, P. (2020a). Length-weight relationship, condition factor and biometric characteristic of albino African catfish (*Clarias gariepinus*) originated from Thailand. *Berita Biologi, 19*(3A), 249–256. (in Indonesian with English abstract).

Iswanto, B., Suprapto, R., & Suwargono, P. (2020b). Heritability and response to selection for growth in the African catfish (*Clarias gariepinus*) treated with a low dietary protein of commercial feed during grow-out phase. *Jurnal Riset Akuakultur, 15*(4), 229–235. (in Indonesian with English abstract).

Iswanto, B., Suprapto, R., Imron, Haryadi, J., Suwargono, P., Pangestika, M. F., & Ilmalizanri. (2021). Zootechnical performance and biometric characteristics of the albino Mutiara African catfish, *Clarias gariepinus*. *Jurnal Riset Akuakultur, 16*(3), 135–144. (in Indonesian with English abstract).

Kechik, I. A. (1995). Aquaculture in Malaysia. In T. U. Bagarinao & E. E. C. Flores (Eds.), *Towards sustainable aquaculture in Southeast Asia and Japan. ADSEA '94 Proceeding* (pp. 125–135). SEAFDEC Aquaculture Department.

Khan, M. M. R., Mollah, M. F. A., & Ahmed, G. U. (2000). Mass production of hybrid magur and its culture potential in Bangladesh. *Aquaculture Research, 31*, 467–472.

Knud-Hansen, C. F., Batterson, T. R., McNabb, C. D., Hadiroseyani, Y., Dana, D., & Eidman, H. M. (1990). Hatchery techniques for egg and fry production of *Clarias batrachus* (Linnaeus). *Aquaculture, 89*(1), 9–19.

Lenormand, S., Slembrouck, J., Pouyaud, L., Subagja, J., & Legendre, M. (1999). *Evaluation of hybridisation in five Clarias species (Siluriformes, Clariidae) of African (C. gariepinus) and Asian origin (C. batrachus, C. meladerma, C. nieuhofii and C. teijsmanni)*. Proceedings of the Mid-Term Workshop of the Catfish Asia Project, 11–15 May 1998, Cantho, Vietnam, pp. 195–209.

Low, B. W., Ng, H. H., & Tan, H. H. (2022). Two new species of blackwater catfishes (Siluriformes: Siluridae and Clariidae) from the Natuna Archipelago, Indonesia. *Raffles Bulletin of Zoology, 70*, 385–396.

Marnis, H., Dewi, R. R. S. P. S., Imron, & Iswanto, B. (2012). Distribution and expression of striped catfish (*Pangasianodon hypophthalmus*) growth hormone gene (PhGH) in the organ of African catfish (*Clarias gariepinus*) transgenic founder. *Indonesian Aquaculture Journal, 7*(1), 11–18.

Marnis, H., Iswanto, B., Suprapto, R., & Imron. (2013). Expression of growth hormone (PhGH) gene and analysis of insuline-like growth factor 1 (IGF 1) production in African catfish (*Clarias gariepinus*) transgenic F-1. *Indonesian Aquaculture Journal, 8*(2), 113–119.

Marnis, H., Iswanto, B., & Suprapto, R. (2014a). *Challenge test of the Mutiara African catfish (Clarias gariepinus) against Aeromonas hydrophila bacteria*. Prosiding Seminar Nasional Perikanan Indonesia. (in Indonesian).

Marnis, H., Iswanto, B., Suprapto, R., Imron, & Dewi, R. R. S. P. S. (2014b). Transmission, expression and growth hormone distribution of striped catfish in African catfish (*Clarias gariepinus*) transgenic F-2. *Jurnal Riset Akuakultur, 9*(2), 179–190. (in Indonesian with English abstract).

Marnis, H., Iswanto, B., Suprapto, R., Imron, & Dewi, R. R. S. P. S. (2015). Performance and zygosity of F-2 transgenic African catfish (*Clarias gariepinus*) containing growth hormone gene of striped catfish (*Pangasianodon hypophthalmus*). *Jurnal Riset Akuakultur, 10*(2), 161–168. (in Indonesian with English abstract).

Marnis, H., Iswanto, B., Suprapto, Imron, & Dewi, R. R. S. P. S. (2016a). Identification of zygosity F2 transgenic African catfish (*Clarias gariepinus*) with growth hormone gene (PhGH) by using realtime-qPCR method. *Jurnal Riset Akuakultur, 11*(1), 39–46. (in Indonesian with English abstract).

Marnis, H., Iswanto, B., Imron, Febrida, S., & Dewi, R. R. S. P. S. (2016b). The ability of fast-growing transgenic African catfish (*Clarias gariepinus*) on predator avoidance. *Indonesian Aquaculture Journal, 11*(1), 15–21.

Marnis, H., Iswanto, B., Febrida, S., Imron, & Dewi, R. R. S. P. S. (2016c). PhGH gene transmission and performance of F3 transgenic African catfish (*Clarias gariepinus*). *Jurnal Riset Akuakultur, 11*(3), 225–234. (in Indonesian with English abstract).

Na-Nakorn, U. (1993). Review of aquaculture genetic researches in Thailand. *Biotropia, 6*, 45–54.

Na-Nakorn, U. (1999). Genetic factors in fish production: A case study of the catfish *Clarias*. In S. Mustafa (Ed.), *Genetics in sustainable fisheries management* (pp. 175–187). Fishing News Books.

Na-Nakorn, U., & Brummet, R. E. (2009). Use and exchange of aquatic genetic resources for food and aquaculture: *Clarias* catfish. *Reviews in Aquaculture, 1*, 214–223.

Ng, H. H., & Hadiaty, R. K. (2011). *Clarias microspilus*, a new walking catfish (Teleostei: Clariidae) from northern Sumatra, Indonesia. *Journal of Threatened Taxa, 3*(3), 1577–1584.

Ng, H. H., & Kottelat, M. (2008). The identity of *Clarias batrachus* (Linnaeus, 1758), with the designation of a neotype (Teleostei: Clariidae). *Zoological Journal of Linnean Society, 153*, 725–734.
Nugraha, S., Ekasari, J., Junior, M. Z., & Widanarni. (2020). Digestive system and growth performance of African catfish larvae *Clarias gariepinus*, (Burchell, 1822) maintained with biofloc technology with the addition of *Chlorella* sp. *Jurnal Iktiologi Indonesia, 20*(3), 297–306. (in Indonesian with English abstract).
Nugroho, E., Putra, S., Syahdan, M. A., Mayadi, L., Budileksono, S., & Zulkifli. (2015). Heterosis of hybrids catfish in West Nusa Tenggara Province. *Jurnal Riset Akuakultur, 10*(1), 33–40. (in Indonesian with English abstract).
Pangestika, M. F., Iswanto, B., Suprapto, R., Suwargono, P., & Ilmalizanri. (2019). BRPI has disseminated more than 8,000 packages of Mutiara African catfish broodstock candidates. *Buletin Penelitian Pemuliaan Ikan, 8*(1), 16. (in Indonesian).
Rahman, M. A., Bhadra, A., Begum, N., Islam, M. S., & Hussain, M. G. (1995). Production of hybrid vigor through cross breeding between *Clarias batrachus* Lin. and *Clarias gariepinus* Bur. *Aquaculture, 138*, 125–130.
Sahoo, S. K., Giri, S. S., Sahu, A. K., & Ayyappan, S. (2003). Experimental hybridization between catfish *Clarias batrachus* (Linn.) x *Clarias gariepinus* (Bur.) and performance of the offspring in rearing operations. *Asian Fisheries Science, 16*, 157–166.
Sudarto. (1993). A review of the fish breeding research and practices in Indonesia. In Main, K. L., & Reynolds, E. (Eds.), *Selective breeding of fishes in Asia and United States*. Proceeding of a Workshop in Honolulu, May 3–7, 1993, pp. 182–189.
Sudarto. (1999). Genetical and zootechnical characterization of the ikan lele (Clariidae) and patin (Pangasiidae) from Southeast Asian regions. Prosiding Seminar Hasil Penelitian Genetika Ikan. 8 Februari 1999, pp. 26–29. (in Indonesian).
Sudarto. (2003). *Systematic revision and phylogenetic relationships among populations of Clariid species in Southeast Asia* [Doctoral dissertasion, University of Indonesia, Depok], 203 pp.
Sudarto, & Pouyaud, L. (2005). Identification key based on morphological characters of the Asian species of the genus *Clarias* (Pisces: Clariidae). *Jurnal Iktiologi Indonesia, 5*(2), 39–47.
Sunarma, A., Nurhidayat, M. A., & Maskur. (2005). *Genetic improvement of African catfish Clarias gariepinus using backcross strategy in Indonesia*. Paper presented on World Aquaculture Society Meeting 2005, Bali, May 9–13, 2005.
Suprapto, R., Iswanto, B., & Imron. (2012). *Effects of various live feeds offered on the growth and survival of the African catfish (Clarias gariepinus) fry*. Prosiding INDOAQUA – Forum Inovasi Teknologi Akuakultur 2013, Makassar, Juni 8–11, 2012, pp. 563–573. (in Indonesian).
Suprapto, R., Alimuddin, Nuryati, S., Imron, Marnis, H., & Iswanto, B. (2017). MHC-II marker potential linked to motile aeromonad septicaemia disease resistance in African catfish (*Clarias gariepinus*). *Indonesian Aquaculture Journal, 12*(1), 21–28.
Suprapto, R., Iswanto, B., Marnis, H., & Haryadi, J. (2020). Screening of MHC-I and MHC-II markers on the African catfish (*Clarias gariepinus*) as a gene code Motile Aeromonas Septicaemia (MAS) disease resistance. *Media Akuakultur, 15*(2), 97–104. (in Indonesian with English abstract).
Teugels, G. G. (1984). The nomenclature of African *Clarias* used in aquaculture. *Aquaculture, 38*, 373–374. (Technical Note).
Teugels, G. G. (1986). A systematic revision of the African species of the genus *Clarias* (Pisces: Clariidae). *Annales Musee Royal de l'Afrique Centrale, 247*, 1–199.
Tripathi, S. D. (1996). Present status of breeding and culture of catfishes in South Asia. *Aquatic Living Resources, 9*, 219–228.
Viveen, W. J. A. R., Richter, C. J. J., Van Oordt, P. G. W. J., Janssen, J. A. L., & Huisman, E. A. (1985). *Practical manual for the culture of the African Catfish (Clarias gariepinus)* (128 pp). Directorate General International Cooperation of the Ministry of Foreign Affairs The Netherlands, Department of Fish Culture and Fisheries of the Agricultural University of Wageningen, Research Group for Comparative Endocrinology, Department of Zoology of the University of Utrecht.

Wu, G. M., Luo, J. R., Chen, K. C., Xian, C. B., Lin, G. G., Wang, Z. X., Lin, Z. P., Luo, J. L., & Pan, K. Q. (1990). Analysis and comparison of morphology and cytology between the F1 hybrid catfish (*Clarias fuscus* x *C. lazera*) and its parents. *Acta Hydrobiologica Sinica, 14*(4), 328–335.

Zheng, W. B., Pan, J. H., & Liu, W. S. (1988). Culture of catfish in China. *Aquaculture, 75*, 35–44.

Zonneveld, N., Rustidja, Viveen, W. J. A. R., & Mudana, W. (1988). Induced spawning and egg incubation of the Asian catfish, *Clarias batrachus*. *Aquaculture, 74*, 41–47.

Development of Pangasius Farming in Indonesia: Domestication and Aquaculture

19

Jadmiko Darmawan, Evi Tahapari, Khairul Syahputra, and Arsad Tirta Subangkit

19.1 Introduction

Pangasius (Pangasiidae) is one of the most economically important fish commodities in Southeast Asia, with a total production of more than 2.1 million tons in 2015, and its production has continued to increase in recent years (Subasinghe, 2017). Indonesia became the third largest pangasius producer in the world, producing 408,539 tons in 2020 (KKP, 2021). The largest cultivation centers of pangasius are dominated by Sumatra (68.07%), Kalimantan (20.23%), and Java (8.48%) islands, and also distributed to some other areas in the country (KKP, 2018).

Pangasius farming has long been carried out traditionally by people who live along the riverside in Sumatra and Kalimantan using submerged and floating cages. The existence of pangasius farming has dramatically developed since the success of artificial spawning through hormonal stimulation induction using the hypophysis of carp in late of 1981 (Hardjamulia et al., 1981). Parallel to the rapid development of cultivation technology, pangasius has spread massively and uncontrollably, and consequently led to the decreasing of genetic quality indicated on the slow growth performance. In addition, other frequent issues faced by the farmers are high mortality rates due to pathogen infections, such as *Aeromonas hydrophila* and *Edwardsiella ictaluri*. Some efforts have been placed to overcome these problems, including genetic improvement of pangasius through domestication of local species and breeding programs (selective breeding and hybridization), as well as innovation in cultivation technologies.

J. Darmawan (✉) · E. Tahapari · K. Syahputra · A. T. Subangkit
Research Center for Fishery, Research Organization for Earth Sciences and Maritime, National Research and Innovation Agency, Cibinong, Indonesia
e-mail: jadm001@brin.go.id

E. Nugroho et al. (eds.), *Aquaculture Development in Archipelago*,
https://doi.org/10.1007/978-981-95-1084-9_19

19.2 Taxonomy and Morphology

Indonesia has a high genetic resource of pangasius covering 15 species. They consist of 13 local species, one introduced species, and one species resulting from hybridization program. The local species of pangasius includes *Pangasius nieuwenhuisii*, *P. humeralis*, *P. lithostoma*, *P. mahakamensis*, *P. djambal*, *P. macronema*, *P. polyuranodon*, *P. nasutus*, *P. kunyit*, *P. rheophilus*, *Pteropangasius micronemus*, *Helicophagus typus*, and *Helicophagus waandersii*. Striped catfish (*Pangasianodon hypophthalmus* Sauvage, 1878) is known as an introduced species, while Pasupati strain of catfish (*Pangasius* sp.) is a hybrid of catfish produced through crossbreeding between *P. hypophthalmus* female and *P. djambal* male (Ferraris, 2007; Gustiano & Pouyaud, 2006, 2008; Pouyaud et al., 1998, 1999, 2004; Tahapari et al., 2014).

In Indonesia, the striped catfish (also referred to *Pangasius hypophthalmus* Sauvage, 1878; *Pangasius sutchi* Fowler, 1937) introduced from Thailand in 1972 is the preferred aquaculture species (Hardjamulia et al., 1981). Taxonomically, striped catfish is classified into the following:

Kingdom:	Animalia
Phylum:	Vertebrate
Class:	Osteichthyes
Order:	Siluriformes
Family:	Pangasiidae
Genus:	*Pangasius*
Subgenus:	Pangasianodon
Species:	*Pangasianodon hypophthalmus* Sauvage, 1878.

In some areas particularly Sumatra and Kalimantan, *P. djambal* belonging to subgenus pangasius is the preferred species for aquaculture due to having a better meat quality with economic price higher compared to the striped catfish.

Morphologically, pangasius has a laterally compressed body which is dominated with ash gray (silver) color and a back of greenish blue. Pangasius is a type of scaleless fish with a wide mouth and equipped with two pairs of barbels as whisker-like sensory organs. Morphometric characteristics presented as a percentage relative to head length (%HL) and standard length (%SL) show that pangasius has head length of <29% SL, prepectoral length of <25% SL, snout width of <30% HL, and anal fin length of <27.5% SL (Gustiano & Pouyaud, 2008). In juveniles, pangasius is recognized with palatal teeth in vomerine and palatine bands and two pairs of barbels. However, the vomerine teeth plate becomes thin and disappear in the larger size (Gustiano & Pouyaud, 2008).

Pangasius has a dorsal fin consisting of one spine and 6–8 rays and an adipose fin, which is a small fat fin located between posterior the dorsal fin and anterior of the caudal fin. Caudal fin is deeply forked and symmetrical, and anal fin with 30–36 rays extends toward the caudal fin. The pectoral fin is composed of 9–11 soft-fin rays and a single leading-edge spine used as a primarily defensive role. Further, pangasius is equipped with the pelvic fin, and interestingly the number of pelvic soft-fin rays can be used to distinguish between *P. hypophthalmus*, *P. djambal,* and its hybrid (known as Pasupati) having 8–9, 6, and 7 fin rays, respectively.

19.3 Distribution of Pangasius in Indonesia

Indonesia's genetic diversity accounts for 13 out of 28 pangasius species worldwide, even though the introduced striped catfish become the main species of pangasius farming. These local species are widely distributed in freshwater particularly in the major rivers located in Sumatra (Way Rarem, Musi, Batanghari, and Indragiri rivers), eastern Java (Brantas and Bengawan Solo rivers), and Kalimantan (Kayan, Berau, Mahakam, Barito, Kahayan dan Kapuas's rivers). However, a significant decrease of fishing in these rivers has been recorded in recent years, which might be affected by habitat changes, environment degradation, and overfishing.

Striped catfish, which is a species of introduced catfish in Indonesia, is mainly found in aquaculture environments, such as ponds and floating net cages, and some have been released and inhabit the waters of large rivers in Sumatra, Java, and Kalimantan, which are the centers of striped catfish production, namely, in the Batanghari, Musi, Komering, Citarum, Kapuas, and Mahakam rivers (Pouyaud et al., 1998). In addition, striped catfish populations found in nature are also the result of restocking development, such as in Wonogiri Reservoir (Kartamihardja et al., 2011) and Malahayu Reservoir (Andriyanto et al., 2012). In addition to Indonesia, striped catfish are also widely found in the Asian region, especially in their native areas such as Vietnam, Thailand, and China.

In the wild, pangasius usually hides in burrows on the riverbank. They will only come out of its hiding place after dark. This is in accordance with its nocturnal nature. In their natural habitat, fish settle more at the bottom of the water than at the surface, therefore they are classified as bottom fish (demersal) (Khairuman, 2010).

19.4 Reproduction

As a fish species that has long been cultivated, pangasius spawning technology has been widely studied. In general, the males can be identified by looking at the genital area, where the urogenital organs are slightly inward in which on the naked eye only two holes are seen at the end of the abdomen, namely, the anus and the urogenital hole. In females, the position of the urogenital organs is outside where three holes are seen at the end of the abdomen by the naked eye, namely, the anus, urine, and genital holes. Males of the striped catfish are recorded to enter the adult phase at the age of 9 months, while females do so at the age of 14 months (Tahapari et al., 2017). In Jambal catfish, males are recorded to enter adulthood at the age of 12 months, while females do so at the age of 3 years (Slembrouck et al., 2005). However, both species can generally be spawned when males are over 2 years old and weigh more than 2 kg, and when females are over 2.5 years old and weigh more than 2.5 kg. Checking the gonadal maturity in males is relatively easier to do than females by gently pressing the abdomen toward the genital organ and gonadal mature male is characterized by the release of thick white sperm. Gonadal maturity in females is evaluated by collecting oocyte samples through cannulation (intraovarian biopsy) and observing the maturity level based on the following four categories (Slembrouck et al., 2005):

- Level I: oocytes are bright and difficult to separate from each other.
- Level II: oocytes are ivory yellow, larger diameter eggs are more homogeneous, most oocytes can be separated.
- Level III: oocytes are ivory yellow, homogeneous in diameter, and all oocytes are easily separated.
- Level IV: oocytes are partly light-colored and partly ivory yellow, intraovarian fluid, and many degenerative oocytes are present.

Spawning of pangasius occurs when the female is at the category III of gonadal maturity level even though diameter of oocytes in several species differs among others. Siamese catfish have oocyte diameters ranging from 0.9 to 1.2 mm (Tahapari et al., 2017), Jambal catfish have diameters of 1.7 to 2.1 mm (Slembrouck et al., 2005), and nasutus catfish have diameters of 1.4 to 1.7 mm (Iswanto, 2011). This oocyte diameter size ultimately also affects the fecundity or number of eggs produced during spawning. Fecundity of Siamese catfish ranges from 28,000 to 320,000 eggs/kg of female (Tahapari et al., 2017), while Jambal catfish have fecundity ranging from 1100 to 23,000 eggs/kg female (Legendre et al., 2000).

In natural environments such as rivers, reservoirs, and lakes, pangasius can spawn naturally by attaching its eggs to the substrate. However, in the aquaculture system, pangasius cannot spawn naturally despite fully developed gonadal maturity. Therefore, hormonal applications are needed to encourage oocyte maturation and ovulation. Striped catfish brood stock can spawn throughout the year, but the proportion of mature gonads and fecundity is influenced by the season. In the rainy season (September–March), the proportion of mature gonads ranges from 50% to 100%, while in the dry season (April–August), it is less than 30%, with no more than half the fecundity observed in the rainy season (Slembrouck et al., 2005).

19.5 Domestication and Genetic Improvement of Patin Fish in Indonesia

Indonesia has a lot of sources of genetic diversity of pangasius where there are 14 out of 28 species of pangasius genus fish in the world, including an introduced Siamese catfish. Abundant sources of genetic diversity can be the main capital in creating new superior varieties through domestication and breeding (selection and hybridization). Domestication is a process of adapting wild species into the cultivated environment so that species can grow and reproduce well (Effendie, 2002), whereas fish breeding is a series of activities to produce superior fish through measurable trait improvement. Fish breeding can be conducted through methods such as selection, hybridization, chromosome manipulation, and genetic engineering.

Domestication programs on pangasius have been carried out and produced superior strains of Jambal catfish (*P. djambal*), which were released as aquaculture species on May 10, 2000, based on the Decree of the Minister of Agriculture Number 265/Kpts/IK.240/5/2000. Jambal catfish has advantages in premium meat quality with white meat color, dense texture, and savoriness. Domestication of local species of catfish in Indonesia have also been carried out on nasutus catfish (*P. nasutus*) and nieuwenhuisii catfish (*P. nieuwenhuisii*). However, both species are still in the characterization stage and are currently at a standstill due to funding constraints and the priority of national research in Indonesia.

Breeding programs have also been carried out on pangasius and have resulted in Pasupati catfish (Patin Super Harapan Pertiwi), which is a hybrid catfish and Perkasa catfish (Patin Super Karya Anak Bangsa), which is a selected family of Siamese catfish based on growth characteristics (Fig. 19.1). Pasupati was released as an aquaculture species on August 2, 2006, based on the Decree of the Minister of Marine Affairs and Fisheries (MMAF) Number KEP.25/MEN/2006 with the advantages of having premium meat quality with white meat color, dense texture, and more savory like Jambal patin. In addition, Pasupati having high fecundities could produce seeds in large quantities (Darmawan & Tahapari, 2019b). Perkasa was released as an aquaculture species on July 12, 2018, based on the Decree of MMAF Number 75/KEPMEN-KP/2018 with the advantages of faster growth, more resistant to bacterial disease (*A. hydrophila*), and a lower feed conversion ratio (FCR) value compared to the existing Siamese catfish (Darmawan et al., 2023; Darmawan & Tahapari, 2019a). In addition to these varieties, a new variety of pangasius has also been introduced under the name of Pustina patin (National Patin Center), which was predicted to have the advantage of fast growth, resistance to diseases, and tolerance to unfavorable environmental conditions. Pustina catfish was launched by MMAF on April 23, 2021, but until this article was written, a decree of the minister of maritime affairs and fisheries had not been issued to determine this variety as a superior variety of pangasius.

Fig. 19.1 Perkasa striped catfish (*Pangasianodon hypophthalmus*). (Source: Ministerial Decree RI Number 75/KEPMEN-KP/2018)

19.6 Development of Patin Fish Cultivation Technology

Along with the industrialization of pangasius commodities in Indonesia, pangasius cultivation technology can generally be divided into four segments, including spawning technology, indoor hatchery larval rearing, outdoor seed rearing, and grow-out. The rapid development of pangasius cultivation technology occurs in all cultivation segments. In the spawning segment, technology has developed rapidly since the successful artificial spawning technique for Siamese catfish was introduced in Thailand in 1966, using hormonal induction with pangasius hypophysis extract (Boonbrahm et al., 1968). In Indonesia, artificial spawning research on pangasius was initiated in 1977 using hypophysis from pangasius (homoplastic) (Hardjamulia et al., 1978), and then at the end of 1981 artificial spawning was successfully carried out using hypophysis gland extract from carp (heteroplastic) (Hardjamulia et al., 1981). The development of pangasius spawning technology is growing after the successful stimulation of spawning using commercial hormone products, including the hormone HCG (human chorionic gonadotropin) and a combination hormone between GnRHa (gonadotropin-releasing hormone analog) and domperidone (anti-dopamine) (Legendre et al., 2000). The spawning method with hormone application in Siamese catfish has subsequently been successfully applied to other pangasius species such as Jambal catfish (Slembrouck et al., 2005) and nasutus catfish (Iswanto, 2011). The spawning technology has also developed with the application of egg-hatching techniques using a hatching funnel (MacDonald Jar system), which provides effectiveness and efficiency of egg hatching, where one funnel can hatch up to more than 300 g of eggs.

The technology of raising larvae and fry has also undergone a very significant development with super-intensive technology in indoor, semi-outdoor, and outdoor hatcheries. Pangasius farming in Indonesia since 2014 has been directed toward fisheries industrialization demanding effective and efficient hatchery technology. Farmers apply high stocking densities of up to 100 fish/l in the larval stage up to 1 inch size and stocking densities of up to 1000 fish/m^2 in the fingerlings stage from 1 inch to 3 inches size. Further, applications of biosecurity, vaccines, probiotics, and antibiotics in hatchery are required to minimize disease infections that are highly potential to occur in super-intensive hatchery activities. Pangasius hatcheries are currently still dominated in Java, especially Bogor Bekasi, Subang, Jakarta, and Pandeglang, although in recent years new hatcheries have emerged in Lampung, Palembang, East OKU, Jambi, Pekanbaru, Medan, and Banjarbaru.

The pangasius grow-out technology may not experience significant changes in development. Increasing pangasius production in Indonesia is mainly influenced by the high adaptability of fish to water conditions, as well as the expansion of cultivation areas (extensification). In addition to freshwater, pangasius grow-out can also be carried out in peat waters and brackish waters. Used shrimp ponds with salinity below 10 ppt are frequently used as pangasius grow-out facilities by the farmers. On grow-out segmentation, feed is the largest component of production costs. Therefore, research related to feed efficiency have been focused including the use of superior seeds, application of probiotics and prebiotics in feed and water, and utilization of feed substitutes that are available around the facilities. Regular monitoring and management of water quality is also included as a critical aspect for pangasius grow-out segmentation. The centers of pangasius production in Indonesia include the islands of Sumatra, Kalimantan, and parts of Java. This distribution is influenced by market demand, as Pangasius is particularly popular among the local population, especially in Sumatra and Kalimantan.

19.7 Pangasius Production in Indonesia

Pangasius is one of the commodities targeted for fisheries industrialization in Indonesia. The world production of pangasius, which is entirely from Asia, has continued to increase significantly in recent years. In 2015, world pangasius production reached 2.1 million tons with the largest producers being Vietnam (1.1 million tons), Indonesia, and Bangladesh at 400,000 tons each (Subasinghe, 2017). This production value continues to increase and will reach 3.2 million tons in 2020. Indonesia, which since 2010 has been listed as the top five pangasius-producing countries in the world, has production centers spread across several regions. The production centers in Indonesia are dominated from the island of Sumatra (68.07%), Kalimantan (20.23%), and Java (8.48%), and the rest are scattered in several other regions (Fig. 19.2).

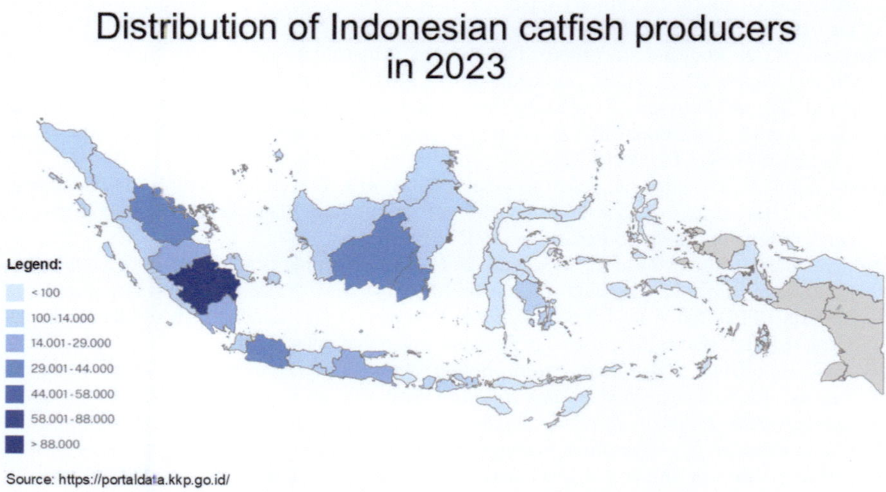

Fig. 19.2 Distribution maps of Indonesian catfish in 2023. (Source: KKP, 2024)

References

Andriyanto, S., Tahapari, E., & Insan, I. (2012). Pendederan Ikan Patin sebagai Pendukung Program Restocking [Striped catfish nursery as a support for restocking programs]. *Prosiding Indoaqua-Forum Inovasi Teknologi Akuakultur, 2012*, 211–216. https://doi.org/10.15578/ma.7.1.2012.20-25

Boonbrahm, M., Tarnchalanukit, W., & Chuapoehuk, W. (1968). Induced spawning by pituitary hormones injection of pond-reared fishes. *Proceedings of the Indo-Pacific Fisheries Council, 13*(2), 162–170. https://openknowledge.fao.org/server/api/core/bitstreams/ed9916fe-1c8b-4f45-94cd-deb399b1b41f/content

Darmawan, J., & Tahapari, E. (2019a). Analisis Produktivitas dan Profitabilitas Budidaya Patin Perkasa (*Pangasianodon hypophthalmus*) dengan Padat Penebaran Berbeda [Productivity and profitability analysis of perkasa striped catfish (*Pangasianodon hypophthalmus*) cultivation with different stocking densities]. *Prosiding Seminar Nasional Tahunan XVI Hasil Penelitian Perikanan Dan Kelautan Tahun, 2019*, 64–67.

Darmawan, J., & Tahapari, E. (2019b). Uji Pendahuluan Tingkat Ketahanan Ikan Patin Pasupati (*Pangasius* sp.) Triploid pada Pengujian Transportasi Statis [Preliminary test of the resistance level of triploid Pasupati striped catfish (*Pangasius* sp.) in static transportation testing]. *Prosiding Seminar Nasional Tahunan XVI Hasil Penelitian Perikanan Dan Kelautan Tahun, 2019*, 307–311.

Darmawan, J., Tahapari, E., & Pamungkas, W. (2023). Effectiveness of anesthesia solution (2-Phenoxyethanol) in the transportation of broodstock of Perkasa Striped Catfish (*Pangasianodon hypophthalmus* Sauvage, 1878). *IOP Conference Series: Earth and Environmental Science, 1137*(1). https://doi.org/10.1088/1755-1315/1137/1/012034

Effendie, M. I. (2002). *Metode biologi perikanan [Fisheries biology methods]* (p. 112). Yayasan Dewi Sri.

Ferraris, C. J. (2007). Checklist of catfishes, recent and fossil (Osteichthyes: Siluriformes), and catalogue of siluriform primary types. *Zootaxa, 1418*(1), 628. https://doi.org/10.11646/zootaxa.1418.1.1. Magnolia Press.

Gustiano, R., & Pouyaud, L. (2006). Diversity of Pangasiid catfishes from Sumatra. *Buletin Plasma Nutfah, 12*(2), 83–88. https://doi.org/10.21082/blpn.v12n2.2006.p83-88

Gustiano, R., & Pouyaud, L. (2008). Systematic revision of the Genera of Pangasiidae (Siluriformes, Ostariophysi). *Indonesian Aquaculture Journal, 3*(1), 13–22. https://doi.org/10.15578/iaj.3.1.2008.13-22

Hardjamulia, A., Djajadiredja, R., Atmawinata, S., & Idris, D. (1981). Fertilization of Siamese catfish (*Pangasius sutchi*) with an injection of the pituitary gland extract of carp (*Cyprinus carpio*). *Buletin Penelitian Perikanan, 1*(2), 183–190.

Hardjamulia, A., Atmawinata, S., & Suseno, D. (1978). Penelitian pembenihan ikan jambal siam (*Pangasius sutchi* F) [*Research on the breeding of Siamese catfish (Pangasius sutchi F)*]. Pewarta LPPD Tahun ke-1 No. 1, Agustus-Desember 1978. Lembaga Penelitian Perikanan Darat (LPPD). Badan Penelitian dan Pengembangan Pertanian, 4 hlm.

Iswanto, B. (2011). *Hibridisasi Antara Betina Patin Siam (Pangasianodon hypophthalmus Sauvage, 1878) dengan Jantan Patin Nasutus (Pangasius nasutus Bleeker, 1863) dalam Upaya Meningkatkan Produktivitas Patin Daging Putih* [Hybridization between female Patin Siam (Pangasianodon hypophthalmus Sauvage, 1878) and male Patin Nasutus (Pangasius nasutus Bleeker, 1863) as an attempt to increase the productivity of white flesh pangasiid catfish]. Thesis. Universitas Brawijaya, Malang, 218 p.

Kartamihardja, E. S., Purnomo, K., Koeshendrajana, S., & Umar, C. (2011). Karakteristik Limnologis, Perkembangan Populasi Dan Perikanan Di Waduk Wonogiri, Jawa Tengah [Limnological characteristics, population and fisheries development in Wonogiri Reservoir, Central Java]. *Jurnal Kebijakan Perikanan Indonesia, 3*(1), 37–50. https://doi.org/10.15578/jkpi.3.1.2011.37-50

Khairuman, H. (2010). *Petunjuk Praktis Budi Daya Patin di Kolam Terpal [Practical instructions for cultivating catfish in Tarpaulin ponds]* (p. 90). AgroMedia Pustaka.

KKP. (2018). *Industri Patin Indonesia Rebut Pasar Global* [Indonesian Patin Industry Captures Global Market]. [online]. http://kkp.go.id/artikel/3163industri-patin-indonesia-rebut-pasar-global. Accessed 10 Nov 2023.

KKP. (2021). *Mendulang Cuan dari Limbah Patin* [Making Money from Patin Waste]. [online]. (https://www.kkp.go.id/bppsdmkp/mendulang-cuan-dari-limbah-patin65c30aee12738/detail.html. Accessed 10 Nov 2023.

KKP. (2024). *Volume Produksi Perikanan Budidaya Pembesaran Komoditas Patin per Provinsi (Ton)* [Production volume of aquaculture for Pangasius Commodity Enlargement per Province (Tons)]. [online]. (https://portaldata.kkp.go.id/portals/data-statistik/prod-ikan/tbl-statis/d/17. Accessed 5 Feb 2025.

Legendre, M., Slembrouck, J., Subagja, J., & Hari Kristanto, A. (2000). Ovulation rate, latency period and ova viability after GnRH-or hCG-induced breeding in the Asian catfish *Pangasius hypophthalmus* (Siluriformes, Pangasiidae). *Aquatic Living Resources, 13*(3), 145–151. https://doi.org/10.1016/S0990-7440(00)00148-0

Pouyaud, L., Gustiano, R., & Legendre, M. (1998). Phylogenetic relationships among pangasiid catfish species (Siluriformes, Pangasiidae) and new insights on their zoogeography. In L. Marc & P. Antoine (Eds.), *The biological diversity and aquaculture of Clariid and Pangasiid Catfishes in South-East Asia: Proceedings of the mid-term workshop of the "Catfish Asia Project", Jakarta (IDN); Can Tho: IRD; Can Tho* (Vol. 1, pp. 49–56) http://www.documentation.ird.fr/hor/fdi:010020346

Pouyaud, L., Teugels, G. G., & Legendre, M. (1999). Description of a new pangasiid catfish from South-East Asia (Siluriformes). *Cybium, 23*(3), 247–258. https://doi.org/10.26028/cybium/1999-233-002

Pouyaud, L., Gustiano, R., & Teugels, G. G. (2004). Contribution to the phylogeny of the Pangasiidae based on mitochondrial 12S Rdna. *Indonesian Journal of Agricultural Science, 5*(2), 45–62. https://doi.org/10.21082/ijas.v5n2.2004.p4562

Slembrouck, J., Subagja, J. D. D., & Legendre, M. (2005). *Petunjuk Teknis Pembenihan Ikan Patin Indonesia, Pangasius djambal [Technical Instructions for Indonesian Striped Catfish Seeding, Pangasius djambal]* (p. 54). Institut de Recherche pour le Développement. https://core.ac.uk/download/pdf/39843279.pdf

Subasinghe, R. (2017). *World Aquaculture 2015: A brief overview* (p. 47). FAO Fisheries and Aquaculture. https://openknowledge.fao.org/handle/20.500.14283/i7546e

Tahapari, E., Darmawan, J., Pamungkas, W., Sularto, & Nurlela, I. (2014). *Petunjuk Teknis Teknologi Pembenihan dan Pembesaran Ikan Patin Pasupati [Technical instructions for Pasupati Patin fish seeding and rearing technology]* (p. 34). Balai Penelitian Pemuliaan Ikan.

Tahapari, E., Darmawan, J., Pamungkas, W., Suharyanto, Iswanto, B., Setyawan, P., Marnis, H., & Imron. (2017). *Naskah Akademik Ikan Patin Siam Tumbuh Cepat Generasi Kedua Hasil Seleksi Famili [Academic manuscript of fast growing siamese catfish second generation result of family selection]* (p. 61). Balai Riset Pemuliaan Ikan.

The Status of Domestication and Aquaculture System on the Tiger Shrimp *Penaeus monodon* in Indonesia

20

Andi Parenrengi, Andi Tenriulo, Samuel Lante, Rosmiati, and Emma Suryati

20.1 Introduction

Tiger shrimp *Penaeus monodon* is recognized as an indigenous crustacean species of Indonesia (Parenrengi et al., 2017) and has been the dominant shrimp species farmed in the eastern hemisphere for many decades, mainly because of its larger final size and faster growth rate compared to other penaeid species (Coman et al., 2005). The marine species *P. monodon*, also referred to as black tiger shrimp, giant tiger prawn, and Asian tiger shrimp, is extensively cultivated for food. The researcher has provided descriptions of taxonomy and morphology, as well as observed the patterns of distribution and reproduction. Some basic and applied research supporting the domestication process has been reported, such as artificial insemination, hormone application, DNA marker, RNAi tool, and transgenic technology (Parenrengi et al., 2021a). Several studies have exposed black tiger shrimp broodstock domestication, including developing domestication techniques of specific pathogen-free (SPF) broodstock in controlled tanks and brackish water ponds (Suwoyo & Sahabuddin, 2017). So far, there has been limited advancement in tiger shrimp aquaculture using traditional and semi-intensive methods within both monoculture and polyculture systems.

A. Parenrengi (✉) · A. Tenriulo · S. Lante · Rosmiati · E. Suryati
Research Center for Fishery, National Research and Innovation Agency, Cibinong, Indonesia
e-mail: andi053@brin.go.id

E. Nugroho et al. (eds.), *Aquaculture Development in Archipelago*,
https://doi.org/10.1007/978-981-95-1084-9_20

20.2 Taxonomy and Morphology

Johan Christian Fabricius published the first description of *P. monodon* in 1798. For a long time, the name was disregarded, but Holthuis (1949) explained which species it applied to and demonstrated that *P. monodon* had to be a species of the Penaeus genus (Fig. 20.1). The complete taxonomic rank is as follows:

Phylum:	Arthropoda
Class:	Malacostraca
Order:	Decapoda
Family:	Penaeidae
Genus:	*Penaeus*
Species:	*Penaeus monodon* Fabricius, 1798

The tiger shrimp is morphologically divided into two parts, namely, the head and the abdomen. The head is called the cephalothorax, covered in thick skin chitin called a carapace. This section consists of a head with five segments and a chest with eight segments. The abdomen contains five pairs of swimming legs (pleopods); on the segment, the 6th swimming legs change shape to become a fan tail (uropod). Tiger prawns are heterosexual animals; that is, they have types of clearly distinguishable male and female sexes. Variety of female tiger prawns can be identified by the thelycum on the 4th and 5th toes. The thelycum is a thin line and will widen after fertilization occurs. The sex of male tiger prawns can be identified by the presence of petasma, namely, the bulge between the first swimming legs (Suyanto & Mujiman, 2005).

Compared to females, males are relatively smaller, averaging 20–25 cm (8–10 inches) in length and 100–170 g (3.5–6.0 oz) in weight, though they can grow up to 33 cm (13 inches) long and weigh between 200 and 320 g (7–11 oz) on average. Transverse bands of red and white alternately decorate the abdomen and carapace. The antennae are brownish gray in color. There are red bordering setae around brown pereiopods and pleopods (Motoh, 1981).

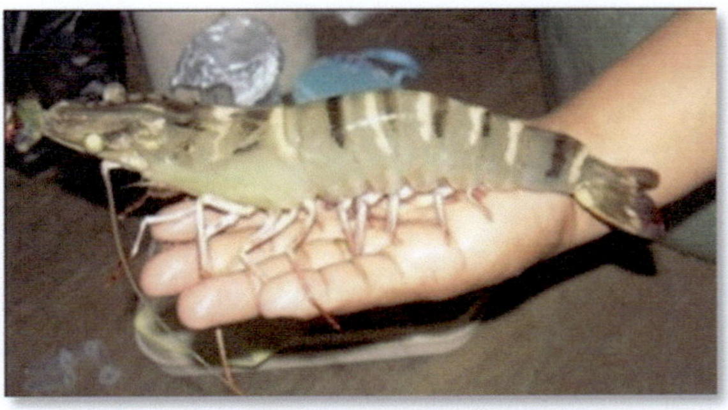

Fig. 20.1 The first generation of domesticated tiger shrimp cultured at a brackish water pond. (Photo: Ministry of Marine Affairs and Fisheries—MMAF)

20.3 Distribution

The Indo-Pacific region, which includes Southeast Asia, northern Australia, and the Pacific Ocean, is where the tiger shrimp is naturally found. It also includes the Arabian Peninsula and the eastern coast of Africa. The northern Gulf of Mexico and Atlantic Ocean areas off the southern United States are habitats for this invasive species (Chan et al., 2021). Shrimp are widely distributed and plentiful. On most beaches, in estuaries, and in rivers and lakes, they can be seen eating close to the seafloor. The species is found naturally throughout the Indo-Pacific region, spanning from the Arabian Peninsula to Southeast Asia, the eastern coast of Africa, northern Australia, and the Sea of Japan. In the Gulf of Mexico's northern seas, it is an invasive species. To avoid predators, certain animals flip off the seafloor and dive into the silt. They typically have a lifespan of 1–7 years. While they can form enormous schools while spawning, shrimp are typically solitary (Suyanto & Mujiman, 2005).

20.4 Reproduction

The sperm of male giant tiger shrimp was stored in sacs located within their closed thelycum until eggs are fully mature, and females have been observed to mate before ovarian development. Although there is limited information on specific mating behaviors, it has been reported that this species mates at night in offshore seas soon after the females have shed their coats. Unlike males, who normally still have hard carapaces during breeding, these animals still have soft carapaces. A male and a female begin to swim parallel to each other during copulation. When the male bends his body, the second pair of pleopods can more easily move forward, enabling the first pair with the petasma to extend downward. This positioning helps the first pair of pleopods separate the petasmal halves to avoid sperm loss during mating. Following that, they align their abdomens with one another. The female shrimp uses the fourth pair of pereiopods to press on the male's petasma, and the male then separates when the female inserts a spermatophore (sperm sac) into her thelycum. Males at carapace lengths of 50, 62, 66, and 72 mm are reported to spawn four times during their lifetimes and most mature individuals copulate more than once (Motoh, 1981).

Although the exact age of sexual maturity is difficult to determine, shrimp males mature when their average carapace size reaches 37 mm, while shrimp females achieve 47 mm. Females are reported to spawn up to four times in their lifetime and have an egg production capacity of 248,000–810,000 at a time. After eggs are fully developed, the eggs are released into the ocean in a greenish white cloud together with any stored spermatophores, where external fertilization takes place. The egg size is 0.27 to 0.31 mm (Motoh, 1981).

20.5 Domestication

Compared to domestication in Australia, which has reached the eighth generation (Coman et al., 2013), domestication of tiger prawns in Indonesia is still very slow. It does not have a sustainable breeding program or continuing finance. Even though the domestication of tiger prawns in Indonesia has reached the sixth generation, the progress of its activities has not continued until the release of domesticated tiger shrimp (Soleh et al., 2012). Domestication of tiger prawns in Indonesia generally uses the basic population from Aceh (Parenrengi et al., 2021a), considering that the best genetic diversity is found in the Aceh population. Sugama et al. (2002) reported on the genetic diversity of tiger prawns spread across Indonesia. The black tiger shrimp from Aceh had the best genetic diversity compared to Aceh, Cilacap, Bali, and other broodstock sources. Genetic characterization of tiger prawn broodstock is the first step in continuing the domestication program. Research support for tiger shrimp domestication in Indonesia has been carried out, including the following:

Artificial Insemination The low rate of natural mating in the tank is one of the key obstacles to domesticating black tiger shrimp. In commercial hatcheries, similar circumstances have occurred. Artificial insemination is a method used to increase the egg's fertility. The success of artificial insemination can produce eggs with a hatchability of around 40.5–61.7% (Lante & Laining, 2016). Furthermore, according to Samuel and Usman (2019), the inseminated broodstock of the F_1 female tiger shrimp had a spawning rate of 67%, a fecundity of 179,257 eggs, an egg fertilization rate of 86.2%, a hatching rate of 59.8%, and a total production of 738,439 nauplii.

Hormone Application The failure of eggs to hatch could be due to low-quality spermatophores from captive male tiger shrimp broodstock. Inducing testosterone is one method that can be used to enhance the quality of spermatophore in captive male tiger shrimp (Haryanti et al., 2015; Mulyaningrum et al., 2021). The injection of 17-methyltestosterone at a dosage of 300 ng/100 g body weight given four times seemed to be sufficient to enhance the performance of captive tiger shrimp reproduction; for example, the spermatophore weight (0.041 g), the gonadosomatic index (7.71%), and the total spermatozoa number (64106 cells/mL) all increased significantly. The rate of hatching increased to 43.14%, while the proportion of defective spermatozoa dramatically decreased by 73% (Rosmiati et al., 2022). The other study results on improved maturation showed that a dose of salmon gonadotropin-releasing hormone (sGnRHa) between 0.1 and 0.2 mL/kg of shrimp weight could be used to enhance the maturation of male tiger shrimp gonads cultured in ponds (Laining et al., 2016). To encourage the development of the spermatophores in male tiger shrimp, the hormone sGnRH-a was administered three times at intervals of 1 week with a dose of 0.2 mL/kg of shrimp weight (Trismawanti et al., 2018).

DNA Markers To support tiger shrimp domestication, the DNA marker has been focused on disease resistance (Tenriulo et al., 2012) and fast growth (Haryanti et al., 2016) (Fig. 20.2). The disease resistance marker was applied to tiger shrimp by microsatellite and sequence-characterized amplified region (SCAR) analysis to select broodstock before breeding. Fast growth has been identified and characterized to get an accurate gene marker corresponding to fast growth (Haryanti et al., 2012a; Wardana et al., 2008). After pond culture and phenotypic prediction, the shrimp will demonstrate the precise degree of fast-growth gene marker on tiger shrimp fry. The fast-growth marker has been applied to the larvae for cultivation. Tonnek and Syafaat (2015) reported that the fast-growing strains of fry could adapt in ponds with quite good growth, making it possible to be further reared as potential superior tiger shrimp broodstock. The traits associated with fast growth in the parent (F1) tiger shrimp were confirmed in the offspring (F2), demonstrating consistency between the male parent, female, sperm, and eggs, which 60% of the F2 fry population samples at the 144 bp level carried the allele linked to fast growth (Haryanti et al., 2016).

RNAi Technology Aquatic animal production by applying RNA interference (RNAi) is a prospective mechanism for improving resistance to viral disease. One of the highly interesting strategies for preventing viral infection in shrimp is the posttranscriptional mechanism known as RNAi, which is produced by the gene-specific constructs of dsRNA (double-stranded RNA) (Itsathitphaisarn et al., 2016). Isolation and characterization of viral protein (VP) of pathogens like WSSV is a basic work for dsRNA application (Hidayani & Tassakka, 2016; Parenrengi et al., 2018a; Tenriulo et al., 2015). The application of dsRNA has been reported to be a valuable strategy for increasing tiger shrimp's survival and immune responses (Binayanti et al., 2019; Mulyaningrum et al., 2018; Parenrengi et al., 2018a, 2019, 2020, 2022).

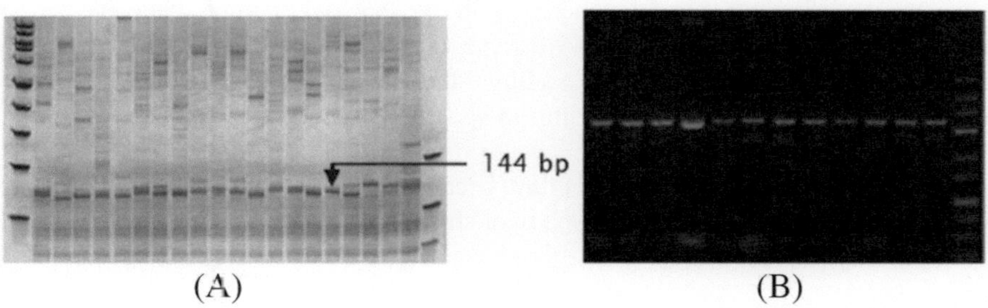

Fig. 20.2 DNA markers for tiger shrimp, *P. monodon* used the single-strand conformation polymorphism (SSCP) method with microsatellite disease resistance (**a**) PmMS-11A locus to indicate fast growth (**b**). (Source: Haryanti et al., 2012b)

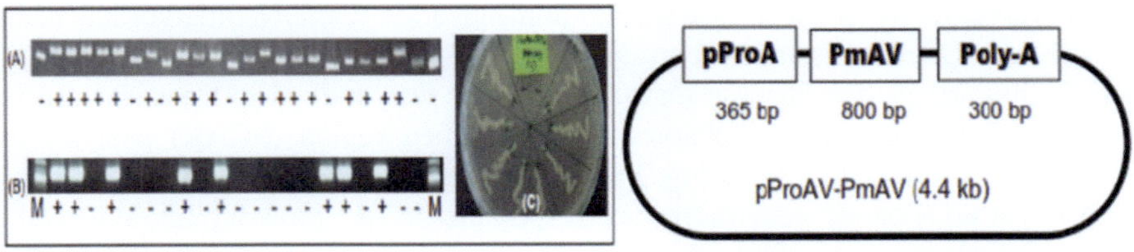

Fig. 20.3 Cracking and PCR assay of gene construct of pProAV-PmAV (**a** and **b**), and plating of bacterial clones (**c**) (left), and map of the schematic structure of gene constructs (right). (Parenrengi et al., 2021b)

Transgenic Technology The novel discovery of promoters and antiviral genes and the success in constructing the all-shrimp gene construct (Fig. 20.3) is an initial milestone toward the development of transgenic technology in tiger shrimp (Parenrengi et al., 2018b). Promoter antiviral genes and gene construct have been analyzed to be active in tiger shrimp larvae (Parenrengi et al., 2014, 2018b). All shrimp gene construct expressions were upregulated, which is a positive indicator. According to the challenge test, the transgenic tiger shrimp survived better against infection of WSSV (95.6%) than the control tiger shrimp (71.1%). In terms of body weight, length increase, appearance, morphology, and activity, the transgenic tiger shrimp at 47 days did not differ significantly from the control group (Parenrengi et al., 2021b). The progress of transgenic tiger shrimp has reached the first generation in cultivation in controlled ponds and tanks (Suwoyo & Sahabuddin, 2017).

20.6 Aquaculture System

Tiger shrimp is the second-largest farmed crustacean species worldwide, following only white-leg shrimp, *Litopenaeus vannamei*. MMAF reported that tiger shrimp production reached 881.6 tons in 2020 and 953.2 tons in 2021 (KKP, 2022). Since 1990, tiger shrimp production in Indonesia has become difficult due to worsening environmental conditions and viral outbreaks. There have been reports of at least 20 viruses that cause diseases in shrimp. A particularly virulent virus that affects shrimp in pond cultures and hatcheries is the WSSV (white spot syndrome virus) (Parenrengi et al., 2017). This sets financial constraints on Australia's farm-raised black tiger shrimp food business. To address these issues, efforts have been made to carefully breed the black tiger shrimp in specific pathogen-resistant lines (Palmer et al., 2021).

The tiger shrimp is accepted in culture because of its quick growth rate and salinity tolerance (Knott et al., 2019). The tiger shrimp aquaculture system in Indonesia is commonly popular with two categories, namely, monoculture and polyculture.

Monoculture The monoculture system is known as only one species cultivated in the pond. The pond is only simply prepared; water enters the pond without being first treated to remove the pathogen and its carrier, a stock is kept at a low stocking density of 1–5 PL/m^2, there is no aeration, and very little food is provided for shrimp. Traditional farmers, with a history of cultivating tiger shrimp, predominantly utilize this species in large pond aquaculture (Taslihan et al., 2015). After 80 days of monoculture rearing, the tiger shrimp were harvested with an average weight between 14.29 and 40.00 g/ind., a survival rate of 27.57%–66.67%, a productivity of 90.1–800.0 kg/ha/season, and an FCR of 0.54 to 2.10 (Mustafa et al., 2021).

Polyculture/Integrated Culture Polyculture of tiger shrimp is generally conducted with seaweed, especially species of *Gracilaria* spp. (Azizah et al., 2018). Polyculture of tiger prawns with milkfish, tilapia, and seaweed has been reported by Mangampa (Mangampa, 2014) in which production reached 118–280.6 kg/ha/season for tiger prawns, 153.8–210 kg/ha/season for milkfish, 200–310 kg/ha/season, and 100–3372 kg dry/ha for seaweeds. The integrated culture concept, called integrated multi-trophic aquaculture (IMTA), is one of the polyculture approaches to help maintain the balance of the ecosystem. This is possible because the various organisms cultivated include omnivores, herbivores, and filter feeders at different stages of their life cycles (Amalia et al., 2022). The IMTA technology has been applied to recycle waste from the aquaculture process produced by the primary cultured species.

The MMAF has introduced a breakthrough program in export-oriented aquaculture, positioning shrimp as a primary commodity. Shrimp is among Indonesia's prominent fishery products with strong export potential. As a major national commodity, shrimp plays a vital role in efforts to boost state revenue and meet the production target of a 250% increase by 2024. The shrimp market is quite large, and Indonesia is among the five leading shrimp producers worldwide, offering a significant chance to drive continued growth in shrimp farming. Shrimp exports are expected to make a major contribution to national economic growth by continuing to optimize the utilization of potential areas.

References

Amalia, R., Rejeki, S., Widowati, L. L., & Ariyati, R. W. (2022). The growth of tiger shrimp (*Penaeus monodon*) and its dynamics of water quality in integrated culture. *Biodiversitas, 23*(1), 593–600. https://doi.org/10.13057/biodiv/d230164

Azizah, I., Rejeki, S., & Ariyati, R. W. (2018). Performa pertumbuhan udang (Penaeus monodon) yang dibudidayakan bersama rumput laut (Gracilaria sp.) dengan padat tebar yang berbeda menerapkan sistem Integrated Multi-Trophic Aquaculture (IMTA) [Growth performance of tiger shrimp (Penaeus monodon) cultivated together with seweed (Gracilaria sp.) with different stocking densities apply the Integrated Multi-Trophic Aquaculture (IMTA)]. *Jurnal Sains Akuakultur Tropis, 2*(2), 1–11.

Binayanti, Anshary, H., Parenrengi, A., & Citra Malina, A. (2019). Use of dsRNA WSSV VP19 with different dose for infection control White Spot Syndrome Virus (WSSV) in vaname shrimp *Litopenaeus vannamei*. *International Journal of Scientific and Research Publications (IJSRP), 9*(8), 9230. https://doi.org/10.29322/ijsrp.9.08.2019.p9230

Chan, T. Y., Muchlisin, Z. A., & Hurzaid, A. (2021). Verification of a pseudocryptic species in the commercially important tiger prawn Penaeus monodon Fabricius, 1798 (Decapoda: Penaeidae) from Aceh Province, Indonesia. *Journal of Crustacean Biology, 41*(1), 1–10. https://doi.org/10.1093/jcbiol/ruaa096

Coman, G. J., Crocos, P. J., Arnold, S. J., Keys, S. J., Preston, N. P., & Murphy, B. (2005). Growth, survival and reproductive performance of domesticated Australian stocks of the giant tiger prawn, *Penaeus monodon*, reared in tanks and raceways. *Journal of the World Aquaculture Society, 36*(4), 464–479. https://doi.org/10.1111/j.1749-7345.2005.tb00394.x

Coman, G. J., Arnold, S. J., Wood, A. T., & Preston, N. P. (2013). Evaluation of egg and nauplii production parameters of a single stock of domesticatedP enaeus monodon(giant tiger shrimp) across generations. *Aquaculture*, 400–401, 125–128. https://doi.org/10.1016/j.aquaculture.2013.03.015

Haryanti, H., Fachruddin, Ardana, I. K., Permana, I. G. N., Mahardika, K., & Moria, B. (2012a). Gen penciri tumbuh cepat sebagai indikator seleksi pada benih udang windu, *Penaeus monodon* [Fast growth characteristic genes as selection indicators in tiger shrimp seeds, *Penaeus monodon*]. *Jurnal Riset Akaukultur, 7*(2), 181–193.

Haryanti, H., Mahardika, K., Fachrudin, F., Wardana, I. K., Permana, I. G. N., & Sembiring, S. B. M. (2012b). Penggunaan gen penyandi tumbuh cepat dalam produksi benih udang windu *Penaeus monodon* [The use of genes encoding fast growth in the production of black tiger shrimp *Penaeus monodon* seeds]. *Jurnal Riset Akuakultur, 7*(3), 345. https://doi.org/10.15578/jra.7.3.2012.345-357

Haryanti, H., Fahrudin, & Sembiring, S. B. M. (2015). Induksi hormon 17α-methyl testosteron terhadap profil spermatogenesis induk jantan udang windu, *Penaeus monodon* [Induction of the hormone 17α-methyl testosterone on the spermatogenesis profile of male tiger shrimp, *Penaeus monodon*]. In *Prosiding Forum Inovasi Teknologi Akuakultur 2015* (pp. 133–139).

Haryanti, H., Sembiring, S. B. M., & Fachrudin. (2016). Konfirmasi sifat tumbuh cepat pada induk (F-1) dan benih turunannya (F-2) udang windu, *Penaeus monodon* hasil budidaya [Konfirmasi sifat tumbuh cepat pada induk (F-1) dan benih turunannya (F-2) udang windu, *Penaeus monodon* hasil budidaya]. *Prosiding Forum Inovasi Teknologi Akuakultur, 2016*, 609–614.

Hidayani, A. A., & Tassakka, A. C. M. (2016). Isolation and characterization of an envelope protein (VP19) of a white spot syndrome virus from diseased vannamei (*Litopenaeus vannamei*) in Indonesia. *AACL Bioflux, 9*(2), 389–395.

Holthuis, L. B. (1949). The identity of Penaeus monodon Fabr. *Proceedings van de Koninklijke Nederlandsche Akademie van Wetenschappen, 52*(9), 1–8.

Itsathitphaisarn, O., Thitamadee, S., Weerachatyanukul, W., & Sritunyalucksana, K. (2016). Potential of RNAi applications to control viral diseases of farmed shrimp. *Journal of Invertebrate Pathology.* https://doi.org/10.1016/j.jip.2016.11.006

KKP. (2022). *Statistic-MMAF*. https://statistik.kkp.go.id/home.php

Knott, D., Fuller, P., Benson, A., & Neilson, M. (2019). *Penaeus monodon: U.S. geological survey, nonindigenous aquatic species database*. https://www.gbif.org/species/165417380

Laining, A., Usman, U., & Syah, R. (2016). Induction of gonadal maturation of pond cultured male tiger shrimp, *Penaeus monodon* with different dosages of gonadotropin releasing hormone analogue against eye stalk ablation. *Indonesian Aquaculture Journal, 11*(1), 23–30. https://doi.org/10.15578/iaj.11.1.2016.23-30

Lante, S., & Laining, A. (2016). Aplikasi inseminasi buatan pada udang windu, *Penaeus monodon* alam menggunakan sumber dan jumlah spermatofor yang berbeda [The application of artificial insemination on wild tiger prawns, *Penaeus monodon* at different sources and amounts of spermatophore]. *Jurnal Riset Akaukultur, 11*(3), 271–280.

Lante, S., & Usman, U. (2019). Pengaruh sumber spermatofora pada inseminasi buatan induk betina udang windu turunan pertama (F-1) terhadap pemijahan, kualitas telur, dan larva turunannya (F-2) [The influence of spermatophore sources on artificial insemination of female tiger shrimp (F-1) on spawning, egg quality, and larvae (F-2)]. *Media Akuakultur, 14*(2), 63. https://doi.org/10.15578/ma.14.2.2019.63-71

Mangampa, M. (2014). Polikultur udang windu (*Penaeus monodon*), bandeng (*Chanos chanos*), nila srikandi (*Oreochromis aureus* x *O. niloticus*), dan rumput laut (*Gracilaria verrucosa*) di tambak tanah sulfat masam (TSM) [Polyculture of tiger shrimp (*Penaeus monodon*), milkfish (*Chanos chanos*), Srikandi tilapia (*Oreochromis aureus* x *O. niloticus*), and seaweed (*Gracilaria verrucosa*) in acidic sulfate soil ponds]. *Prosiding Forum Inovasi Teknologi Akuakultur*, 1–13.

Motoh, H. (1981). *Studies on the fisheries biology of the giant tiger prawn, Penaeus monodon in the Philippines*. SEAFDEC Aquaculture Department.

Mulyaningrum, S. R. H., Parenrengi, A., Tampangallo, B. R., & Trismawanti, I. (2018). Respons imun udang windu *Penaeus monodon* terhadap vaksin dsRNA VP-24 pada dosis berbeda [Immune response of black tiger prawn *Penaeus monodon* to VP-24 dsRNA vaccine at different doses]. *Jurnal Riset Akuakultur, 13*(1), 77–84. https://doi.org/10.15578/jra.13.1.2018.77-84

Mulyaningrum, S. R. H., Parenrengi, A., Lante, S., Tenriulo, A., Daud, R., & Usman. (2021). Effect of different doses of 17α-methyl testosterone (17-α MT) hormone on male broodstock of tiger shrimp *Penaeus monodon* sperm quality. *IOP Conference Series: Earth and Environmental Science, 777*(012024). https://doi.org/10.1088/1755-1315/777/1/012024

Mustafa, A. A., Asaad, A. I. J., & Linthin, D. (2021). Performa budidaya udang windu (*Penaeus monodon*) pada musim kemarau di tambak Kecamatan Marusu Kabupaten Maros [Performance of tiger prawn (*Penaeus monodon*) cultivation during the dry season in ponds in Marusu District, Maros Regency]. *Media Akuakultur, 16*(1), 45–56. https://doi.org/10.15578/ma.16.1.2021.45-56

Palmer, P. J., Rao, M., & Cowley, J. A. (2021). Reduced transmission of IHHNV to Penaeus monodon from shrimp pond wastewater filtered through a polychaete-assisted sand filter (PASF) system. *Aquaculture, 535*. https://doi.org/10.1016/j.aquaculture.2021.736359

Parenrengi, A., Tampangallo, B. R., & Tenriulo, A. (2014). Analysis of immune responses on transgenic tiger shrimp (*Penaeus monodon*) against pathogenic bacterium *Vibrio harveyi*. *Indonesian Aquaculture Journal, 9*(1), 23–32. https://doi.org/10.15578/iaj.9.1.2014.23-32

Parenrengi, A., Alimuddin, A., & Tenriulo, A. (2017). Characteristic of a viral protein (VP-15) of white spot syndrome virus (WSSV) isolated from infected tiger shrimp *Penaeus monodon* (Fabricius, 1798). *Indonesian Aquaculture Journal, 12*(2), 1–9. https://doi.org/10.15578/iaj.12.2.2017.67-75

Parenrengi, A., Mulyaningrum, S. R. H., Tenriulo, A., & Nawang, A. (2018a). Gen penyandi viral protein 15 (VP-15) WSSV dan aplikasinya sebagai vaksin rekombinan pada udang windu [The WSSV viral protein 15 (VP-15) coding gene and its application as a recombinant vaccine in tiger shrimp]. *Jurnal Riset Akuakultur, 13*(1), 57. https://doi.org/10.15578/jra.13.1.2018.57-65

Parenrengi, A., Syah, R., Tenriulo, A., & Alimuddin. (2018b). Analysis of promoter activity on tiger shrimp *Penaeus monodon* using EGFP (enhanced green fluorescent protein) as a marker gene. *AACL Bioflux, 11*(6), 1937–1946.

Parenrengi, A., Tenriulo, A., Mulyaningrum, S. R. H., Lante, S., & Nawang, A. (2019). Pengaruh aplikasi dsRNA VP-15 in vitro dan in vivo terhadap sintasan dan respons imun udang windu *Penaeus monodon* [Effect of application of dsRNA VP-15 in vitro and in vivo on survival and immune response of tiger shrimp *Penaeus monodon*]. *Jurnal Riset Akuakultur, 14*(4), 213. https://doi.org/10.15578/jra.14.4.2019.213-223

Parenrengi, A., Tenriulo, A. A., Lante, S., Nawang, A., Sulaeman, S., & Masruri, M. H. H. (2020). A preliminary experiment on production of dsRNA by in-vivo and its application to tiger shrimp *Penaeus monodon* larvae for survival enhancement. *IOP Conference Series: Earth and Environmental Science, 584*(1), 012051. https://doi.org/10.1088/1755-1315/584/1/012051

Parenrengi, A., Lante, S., Tenriulo, A., Suryati, E., Rosmiati, R., Nawang, A., & Sahabuddin, S. (2021a). *Pemantapan sistem produksi larva, pengembangan teknologi RNAi, dan peningkatan kualitas serta pembesaran calon induk udang windu Penaeus monodon [Strengthening the larval production system, developing RNAi technology, and improving the quality and enlargement of prospective tiger shrimp broodstock Penaeus monodon]*. Technical report of Research Institute for Brackishwater Aquaculture and Fisheries Extension.

Parenrengi, A., Tenriulo, A., Alimuddin, A., & Sukenda, S. (2021b). Enhancement of tiger shrimp *Penaeus monodon* resistance to white spot syndrome virus by overexpression of antiviral gene. *International Journal of Agriculture and Biology, 25*(2), 277–284. https://doi.org/10.17957/IJAB/15.1667

Parenrengi, A., Tenriulo, A., Suryati, E., Rosmiati, R., Lante, S., Azis, A. A., & Alimuddin, A. (2022). Application of dsRNA VP15-WSSV by oral vaccination to increase survival rate and response immunes of tiger shrimp *Penaeus monodon*. *Indian Journal of Animal Research, 56*(7), 893–898. https://doi.org/10.18805/IJAR.BF-1460

Rosmiati, R., Parenrengi, A., Suryati, E., Lante, S., Harlina, H., Tenriulo, A., & Sahabuddin, S. (2022). Improved reproductive performance of captive male broodstock of tiger shrimp *Penaeus monodon* with 17α-methyltestosterone induction. *Israeli Journal of Aquaculture – Bamidgeh, 74*, 1–10. https://doi.org/10.46989/001c.31616

Soleh, M., Priyoutomo, T. P., Anindiastuti, Adiwijaya, D., Maisoni, A. F., Taslihan, A., Erlina, A., Susanto, A., Hardanu, W., Mardiyanta, A., & Sumantri, I. (2012). *Domestication of Tiger Shrimp Penaeus monodon*.

Sugama, K., Haryanti, H., Benzie, J. A., & Ballment, E. (2002). Genetic variation and population structure of the giant tiger prawn, *Penaeus monodon* in Indonesia. *Aquaculture, 205*(1–2), 37–48. https://doi.org/10.1016/S0044-8486(01)00662-7

Suwoyo, H. S., & Sahabuddin, S. (2017). Performa pertumbuhan calon induk udang windu *Penaeus monodon* transfeksi pada generasi yang berbeda [Growth performance of transfected *Penaeus monodon* tiger shrimp broodstock candidates in different generations]. *Jurnal Ilmu Dan Teknologi Kelautan Tropis, 9*(1), 185–200.

Suyanto, R. S., & Mujiman, A. (2005). *Budidaya Udang Windu [Tiger shrimp culture]*. Penebar Swadaya. http://perpus.tasikmalayakab.go.id/opac/detail-opac?id=7954

Taslihan, A., Callinan, R., Torribio, J.-A., Sumiarto, B., & Nitimulyo, K. H. (2015). Cluster model for extensive giant tiger shrimp (*Penaeus monodon* Fab.) to prevent transmission of white spot syndrome virus. *Indonesian Aquaculture Journal, 10*(1), 65–70.

Tenriulo, A., Parenrengi, A., & Alimuddin. (2012). Analisis marker DNA mikrosatelit udang windu *Penaeus monodon* tahan penyakit dari beberapa populasi alam [Analysis of disease-resistant tiger shrimp *Penaeus monodon* microsatellite DNA markers from several natural populations]. *Prodising Seminar Nasional Tahunan IX Hasil Penelitian Perikanan Dan Kelautan Tahun 2012, UGM*.

Tenriulo, A., Tampangallo, B. R., Parenrengi, A., & Dewi, R. A. (2015). Isolasi dan karakterisasi gen penyandi protein VP-24 WSSV pada udang windu (*Penaeus monodon*) untuk pengembangan teknologi RNAi [Isolation and characterization of the WSSV VP-24 protein coding gene in tiger shrimp (*Penaeus monodon*) for the development of RNAi technology]. *Prosiding Forum Inovasi Teknologi Akuakultur*, 593–598.

Tonnek, S., & Syafaat, M. N. (2015). Pertumbuhan larva udang windu strain cepat tumbuh dan seleksi calon induk asal tambak [Growth of fast growing strains of tiger shrimp larvae and selection of potential broodstock from ponds]. *Forum Inovasi Teknologi Akuakultur, 2015*, 979–984.

Trismawanti, I., Nawang, A., & Laining, A. (2018). Stimulasi pematangan spermatofor udang windu (Penaeus monodon) apkiran tanpa ablasi [Stimulation of spermatophore maturation in tiger shrimp (Penaeus monodon) unablation]. *Media Akuakultur, 13*(2), 67–74. https://doi.org/10.15578/ma.13.2.2018.67-74

Wardana, I. K., Muzaki, A. A., Permana, I. G. N., Haryanti, H., Fahruddin, Permana, I. G. N., & Harya. (2008). Selektif breeding udang windu *Penaeus monodon* dengan karakter pertumbuhan dan SPF (Specific pathogen free) [Selective breeding of *Penaeus monodon* tiger shrimp with growth characteristics and SPF (Specific pathogen free)]. *Jurnal Riset Akuakultur, 3*(3), 301–312.

Domestication of Blue Swimming Crab, *Portunus pelagicus* (Linnaeus, 1758)

21

Gunarto Gunarto, Herlinah Herlinah, and Sulaeman Sulaeman

21.1 Introduction

Blue swimming crab is a marine fishery commodity and the third main export commodity after shrimp, tuna-skipjack-cob, cuttlefish, and seaweed in Indonesia. Besides that, it is a source of income for fishermen and coastal communities. Blue swimming crab is marketed for export in fresh and processed form in cans to Singapore, Japan, and the United States (Oniam & Arkronrat, 2022). In 2019, the export volume of blue swimming crabs amounted to 2749 tons, with a total export value exceeding USD259 million (Asosiasi Pengelolaan Rajungan Indonesia, 2019). To fulfill the market, the need for those swimming crabs exported is mostly caught from the wild. This has caused the crab population in the wild to decrease, and the size of the crabs caught is getting smaller (Madduppa et al., 2021). This condition is an indication of overfishing, as occurs in the Sulawesi Sea and Java Sea (Zairion et al., 2015). Therefore, the aquaculture of swimming crabs should be enhanced in both hatchery and grow-out phases. Additionally, the domestication of this species is crucial for acquiring high-quality broodstock and achieving improved growth rates in subsequent generations.

G. Gunarto (✉) · S. Sulaeman
Research Center for Fishery, National Research and Innovation Agency, Cibinong, Indonesia
e-mail: guna019@brin.go.id

H. Herlinah
Research Center for Fishery, National Research and Innovation Agency, Cibinong, Indonesia

Faculty of Vocational Studies, Hasanuddin University, Makassar, Indonesia

E. Nugroho et al. (eds.), *Aquaculture Development in Archipelago*,
https://doi.org/10.1007/978-981-95-1084-9_21

21.2 Swimming Crab Distribution

Swimming crab, especially blue swimming crab, has a blue base color with white patches on male crabs, while the female crab has a dirty-green base color with dirty-white patches and has five pairs of legs consisting of one pair of claws functioning as holders, three pairs of legs as walking legs, and one pair of legs serving as oars for swimming. Some species of swimming crabs most caught in Indonesia consist of several species including *Portunus pelagicus* Linnaeus, 1758; *P. sanguinolentus* J. F. W. Herbst, 1783; *P. sanguinus*; *P. trituberculatus* Hiers; *P. gladiator*; *P. hastatoides*; *Thalamita crenata* Latr.; *Thalamita danae* Stimpson; *Charybdis cruciata*; *Charybdis natator* Herbst; and *Podophthalmus vigil* Fabr. The distribution area covers tropical waters, especially in Southeast and East Asia or the eastern Indian Ocean and the western Pacific Ocean (Lai et al., 2010). The sandy seagrass meadow area is a nursery of swimming crabs. In Indonesia, crabs inhabit the waters of the sand bottom substrate at 2–4 miles from the coastline in the East Lampung waters (Radifa et al., 2020). Crabs can be found on various substrate fractions, including sandy mud, and gravel beaches but the most preferred is the sandy bottom (Zhu et al., 2022). In the waters of Pangkep Regency, South Sulawesi, the maximum carapace width recorded for male crabs is 183.5 mm, whereas the largest female crabs measure 171.5 mm. However, the predominant size of crabs captured ranges from 111.5 to 123.5 mm, with the fishing season lasting from May to November and peak catches observed in June and September (Ihsan et al., 2015). Additionally, in Central Kalimantan, the crab fishing season closely aligns with that of Southern Sulawesi, occurring from May to October (Hidayat et al., 2020). According to Hamid (2017), the size of swimming crabs captured is significantly influenced by the depth of the water. Specifically, crabs caught at depths of 14–31 m are larger than those harvested from shallower depths of 1–4 m.

Classification:

Phillum:	Arthropoda
Class:	Crustacea
Subclass:	Malacostraca
Ordo:	Decapoda
Sub Ordo:	Brachyura
Familia:	Portunidae
Genus:	*Portunus*
Species:	*Portunus pelagicus*

The morphological shape of the blue swimming crab is different from that of the mud crab, *Scylla* spp. The swimming crab has a slimmer body shape with longer claws and an interesting color on its carapace. Blue swimming crabs have longer legs and are also flatter when compared to mud crabs. The end spines on both sides of the carapace are relatively longer on the swimming crab. The swimming crab broodstock has a longer cheliped, and the carapace has nine teeth on the right and left side of the eye. The female has a dilated abdomen and a bright speckled

Fig. 21.1 The female of the swimming crab, *Portunus pelagicus,* and its color (top left and bottom left), and the male of the swimming crab, *P. pelagicus* crab, and its color (top right and bottom right). (Source: Ruliaty, 2017)

carapace color. Meanwhile, the male has a narrowed abdomen shape with a mottled and dull blue carapace color (Fig. 21.1). Crabs live in saltwater environments with salinity levels exceeding 25 parts per thousand (ppt) and cannot survive without water for extended periods, typically lasting only a few hours.

21.3 Domestication

Domestication of swimming crab was initiated by obtaining hatchery seed production. The environment and feed are important factors in determining the survival rate of crab larvae. At the initial stage of the first 24 h, the crab larvae will eat a lot of rotifers, while after that the larvae will prefer to eat *Artemia* nauplii (Ikhwanuddin et al., 2011; Redzuari et al., 2012). Culturing larvae under conditions of food scarcity can elevate the incidence of cannibalism, particularly when the larvae have reached the megalopa stage. The rate of larval metamorphosis is significantly influenced by the temperature of the water in which the larvae are cultured, in addition to the quality and quantity of the available feed. The optimal water temperature for culturing crab larvae ranges from 30 °C to 32 °C, and the sufficiency of feed, both in quality and quantity, is essential for their growth. Excessive nutrients from feed derived from *Nannochloropsis* sp., consumed by rotifers (*Brachionus plicatilis*), may lead to the development of abnormal megalopa morphology. Conversely, a deficiency in essential nutrients can result in increased mortality rates among megalopa (Dan et al., 2016). Therefore, it is recommended to provide only sufficient feed in the form of *Artemia* nauplii for crab larvae culture, without the need for supplementation with *Nannochloropsis* spp. During molt initiation, larvae require elevated levels of total sugars in the hemolymph, which subsequently decrease following the release of the old cuticle (apolysis) at the conclusion of the molting process. This necessitates a substantial amount of total sugars for the apolysis process. Consequently, glycogen concentrations in the hepatopancreas increase from initially low levels at the onset of molting, reaching their peak during the late stages of the molting process, just prior to apolysis (Sugumar et al., 2013).

Several studies showed that crablet production in hatcheries is often inconsistent, typically resulting in a low percentage of successful crablet yields. In India, crablet production in hatcheries reached 10.3 ± 5.76% from the zoea-1 stage to Crab-1 stage with a stocking density of 50 larvae/l (Maheswarudu et al., 2008). Furthermore, utilizing mulberry (*Morus alba*) leaf extract at a concentration of 4 mg per 100 mg of feed has been shown to decrease mortality associated with molting syndrome in both zoea larvae and the megalopa stage (Fujaya et al., 2013). Susanto (2007) reported that the highest survival rate of zoea-1 larvae to the D-1 crablet stage was 23% when reared at a salinity of 32–34 ppt, whereas survival rates were less than 23% when zoea-1 larvae were cultured at salinities below 32 ppt. In addition, the highest crablet production reaching 23.11% was also reported by Efrizal (2017).

A domestication program for crabs has been initiated by a research team from Hasanuddin University in Makassar, Indonesia. As a result, the fourth generation (G4) of domesticated blue swimming crab (*Portunus pelagicus*) demonstrated a faster growth rate in carapace width (109 mm) compared to the G1 (103 mm) after 100 days of culture in brackish water (Hidayani et al., 2019). In Thailand, it has been reported that crablets with a carapace width of 1.0–2.0 cm, when stocked in a brackish water pond, begin to mature their gonads at 90 days of culture for G1, while G2 and G3 reach gonadal maturity at 60 days of culture. Furthermore, after 150 days of culture, 20–50% of the female crabs in the population are observed to spawn, producing eggs in yellow, brown, and black colors (Oniam & Arkronrat, 2022).

21.4 Reproduction

A study of swimming crab populations in the waters of Bone Regency, South Sulawesi, indicated that female crabs with a carapace width ranging from 69.36 to 73.97 mm exhibited the first signs of gonadal maturation. Additionally, the sex ratios of males to females were found to be unbalanced. The spawning season occurs throughout the year, with peaks observed in May and December (Kembaren et al., 2012). In the waters of Lampung Province, crabs spawn at a carapace width ranging from 111.0 to 155.9 mm, with fecundity levels between 229,468 and 2,236,355 eggs. As suggested, the number of eggs is closely related to the carapace width and the weight of the crab individual (Zairion et al., 2015). Like the waters of Bone Regency in South Sulawesi Province, crabs in the waters of Pati Regency, Central Java, reach their first gonadal maturation at a carapace width ranging from

Fig. 21.2 The life cycle of the blue swimming crab (*Portunus pelagicus,* Linnaeus). (Source: Cambell, 2017)

61 to 70 mm. Typically, crabs exhibit a carapace width ranging from 64 to 164 mm and body weights between 23 and 333 g (Iksanti et al., 2022).

Swimming crabs inhabit estuarine environments from the juvenile stage until spawning, after which they migrate to seawater with higher salinity (30–34 ppt) to hatch their eggs into the zoea-1 stage. The larvae will reside in the sea as plankton, developing into zoea-2, zoea-3, and finally zoea-4 stages. Within 4–5 days, they metamorphose into the megalopa stage, which subsequently transforms into a crablet, resembling the adult form of the swimming crab. Zoea-1 larvae can develop into the crablet stage within 20 days. Upon reaching a young stage, the crabs return to the estuary to grow to adulthood and mate in estuarine waters. The crab often remains submerged in the sand, with only its eyes visible, enabling it to stay hidden while it patiently awaits the arrival of fish and various invertebrates to capture as prey. The life cycle of crabs is depicted in Fig. 21.2.

21.5 The Development of Swimming Crab Aquaculture Technology

In Indonesia, the aquaculture of blue swimming crabs begins with seed production in hatcheries, followed by grow-out phases in brackish water ponds. This process has been implemented by the Institute for Brackish Water Aquaculture in Takalar, South Sulawesi, and the Institute of Brackish Water Aquaculture in Jepara, Central Java, under the coordination of the Directorate General of Aquaculture within the Ministry of Marine Affairs and Fisheries (MMAF). In hatchery facilities, the swimming crab produces a greater number of crablets compared to the mud crab (*Scylla* spp.). This disparity is attributed to the higher vitality of swimming crab larvae, which exhibit greater resistance to disease than mud crab larvae, resulting in a higher survival rate of swimming crab larvae as they develop into the crablet stage. The fundamental attributes required for a crab hatchery include effective broodstock management, larval rearing, and the management of water and feed until the larvae develop into crablets. For grow-out, crablets at D-5 to D-10 are stocked in selected ponds that meet specific criteria conducive to rapid growth, such as those with sandy substrates. As previously indicated, the growth of crabs in sandy substrates is faster than in clay substrates (Oniam et al., 2018). To achieve optimal growth of crabs, it is crucial to stock healthy crablets and ensure that the pond is properly prepared. This preparation includes drying the pond, repairing the dikes, eliminating competing species, topping up the water, fertilizing, and eradicating pests. With proper water and feed quality management, it is expected that the crab culture has reached a size of 100–150 g/individual and can be harvested for marketing in the third to the fourth month of rearing period.

Crab cultures on the island of Panjang, Jepara Regency, Central Java province, have been carried out since 2016, with swimming crab seeds sourced from either the wild or hatcheries. Crab grow-out is conducted using green nets installed along the coast, in ponds, or in pond compartments on the beach, where water is minimally impacted by freshwater sources, maintaining salinity levels in the range of 30–33 ppt. Economically, crab culture on Panjang Island can enhance farmers' income. In addition to being sold in live, fresh form due to the high volume of tourists seeking crab dishes, crab meat is also processed into products such as *petis* and chips, resulting in product diversification from swimming crab culture.

Survival rates of crabs cultivated in floating cages or ponds have not been extensively documented. A survival rate of 90% is achieved after 90 days of culture in fiberglass tanks, with each tank featuring a different substrate at the bottom: sandy loam, sand, and clay. Furthermore, higher growth rates are observed in crab cultures using sandy loam and sand substrates at the bottom of the tanks. In contrast, clay substrates yield the least favorable growth rates for crabs. Cannibalism and poor feed quality are common factors contributing to low survival rates in crab culture within ponds that have sandy or sandy clay bottom substrates. Therefore, it is recommended to maintain a stocking density of 2 crablets per square meter for cultivation in these ponds (Oniam & Arkronrat, 2022).

Moreover, the quality of the broodstock is the most critical factor among several determinants of successful crab culture in ponds. High-quality broodstock significantly influences the vitality of the produced larvae and subsequently affects the growth rate of the crablet offspring. Thus, the domestication of crabs is an important program aimed at producing high-quality broodstock, as domesticated biota become better adapted to the cultured environment. Ultimately, the success of biota aquaculture is largely determined by effective domestication activities, which can enhance the reproductive capacity of the broodstock and increase the growth rates of offspring.

References

Asosiasi Pengelolaan Rajungan Indonesia. (2019). *Indonesian blue swimming crab, fishery improvement projects*. https://www.apri.or.id/fip/

Cambell, T. I. (2017). *Dietary composition of the blue swimming crabs, Portunus armatus, and life history characteristic of related species*.

Dan, S., Ashidate, M., & Hamasaki, K. (2016). Improved method for culturing the swimming crab *Portunus trituberculatus* larvae to prevent mass mortality during seed production. *Fisheries Science, 82*(1), 113–126. https://doi.org/10.1007/s12562-015-0935-y

Efrizal, E. (2017). Effects of stocking density on survival rate and larval development of blue swimming crab, *Portunus pelagicus* (Linnaeus, 1758) under laboratory conditions. *AACL Bioflux, 10*(2), 217–226.

Fujaya, Y., Trijuno, D. D., Nikhlani, A., Cahyono, I., & Hasnidar, H. (2013). The use of mulberry (*Morus alba*) extract in the mass production of blue swimming crab (*Portunus pelagicus* L.) larvae to overcome the mortality rate due to molting syndrome. *Aquatic Science and Technology, 2*(1), 1. https://doi.org/10.5296/ast.v2i1.4048

Hamid, A., Wardiatno, Y., Lumban Batu, D., & Riani, E. (2017). Pengelolaan rajungan (*Portunus pelagicus*) yang berkelanjutan berdasarkan aspek bioekologi di teluk Lasongko, Sulawesi Tenggara [Sustainable management of crabs (*Portunus pelagicus*) based on bioecological aspects in Lasongko Bay, Southeast Sulawesi]. *Jurnal Kebijakan Perikanan Indonesia, 9*, 41–50. http://ejournal-balitbang.kkp.go.id/index.php/jkpi

Hidayani, A., Fujaya, Y., Tassakka, A., & Kalsum, U. (2019). Domestication of blue swimming crab for sustainability: Growth and genetic change. *European Journal of Sustainable Development, 8*(4), 254. https://doi.org/10.14207/ejsd.2019.v8n4p254

Hidayat, R., Gumiri, S., & Neneng, L. (2020). Studi Bioekologi dan Pola Distribusi Rajungan di Perairan Laut Jawa Kabupaten Sukamara [Study of bioecology and distribution patterns of crabs in Java Sea Waters, Sukamara Regency]. *Anterior Jurnal, 19*(2), 32–41. https://doi.org/10.33084/anterior.v19i2.1243

Ihsan, I., Wiyono, E. S., Wisudo, S. H., & Haluan, J. (2015). Alternatif pengelolaan perikanan rajungan (*Portunus pelagicus*) di perairan Kabupaten Pangkep, Sulawesi Selatan [Alternative management of crab fisheries (*Portunus pelagicus*) in the waters of Pangkep Regency, South Sulawesi]. *Jurnal Kebijakan Perikanan Indonesia, 7*(1), 25. https://doi.org/10.15578/jkpi.7.1.2015.25-36

Ikhwanuddin, M., Adila, T. N., Azra, M. N., Hii, Y. S., & Talpur, A. D. (2011). Determination of live prey ingestion capability of blue swimming crab, *Portunus pelagicus* (Linnaeus, 1758). *Larvae., 3*(6), 570–575.

Iksanti, R. M., Redjeki, S., & Taufiq-Spj, N. (2022). Aspek Biologi Rajungan (*Portunus pelagicus*) Linnaeus, 1758 (Malacostraca: Portunidae) Ditinjau dari Morfometri dan Tingkat Kematangan Gonad di TPI Bulu, Jepara [Biological Aspects of Crab (*Portunus pelagicus*) Linnaeus, 1758 (Malacostraca: Portunidae) Reviewed from Morphometry and Gonad Maturity Level in TPI Bulu, Jepara]. *Journal of Marine Research, 11*(3), 495–505. https://doi.org/10.14710/jmr.v11i3.31258

Kembaren, D. D., Ernawati, T., & Suprapto. (2012). Biologi Dan Parameter Populasi Rajungan (*Portunus pelagicus*) di Perairan Bone dan Sekitarnya [Biology and Parameters of Crab Population (*Portunus pelagicus*) in Bone Waters and Surrounding Areas]. *Jurnal Penelitian Perikanan Indonesia, 18*(4), 273–281. https://doi.org/10.15578/jppi.18.4.2012.273-281

Lai, J. C. Y., Ng, P. K. L., & Davie, P. J. F. (2010). A revision of the Portunus pelagicus (Linnaeus, 1758) species complex (Crustacea: Brachyura: Portunidae), with the recognition of four species. *Raffles Bulletin of Zoology, 58*(2), 199–237.

Madduppa, H., Martaulina, R., Zairion, Z., Renjani, R. M., Kawaroe, M., Anggraini, N. P., Subhan, B., Verawati, I., & Sani, L. M. I. (2021). Genetic population subdivision of the blue swimming crab (*Portunus pelagicus*) across Indonesia inferred from mitochondrial DNA Implication to sustainable fishery. *PLoS One, 16*(2 February), 1–14. https://doi.org/10.1371/journal.pone.0240951

Maheswarudu, G., Jose, J., Manmadhan Nair, K. R., Arputharaj, M. R., Ramakrishna, A., Vairamani, A., & Ramamoorthy, N. (2008). Evaluation of the seed production and grow out culture of blue swimming crab *Portunus pelagicus* (Linnaeus, 1758) in India. *Indian Journal of Marine Sciences, 37*(3), 313–321.

Oniam, V., & Arkronrat, W. (2022). Growth rate and sexual performance of domesticated blue swimming crab, *Portunus pelagicus* (Linnaeus, 1758) in Earthen Ponds. *Trends in Sciences, 19*(20). https://doi.org/10.48048/tis.2022.6235

Oniam, V., Taparhudee, W., & Yoonpundh, R. (2018). Impact of different pond bottom soil substrates on blue swimming crab (*Portunus pelagicus*) culture. *Walailak Journal of Science and Technology, 15*(4), 325–332. https://doi.org/10.48048/wjst.2018.3914

Radifa, M., Wardiatno, Y., Simanjuntak, C. P. H., & Zairion, Z. (2020). Preferensi Habitat dan Distribusi Spasial Yuwana Rajungan (*Portunus pelagicus*) di Perairan Pesisir Lampung Timur, Provinsi Lampung [Habitat preferences and spatial distribution of crab Yuwana (Portunus pelagicus) in the coastal waters of East Lampung, Lampung Province]. *Jurnal Pengelolaan Sumberdaya Alam Dan Lingkungan (Journal of Natural Resources and Environmental Management), 10*(2), 183–197. https://doi.org/10.29244/jpsl.10.2.183-197

Redzuari, A., Azra, M. N., Abol-Munafi, A. B., Aizam, Z. A., Hii, Y. S., & Ikhwanuddin, M. (2012). Effects of feeding regimes on survival, development and growth of blue swimming crab, *Portunus pelagicus* (Linnaeus, 1758) larvae. *World Applied Sciences Journal, 18*(4), 472–478. https://doi.org/10.5829/idosi.wasj.2012.18.04.313

Ruliaty, L. (2017). *Teknik produksi benih rajungan (Portunus pelagicus).* Direktorat Jenderal Perikanan. ISBN/ISSN 978-602-61170-5-2.

Sugumar, V., Vijayalakshmi, G., & Saranya, K. (2013). Molt cycle related changes and effect of short term starvation on the biochemical constituents of the blue swimmer crab *Portunus pelagicus*. *Saudi Journal of Biological Sciences, 20*(1), 93–103. https://doi.org/10.1016/j.sjbs.2012.10.003

Susanto, B. (2007). Pertumbuhan, Sintasan dan Keragaan Zoea Sampai Megalopa Rajungan (*Portunus pelagicus*) Melalui Penurunan Salinitas [Growth, survival and extinction of Zoea to Megalopa Crab (*Portunus pelagicus*) through salinity decline]. *Jurnal Perikanan Universitas Gadjah Mada, 9*(1), 154–160.

Zairion, Wardiatno, Y., Boer, M., & Fahrudin, A. (2015). Reproductive biology of the blue swimming crab *Portunus pelagicus* (Brachyura: Portunidae) in east Lampung waters, Indonesia: Fecundity and reproductive potential. *Tropical Life Sciences Research, 26*(1), 67–85.

Zhu, B., Zhang, H., Liu, D., Lu, Y., & Wang, F. (2022). Importance of substrate on welfare in swimming crab (*Portunus trituberculatus*) culture: A territorial behavior perspective. *Aquaculture Reports, 24*(March), 101113. https://doi.org/10.1016/j.aqrep.2022.101113

The Status of Domestication and Aquaculture System on *Gracilaria verrucosa* in Indonesia

22

Rosmiati, Emma Suryati, Andi Parenrengi, Andi Tenriulo, and Samuel Lante

22.1 Introduction

Gracilaria verrucosa, commonly known as seaweed, is a species of red algae (Rhodophyta) found in shallow tropical and subtropical marine environments. This seaweed holds significant economic value and is particularly abundant in Indonesia, especially in regions such as northern Java, West Nusa Tenggara, South Sulawesi, and Lampung (Nurman et al., 2021; Rejeki et al., 2018; Suhartini et al., 2023). *Gracilaria verrucosa* is recognized as an export commodity (Fatmawati et al., 2022) and is valued for its high fiber content, making it suitable for food products (Darmanto et al., 2022). Additionally, this species of seaweed offers significant economic benefits in various sectors, including gelatine production, human food, pet food, medicine, biofuels, and fertilizers (Nasmia et al., 2021; Torres et al., 2014). Its high growth rate, ease of acquisition, and low cost contribute to the strong demand for *Gracilaria* in gelatine production (Ak et al., 2011). *Gracilaria verrucosa* can be cultivated in shallow ponds or harvested from the wild (Rustam & Wamnebo, 2021). However, cultivation methods pose challenges, as the stocking techniques employed by seaweed cultivators often lead to poor production outcomes. Factors contributing to this issue include pests, such as triptan, which prey on the seaweed, and contamination from mud at the bottom of the ponds, negatively impacting quality. Addressing these challenges will be crucial for the future of *Gracilaria verrucosa* cultivation.

Rosmiati (✉) · E. Suryati · A. Parenrengi · A. Tenriulo · S. Lante
Research Center for Fishery, National Research and Innovation Agency, Cibinong, Indonesia
e-mail: rosm005@brin.go.id

E. Nugroho et al. (eds.), *Aquaculture Development in Archipelago*,
https://doi.org/10.1007/978-981-95-1084-9_22

22.2 Taxonomy, Morphology, and Habitat of *Gracilaria verrucosa*

Gracilaria verrucosa is a type of seaweed commonly found in Indonesia, where it is widely cultivated in ponds. *Gracilaria verrucosa* can exist as unicellular or multicellular colonies and typically inhabits shallow, sandy, rugged, tidal, clear areas. *G. verrucosa* is usually attached to dead corals, pieces of coral, or various natural and artificial substrates (Afrianto & Liviawati, 1993). Morphologically, *G. verrucosa* lacks distinct roots, stems, and leaves, making it difficult to differentiate among these structures. Despite the similarity in appearance, the plants exhibit various branching patterns, which can be dichotomous, alternate, pinnate, or take on other forms (Aslan, 1998). According to Skriptsova and Choi (2009), adult *G. verrucosa* typically presents colors ranging from brown to dark olive or dark red. The thalli are upright and cylindrical, reaching lengths of 25–30 cm, and are anchored to the substrate by a discoid holdfast measuring 1–4 mm in diameter, from which multiple thalli can arise. The central axis has a diameter of 1.5–2 mm and usually features 3–4 sequences of lateral branches of varying sizes at irregular intervals, arranged in an alternate to unilateral pattern (Fig. 22.1). The branches are often slightly narrowed at their base. The cortex contains one to three layers of well-pigmented cells, with the outer cortical cells being radially elongated and elliptical, measuring 4.5–7 mm in diameter. Subcortical cells range from 12 to 25 mm in diameter. The medulla consists of large isodiametric cells, measuring 150–360 mm in diameter. Ruptures in medullary cells may occur on older branches, resulting in a partially hollow thallus. According to Dawes (1981), the taxonomy of *G. verrucosa* is as follows:

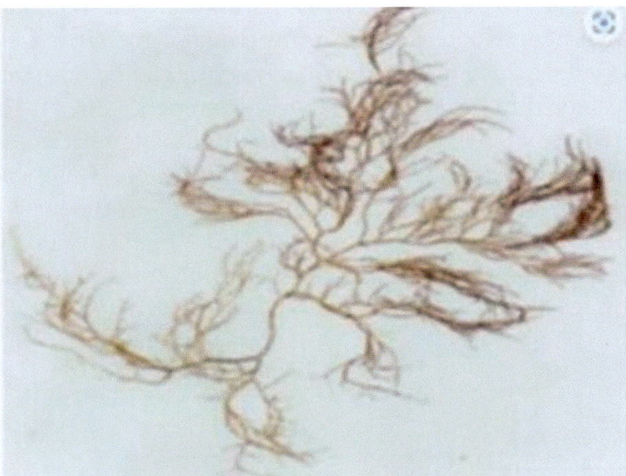

Fig. 22.1 *Gracilaria verrucosa*. (Source: Amalia, 2013)

Division: Rhodophyta (red algae)
Class: Rhodophyceae
Order: Gracilariales
Family: Gracilariaceae
Genus: *Gracilaria*
Species: *Gracilaria verrucosa*

22.3 Distribution

This genus *Gracilaria* is widely distributed in tropical and temperate seas, with notable occurrences in regions such as the open waters around the Korean Peninsula, Chile, China, Taiwan (Choi et al., 2006), Venezia, and India (Padhi et al., 2011). *Gracilaria verrucosa* is a typical marine red alga that thrives on pebbles or rocks in shallow littoral zones (Lee et al., 2009). *Gracilaria verrucosa* exhibits optimal growth in a temperature range of 17–30 °C and salinity levels between 15 and 30‰. Its ribbonlike structure allows it to be less affected by tidal flows and the deposition of particulate matter. Like other members of the Rhodophyceae class, *G. verrucosa* can survive in murky and deep waters, often demonstrating better growth than other seaweed species under similar conditions (Choi et al., 2006).

22.4 Reproduction

Generally, *Gracilaria* species are primarily characterized by a triphasic sexual life cycle that involves alternation of generations. The life cycle of most red algae, including *Gracilaria verrucosa*, is diplohaplontic, featuring haploid gametophytes that alternate with diploid sporophytes (Mantri et al., 2009). Variations in reproductive success have been observed at both the haploid and diploid stages. Differences in the ability to spread in haploid and diploid life cycle stages and different survival strategies under artificial environmental conditions have also been investigated. Research has investigated differences in the ability to disperse during the haploid and diploid life cycle stages, as well as varying survival strategies under artificial environmental conditions. The success of male fertilization is influenced by factors such as the distance traveled by spermatia, although competition among males and female choice have not been demonstrated using microsatellite DNA markers. Understanding the reproductive biology of *Gracilaria* is essential for effective resource management and cultivation practices.

The sexual life cycle in *Gracilaria* involves a cyclic alternation between diploid and haploid phases, with meiosis facilitating the transition from the diploid to the haploid state, while fertilization reconstitutes the diploid genome. The typical life history of *Gracilaria* follows an archetype known as the "Polysiphonia type," which is triphasic (Fig. 22.2). In this cycle, gametophytes and sporophytes are

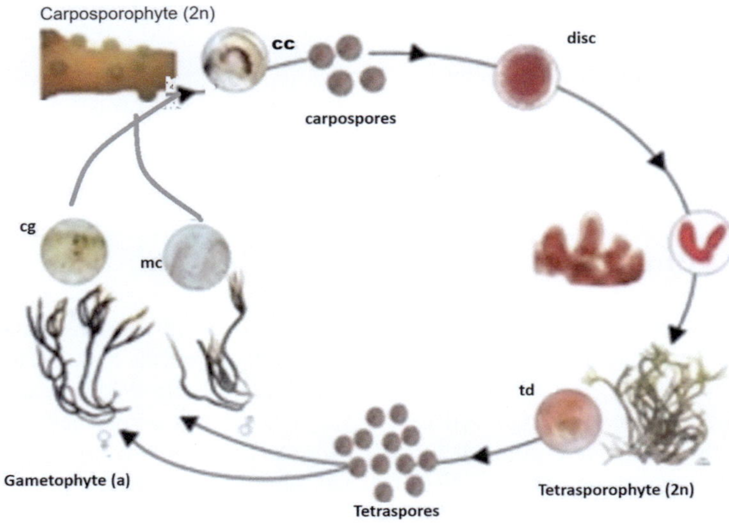

Fig. 22.2 Reproductive process in the red algal genus *Gracilaria*

morphologically identical and independent, while the gametophytes are dioecious, meaning individual plants are either male or female.

22.5 Domestication

Understanding the life cycle of various species within the genus *Gracilaria* and recognizing the optimal conditions for each phase have facilitated the domestication and development of a valuable industry. In the cultivation of tropical seaweeds, biomass production generally does not require a hatching phase, as gametes or spores are rarely involved in mass cultivation. *Gracilaria verrucosa* is primarily cultivated through the vegetative propagation of thalli fragments. Maintaining genotypes of commercial interest by controlling the life cycle of *Gracilaria verrucosa* is crucial for enhancing the long-term success of seaweed production (Hayashi et al., 2020).

However, several challenges persist in seaweed production, including disease outbreaks, biofouling, pests, and low seed quality. Conventionally, procuring seaweed seeds is relatively easy and cost-effective, as they can be sourced from natural environments or seedling gardens. The provision of superior seeds can also be achieved through generative methods, although this approach tends to be time-consuming and costly (Parenrengi et al., 2006). Repeated use of seedlings has been identified as a significant factor contributing to declines in both quality and quantity, making cultivars more susceptible to disease (Yow et al., 2011). To enhance the productivity of *Gracilaria verrucosa* cultivation in ponds, it is essential to utilize quality seedlings that can adapt and thrive optimally in their specific maintenance environments.

22.6 Development of Cultivation Technology

Gracilaria verrucosa is one of Indonesia's most widely cultivated seaweed species, particularly in pond environments. *G. verrucosa* can be cultivated using simple low-cost culture technology, which has a high market absorption (Hernanto et al., 2015). The cultivation method can be adapted to the specific conditions and characteristics of the cultivation site. However, challenges remain, particularly regarding the cultivation methods themselves. There are two primary cultivation technologies for *G. verrucosa*: horizontal and vertical methods. According to Akbarurrasyid et al. (2021), the horizontal cultivation method includes the following technique:

1. Bottom-method: Stocking seeds directly at the bottom of the pond.
2. Off-bottom method: Tying seedlings to a rope, which is then secured to a stake or raft.
3. Floating rack method: Utilizing a bamboo construction that is anchored with ballast to prevent it from being washed away by currents.
4. Longline method. Similar to the floating rack method, but using plastic bottles as buoyancy aids instead of bamboo.

The area of the farm significantly influences the horizontal cultivation method, which in turn impacts production. One of the disadvantages of the horizontal method is that it can disrupt traffic in the surrounding waters due to the use of long transverse ropes. Additionally, this method does not optimize the use of the water column, resulting in less-than-optimal production. In contrast, vertical cultivation methods, such as the use of stacked net bags, can maximize the cultivated area and enhance production (Agus Nadlir et al., 2019). Widowati et al. (2015) note that while the horizontal longline method allows for cultivation along a single line, vertical cultivation enables multiple rows to be cultivated within the same column. This approach significantly increases seaweed production.

The vertical method employs ropes to bind seaweed seeds in an upright position (perpendicular to the water surface), utilizing the water column to its full potential, typically to depths of 3–5 m, depending on water clarity. Research indicates that vertical cultivation can enhance productivity by at least 420% compared to the longline method, which primarily utilizes the surface water column.

Furthermore, the spacing of seedlings plays a crucial role in seaweed growth. Adequate planting distance allows for better nutrient distribution in the water, accelerates diffusion, and potentially enhances growth. Field surveys have shown that the highest seaweed production occurs at a planting distance of 20 cm with seedling weights of 50 g, yielding approximately 2423.7 g/m^2 (Fanni et al., 2021). Proper spacing is essential; if seedlings are planted too closely, competition among sea plants can hinder growth, while excessive spacing may allow for unwanted phytoplankton growth.

22.7 Production of *Gracilaria verrucosa* in Indonesia

Seaweed has emerged as one of the leading commodities in aquaculture due to its high economic value, comprehensive industrial applications, and significant labor absorption capacity. Cultivating seaweed is supported by accessible cultivation technologies, a relatively short planting period of approximately 45 days, and low production costs. Given these advantages, seaweed cultivation is poised for a promising future, contributing to economic growth, increasing employment opportunities, and alleviating poverty. The vast potential of seaweed in Indonesia is not solely attributed to its economic benefits; it is also enhanced by the country's rich marine territory, which offers extensive coastal waters conducive to seaweed farming (Yuliana, 2022).

One of the most sought-after species in the global market is *Gracilaria verrucosa*. The rising demand for processed macroalgae extracts underscores the versatility of seaweed, as it serves as a key ingredient in various sectors, including food, cosmetics, pharmaceuticals, and other industries such as paper, textiles, photography, shoe polish, toothpaste, fish and meat canning, and fertilizer production. Despite this, domestic production of *G. verrucosa* has not yet met the high market demand, particularly in the agar industry.

In 2018, Indonesia recorded a national production of cultivated seaweed at 10.18 million tons, with a continuous upward trend. Between 2014 and 2019, Indonesian seaweed exports grew at an average rate of 6.53% per year. In 2019, the export value of Indonesian seaweed reached USD324.84 million, reflecting an increase of 11.31% compared to USD291.83 million in 2018 (MMAF, 2020). According to 2021 data, Indonesia ranked as the second-largest seaweed producer in the world, trailing only China, with a production volume of 9.1 million tons (Muflih, 2022). Furthermore, seaweed production in Indonesia increased by 6.89% in 2021, with a total value of IDR 28.48 trillion, compared to IDR 26.65 trillion in 2020. Comprehensive data on Indonesian seaweed production from 2011 to 2021 is illustrated in Fig. 22.3.

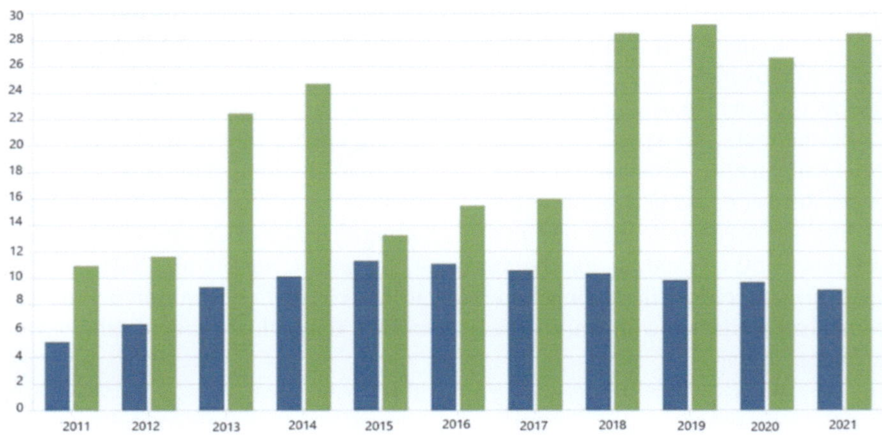

Fig. 22.3 Seaweed production in Indonesia in 2011–2021

References

Afrianto, & Liviawati. (1993). *Budidaya rumput laut dan cara pengolahannya [Seaweed cultivation and how to process it]* (2nd ed.). Bharata.

Agus Nadlir, Susilowati, T., Adi, K., Harwanto, D., Haditomo, A. H. C., & Windarto, S. (2019). Production performance of *Gracilaria verrucosa* using verticulture method with various wide planting area in Karimunjawa. *Omni-Akuatika, 15*(1), 47–58.

Ak, I., Çetin, Z., Cirik, Ş., & Göksan, T. (2011). *Gracilaria verrucosa* (Hudson) papenfuss culture using an agricultural organic fertilizer. *Fresenius Environmental Bulletin, 20*(8a), 2156–2162.

Akbarurrasyid, M., Prietoyo, A., Astiyani, W. P., & Mustia, D. A. (2021). Teknologi budidaya rumput laut *Gracilaria verrucosa* menggunakan kantong jaring bersusun dengan bobot awal bibit berbeda di Karawang [Technology of seaweed culture (*Gracilaria verrucosa*) using multilevel net bags with different initial seed weight in Karawang]. *Maspari Journal, 13*(2), 117–128. https://doi.org/10.56064/maspari.v13i2.14036

Amalia, D. R. N. (2013). Efek temperature terhadap pertumbuhan *Gracilaria verrucosa* [Effect of temperature on growth of *Gracilaria verrucosa*]. In *Degree Thesis*.

Aslan, L. M. (1998). *Budidaya Rumput Laut* (1st ed.). [Seaweed Cultivation]. Kanisius.

Choi, H. G., Kim, Y. S., Kim, J. H., Lee, S. J., Park, E. J., Ryu, J., & Nam, K. W. (2006). Effects of temperature and salinity on the growth of *Gracilaria verrucosa* and *G. chorda*, with the Potential for mariculture in Korea. *Journal of Applied Phycology, 18*(3–5), 269–277. https://doi.org/10.1007/s10811-006-9033-y

Darmanto, Y. S., Kurniasih, R. A., Romadhon, R., Riyadi, P. H., & Anggraeni, N. (2022). Characteristic of analog rice made from arrowroot (*Maranta arundinacease*) and seaweed (*Gracilaria verrucosa*) flour fortified with fish collagen. *Food Research, 6*(5), 370–379. https://doi.org/10.26656/fr.2017.6(5).473

Fanni, N. A., Rahayu, A. P., & Prihatini, E. S. (2021). Seaweed (*Gracilaria verrucosa*) production based on different spacings and seed weights in pond of Tlogosadang Village, Paciran Sub-District, Lamongan District [Produksi rumput laut (*Gracilaria verrucosa*) berdasarkan perbedaan jarak tanam dan bobot benih yang berbeda di kolam Desa Tlogosadang, Kecamatan Paciran, Kecamatan Lamongan]. *Jurnal Ilmu Pertanian Indonesia, 26*(2), 177–183. https://doi.org/10.18343/jipi.26.2.177

Fatmawati, F., Sibero, M. T., Trianto, A., Wijayanti, D. P., Sabdono, A., Pringgenies, D., & Radjasa, O. K. (2022). The influence of fermentation using marine yeast Hortaea werneckii SUCCY001 on antibacterial and antioxidant activity of *Gracilaria verrucosa*. *Biodiversitas Journal of Biological Diversity, 23*(10), 5258–5266. https://doi.org/10.13057/biodiv/d231035

Hayashi, L., de Cantarino, S. J., & Critchley, A. T. (2020). Challenges to the future domestication of seaweeds as cultivated species: Understanding their physiological processes for large-scale production. *Advances in Botanical Research, 95*, 1–27. https://doi.org/10.1016/bs.abr.2019.11.010

Hernanto, A. D., Rejeki, S., & Ariyati, R. W. (2015). Pertumbuhan budidaya rumput laut (*Eucheuma cottoni* dan *Gracilaria* sp.) dengan metode long line di Perairan Pantai Bulu Jepara [Growth of seaweed culture (*Eucheuma cottoni* and *Gracilaria* sp.) with long line method in Bulu Sea Jepara]. *Journal of Aquaculture Management and Technology, 4*(2), 60–66. http://ejournal-s1.undip.ac.id/index.php/jfpik

Lee, H.-J., Dang, H.-T., Kang, G.-J., Yang, E.-J., Park, S.-S., Yoon, W.-J., Jung, J. H., Kang, H.-K., & Yoo, E.-S. (2009). Two enone fatty acids isolated from *Gracilaria verrucosa* suppress the production of inflammatory mediators by down-regulating NF-κB and STAT1 activity in lipopolysaccharide-stimulated RAW 264.7 cells. *Archives of Pharmacal Research, 32*(3), 453–462. https://doi.org/10.1007/s12272-009-1320-0

Mantri, V. A., Thakur, M. C., Kumar, M., Reddy, C. R. K., & Jha, B. (2009). The carpospore culture of industrially important red alga *Gracilaria dura* (Gracilariales, Rhodophyta). *Aquaculture, 297*(1–4), 85–90. https://doi.org/10.1016/j.aquaculture.2009.09.004

MMAF. (2020). Laporan tahunan Kementerian Kelautan dan Perikanan 2020 [Annual report of the Ministry of Maritime Affairs and Fisheries 2020]. *Laporan Tahunan KKP*. https://kkp.go.id/an-component/media/upload-gambar-pendukung/zzzz/LaporanTahunanKKP2020.pdf

Muflih, M. A. (2022). Strategi KKP genjot produktivitas rumput laut untuk tingkatkan devisa [KKP strategy to boost seaweed productivity to increase foreign exchange]. *Pantau.*. https://www.pantau.com/nasional/133267/strategi-kkp-genjot-produktivitas-rumput-laut-untuk-tingkatkan-devisa

Nasmia, Rosyida, E., Masyahoro, A., Putera, F. H. A., & Natsir, S. (2021). The utilization of seaweed-based liquid organic fertilizer to stimulate *Gracilaria verrucosa* growth and quality. *International Journal of Environmental Science and Technology, 18*(6), 1637–1644. https://doi.org/10.1007/s13762-020-02921-8

Nurman, S., Saiful, S., Ginting, B., Rahmi, R., Marlina, M., & Wibisono, Y. (2021). Synthesis of polyurethane membranes derived from red seaweed biomass for ammonia filtration. *Membranes, 11*(9), 1–14. https://doi.org/10.3390/membranes11090668

Padhi, S., Swain, P. K., Behura, S. K., Baidya, S., Behera, S. K., & Panigrahy, M. R. (2011). Cultivation of *Gracilaria verrucosa* (Huds) papenfuss in Chilika Lake for livelihood generation in coastal areas of Orissa State. *Journal of Applied Phycology, 23*(2), 151–155. https://doi.org/10.1007/s10811-010-9592-9

Parenrengi, A., Sulaeman, S., Suryati, E., & Tenriulo, A. (2006). Karakterisasi genetika rumput laut *Kappaphycus alvarezii* yang dibudidayakan di Sulawesi Selatan [Genetic characterization of *Kappaphycus alvarezii* seaweed cultivated in South Sulawesi]. *Jurnal Riset Akuakultur, 1*(1), 1–11. https://doi.org/10.15578/jra.1.1.2006.1-11

Rejeki, S., Ariyati, R. W., Widowati, L. L., & Bosma, R. H. (2018). The effect of three cultivation methods and two seedling types on growth, agar content and gel strength of *Gracilaria verrucosa*. *Egyptian Journal of Aquatic Research, 44*(1), 65–70. https://doi.org/10.1016/j.ejar.2018.01.001

Rustam, R., & Wamnebo, M. I. (2021). Feasibility study of combined *Gracilaria verrucosa* with giant tiger prawn *Penaeus monodon* farming. *AACL Bioflux, 14*(6), 3197–3203.

Skriptsova, A. V., & Choi, H.-G. (2009). Taxonomic revision of Gracilaria "verrucosa" from the Russian Far East based on morphological and molecular data. *Botanica Marina, 52*(4), 331–340. https://doi.org/10.1515/BOT.2009.008

Suhartini, S., Indah, S. H., Rahman, F. A., Rohma, N. A., Rahmah, N. L., Nurika, I., Hidayat, N., & Melville, L. (2023). Enhancing anaerobic digestion of wild seaweed *Gracilaria verrucosa* by co-digestion with tofu dregs and washing pre-treatment. *Biomass Conversion and Biorefinery, 13*(5), 4255–4277. https://doi.org/10.1007/s13399-022-02507-z

Torres, F. A. E., Passalacqua, T. G., Velásquez, A. M. A., de Souza, R. A., Colepicolo, P., & Graminha, M. A. S. (2014). New drugs with antiprotozoal activity from marine Algae: A review. *Revista Brasileira de Farmacognosia, 24*(3), 265–276. https://doi.org/10.1016/j.bjp.2014.07.001

Widowati, L. L., Rejeki, S., Yuniarti, T., & Ariyati, R. W. (2015). Efisiensi produksi rumput laut *E. Cotonii* dengan metode budidaya long line vertikal sebagai alternatif pemamfaatan kolom air [Efficiency of *E. cotonii* seaweed production with vertical long line cultivation method as an alternative to water column utilization]. *Jurnal Saintek Perikanan, 11*(1), 47–56. file:///C:/Users/DELL/Downloads/10564-23969-1-SM.pdf

Yow, Y.-Y., Lim, P.-E., & Phang, S.-M. (2011). Genetic diversity of *Gracilaria changii* (Gracilariaceae, Rhodophyta) from west coast, Peninsular Malaysia based on mitochondrial cox1 gene analysis. *Journal of Applied Phycology, 23*(2), 219–226. https://doi.org/10.1007/s10811-010-9535-5

Yuliana, E. (2022). *Pertumbuhan dan kualitas rumput laut Gracilaria verrucosa dengan jarak ikatan berbeda di tambak udang supra intensif* [Growth and quality of *Gracilaria verrucosa* seaweed with different bond distances in intensive supra shrimp ponds]. Universitas Hasanuddin.

The Hatchery Technology of Milkfish (*Chanos chanos* Forsskal) to Support Cultivation and Development of Sustainable Fisheries Industry

23

Tony Setia Dharma, Gusti Ngurah Permana, Muhammad Marzuqi, Gigih Setia Wibawa, and Zafran Zafran

23.1 Introduction

Milkfish (*Chanos chanos* Forsskal) is a widely recognized consumption fish that can be cultivated in ponds. It holds significant economic value due to the consistently increasing demand for milkfish each year. This demand arises from local consumption and its use as live bait for tuna fishing and broodstock needs in hatchery activities and the export market. The growing need for milkfish for export presents a positive business opportunity. As an essential aquaculture commodity, milkfish has many advantages, including its appealing taste, relatively affordable price, resistance to disease, noncannibalistic nature, ability to be cultivated in polyculture with other species, and euryhalinity, which allows it to thrive in various habitats such as brackish water, seawater, and freshwater.

The domestic market shows a high demand for milkfish seeds, and exports to several countries are rising. Therefore, it is crucial to maintain the quality of milkfish seeds produced by household-scale hatchery. This focus should include effective feed management and a controlled environment to ensure that the quality of seeds and fish meets export standards. The role of milkfish broodstock is fundamental for seed production. Attention must be given to selecting high-quality milkfish individuals, choosing the right seeds, employing effective growth methods, and maintaining suitable water conditions to produce seeds and fish that grow rapidly and sustainably.

T. S. Dharma (✉) · G. N. Permana · M. Marzuqi · Z. Zafran
Research Center for Fishery, National Research and Innovation Agency, Cibinong, Indonesia
e-mail: tony005@brin.go.id

G. S. Wibawa
Research Center for Marine Cultivation and Fisheries Extension, Marine Fisheries Research and Human Resources Agency, Marine and Fisheries Ministry, Cibinong, Indonesia

E. Nugroho et al. (eds.), *Aquaculture Development in Archipelago*,
https://doi.org/10.1007/978-981-95-1084-9_23

In addition to being a popular consumption fish, milkfish has gained demand in recent years as live bait for tuna (*Thunnus* sp.) and skipjack (*Katsuwonus pelamis*) fishing. Efforts in milkfish cultivation and development have encompassed various aspects, including collecting and documenting natural and artificial hatchery technologies and enlargement methods. These initiatives aim to address the shortage of milkfish seeds and contribute to developing a sustainable national milkfish cultivation industry.

23.2 Taxonomy and Morphology

Milkfish, known scientifically as *Chanos chanos*, is called "Milkfish" in English and "Bale Bolu" by the Makassar people. It was first discovered by Dane Forsskal in 1925 in the Red Sea. According to Sudrajat (2008), the taxonomy and classification of milkfish are as follows:

Kingdom:	Animalia
Phylum:	Chordata
Subphylum:	Vertebrata
Class:	Osteichthyes
Order:	Gonorynchiformes
Family:	Canidae
Genus:	*Chanos*
Species:	*Chanos chanos*

Identification was conducted by sending samples of G1 fish obtained from domestication (Fig. 23.1). The fish samples were determined to belong to the *Chanos chanos* Forsskal species (Blecker).

Fig. 23.1 Milkfish (*Chanos chanos* Forsskal). Remarks: Eyes (**a**), mouth (**b**), operculum (**c**), pectoral fin (**d**), linea lateral (**e**), abdominal fin (**f**), dorsal fin (**g**), anal fin (**h**), caudal fin (**j**). (Tony, 2016)

23.3 Distribution

Milkfish is an euryhaline fish found in freshwater, brackish water, and seawater. During its development period, milkfish like to live in brackish water or estuaries. When they reach maturity, milkfish will return to the sea to breed. The growth of milkfish (*Chanos chanos*) is relatively fast, with a rate of 1.1% to 1.7% of body weight per day (Sudrajat, 2008). They can reach an average weight of 0.60 kg at 5–6 months when kept in ponds. Milkfish have a wide distribution area, primarily found in the tropical seas of the Indo-Pacific, with a dominant presence around Asia. Their habitat includes the waters of Burma, Thailand, Vietnam, the Philippines, Malaysia, and Indonesia. Milkfish in the Central Pacific Ocean are also spread around the Hawaiian Islands. In Indonesia, they are commonly found in the eastern waters of Sumatra, Java, Kalimantan, Sulawesi, Maluku, Irian, Nusa Tenggara, and Bali. Milkfish exhibit a diurnal feeding habit, primarily taking food from the top layer of the seabed in their natural habitat. Their diet consists of microscopic plants such as plankton, tiny shrimp, microorganisms, and other multicellular plants. The food intake of milkfish is adjusted according to the size of their mouths. During the larval stage, milkfish are classified as carnivorous. As fry, they become omnivorous, and at the juvenile stage, they are considered herbivores, capable of consuming artificial feed in the form of pellets. As adults, milkfish revert to omnivores, feeding on algae, zooplankton, soft benthos, and artificial pellets (K. Imsland et al. 2008; Setiadharma. et al. 2014).

23.4 Reproduction

Candidates for good milkfish breeders possess specific characteristics: they are typically more than 4 to 5 years old, weigh at least around 4 kg, and have a total length of 60–65 cm. Milkfish can spawn daily within a population, and spawning occurs partially in concrete tanks or ponds. Both natural and artificial spawning methods are employed. The signs that milkfish are ready to spawn include reduced appetite and increased activity, where they often chase each other. At night, there are noticeable ripples and splashes of water during mating. Generally, mating begins at midnight, with maximum production occurring close to the new moon.

Distinguishing between male and female parent candidates remains challenging. However, examination of the genitals reveals that male milkfish have two small protrusions (papillae) that are externally visible, located in the form of the outer anal membrane and a discharge hole at the end. The vasa differentia of the testes merges approximately 5–10 mm deep from the opening of the discharge. In contrast, female milkfish exhibit three small protrusions (papillae) in the anal area, with the anal canal parallel to the genital pore. The third opening is the posterior entrance of the genital pore, and the two oviducts merge into a wide channel that ends at the genital pore. The diameter of the eggs can serve as an indicator of the maturity level of the gonads of the broodstock, and a microscope can be employed to measure egg diameter.

Males that are ready to spawn typically possess third-order sperm, indicating a substantial quantity of sperm during stripping. Selected broodstock are then mass-mounted in concrete tanks with a volume of 200 tons, housing 60 individuals with a female-to-male ratio of 2:1. The gonad maturation of prospective milkfish broodstock can be stimulated through hormone implantation. The hormone administered consists of LHRH-analog pellets at 50 μg/kg for both male and female candidates.

The hormone implantation schedule includes the following stages: in the first month, LHRH-a is used; the second month involves both LHRH-a and 17α methyl testosterone; and the third month again utilizes LHRH-a. This hormone implantation can be performed continuously until eggs with a diameter exceeding 750 μm are obtained. Milkfish spawn en masse and naturally within a single population kept in the tank. This mass spawning increases the likelihood of daily spawning events, as milkfish are known to be partial spawners, releasing their eggs over multiple spawning events rather than all at once. After spawning, viable eggs float and enter the egg collector at the mother tank's top outlet.

The harvested eggs are then incubated in a fiberglass tank of known volume for easier division. After 6 h in the incubator, the eggs are sorted, beginning with separating good eggs from damaged ones. The process involves removing aeration and stirring the eggs in a unidirectional and regular rotation, allowing damaged eggs to settle at the bottom of the fiberglass tank while the good eggs remain afloat. Damaged eggs at the bottom are gently removed using a small hose. The good eggs are then aerated, counted through random sampling, and divided according to the volume of water (Figs. 23.2, 23.3, 23.4, and 23.5).

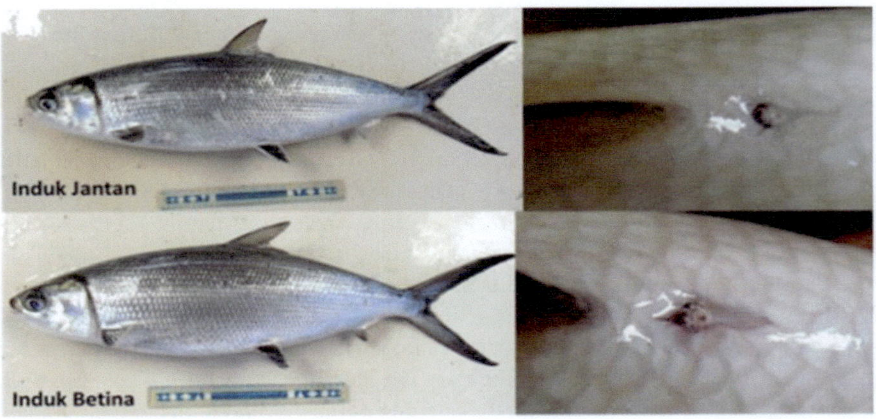

Fig. 23.2 Male and female into broodfish milkfish (*Chanos chanos* Forsskal).

Fig. 23.3 Implantasi hormon pada induk. (Priyono et al. 2012; Setiadharma et al. 2011; 2012 and 2014; Anonymous, 2016)

Fig. 23.4 Construction of main tank and egg collector.

23.5 Domestication

Domestication is a process of adaptation to the cultivation environment across generations. This process is a crucial step toward breeding and encompasses exploration, collection, documentation, characterization, and mastery of natural and artificial hatchery technologies and their enlargement. Mastery of seed technology includes gonad maturation, spawning, egg care, larval treatment, seed treatment, and disease control. Effective domestication can ensure a sustainable and stable supply of fish for consumption from cultivation-based production. Based on the level of perfection, domestication can be categorized into four levels: (1) Perfect domestication, where the entire life cycle of the fish can be maintained within the aquaculture system; (2) Almost perfect domestication, where the entire life cycle can be maintained but with low success rates; (3) Imperfect domestication, where only part of the life cycle can be maintained in the cultivation system; and (4) Undomesticated, where the entire life cycle cannot be maintained in the cultivation system.

Currently, milkfish cultivators face challenges related to the origin of milkfish broodstock and the fast-growing seeds resulting from domestication. Prospective broodstock sourced from the wild are generally not well-adapted to new environments, making them weak and susceptible to high mortality rates. The history and preparations for the domestication of milkfish began at the Kamal Experimental Pond Installation in 1984, coordinated by the Bogor Fisheries Research Institute

Fig. 23.5 Egg incubation and counting.

(LPPD). This effort continued at the Gondol Marine Cultivation Research and Development Center (BBPBPL) in 1985. The initial broodstock (G0) was successfully spawned in 1987, leading to the development of natural and semi-artificial seed production techniques and nursery and rearing methods in 1988. In 1998, seeds from G0 were designated as potential G1 broodstock, and this process continued until 2005. In 2010, 1000 first-generation milkfish (G1) broodstock were produced. These domesticated milkfish were then ready for breeding to obtain seeds and second-generation parent candidates (G2) by 2015.

The results of this research, in the form of milkfish hatchery technology and biological products, have been utilized for enlargement in community-owned ponds throughout Indonesia and have contributed to the development of milkfish cultivation through dissemination programs. The Jakarta Cultivation Research and Development Center has further applied these cultivation practices. As a result of these activities, a comprehensive technology package has been produced to develop milkfish cultivation, including hatchery and enlargement techniques implemented in local communities.

23.6 Development of Aquaculture Technology

Milkfish cultivation technology, encompassing hatchery, nursery, and enlargement phases, is straightforward and cost-effective, with a broad market demand. Milkfish hatchery technology was initially conducted in indoor tanks, with a 22- to 25-day harvest period. The feeding regimen relied on chicken eggs and artificial feed, producing relatively high production costs. As the demand for milkfish seeds increased, fostering synergy between research and cultivators to make seeds more efficient and at lower costs became essential. In the 1990s, outdoor hatcheries without roofs were introduced, utilizing full rotifer feed, which led to estimated production cost savings of up to 50% and continued until 2010. However, by 2011, concerns arose regarding the slow growth of milkfish seeds from Bali. In response, improvements in cultivation technology were implemented, including using domesticated broodstock, providing high-quality prospective broodstock, enhanced seed production management, and better rearing practices within the community sustainably. The milkfish cultivation segment and its distribution are presented in Table 23.1, while the relationship between milkfish productivity and survival across various cultivation segments is illustrated in Table 23.1 and Figs 23.6, 23.7, 23.8, 23.9, and 23.10.

Table 23.1 Business segment on milkfish cultivation

Segment of business	Eggs quality (%)	Remark
Eggs	0–95	Fertility (%)
Eggs incubation	≥80	22–24 jam
Survival rate (%)		
Seed production	86–89	Size 1.2–1.4 cm
Nursery 1	55–80	4–6 cm, 1 month
Nursery 2	60–80	8–12 cm, 2 months
Grow-out	70–85	Size of harvest 150–450 g

Source: Prijono et al. (2011)

Fig. 23.6 Harvest of milkfish in ponds

Fig. 23.7 Packing and transport of milkfish logs.

Fig. 23.8 Milkfish harvest in ponds.

Fig. 23.9 Development of hatchery of milkfish in the Bali community.

Fig. 23.10 Local and export market of milkfish seeds in the community.

References

Anonymous (2016). Naskah akademik rilis ikan bandeng unggul (Chanos chanos Forsskal) hasil domestikasi. *Balai Besar Penelitian dan Pengembangan Budidaya Laut (BBPPBL) Gondol*. Bali KKP. 66 hlm

K. Imsland, S. Gustavsson, A. Foss, J. Amason, I. Amarso, A. F. Jhonsson, H.Smaradottir, and H. Thararensen (2008), Effects of Reduced Salinity on Growth, Feed Confersion Efficiency and Blood Physiology of Juvenile Atlantic Halibut (Hippoglossus hippoglossus L). *Aquaculture* 274 254-259.

Priyono.A, T.Aslianti,T. Setiadharma dan I.N.A. Giri. (2012), Petunjuk Teknis Perbenihan Ikan Bandeng (Chanos chanos Forsskal). Balai Besar Penelitian dan Pengembangan Budidaya Laut. Badan Penelitian dan Pengembangan Kelautan dan Perikanan. *Kementerian Kelautan dan Perikanan*. ISBN: 978-979-17440-6-5. 45 hlm

Setiadharma T., A.A. K. Alit., Priyono A., dan Wibawa G S. (2011). Performa Benih Ikan Bandeng (Chanos-Chanos Forsskall) Pada Pembesaran Di Bak Dan Hafa Di Tambak Dengan Perbedaan Kepadatan. BBPPBL. *Laporan Hasil Penelitian*

Setiadharma,T, A. Priyono, Aslianti, T, Setiadi. I, wibawa,G.S. (2012). Teknik Pendederan ikan Bandeng, Chanos chanos Forskall). Makalah. *Disampaikan Dalam Acara Diseminasi dii Kabupaten Gresik Jawa Timur pada tanggal* 17 April 2012. 15 hlm

Setiadharma T, Setiadi, I dan Wibawa, S.W. (2014). Pengelolaan induk bandeng, chanos chanos forskall hasil seleksi. Laporan teknis akhir kegiatan. Balai besar penelitian dan pengembangan budidaya laut. Badan penelitian dan pengembangan kelautan dan perikanan. *Kementerian kelautan dan perikanan*.15 Hlm.

Sudradjat, A. (2008). Budidaya 23 Komoditas laut Menguntungkan. *Penebar Swadaya*, Jakarta.

24 Prospect of Kuwe Fish (*Gnathanodon speciosus* Forsskal) Hatchery Supports to the Aquaculture Development

Tony Setia Dharma, Jhon Harianto Hutapea,
Ketut Mahardika, and April Supii

24.1 Introduction

The golden trevally (*Gnathanodon speciosus*), also known as the leopard catfish, yellow penal, or yellow simba, belongs to the family Carangidae. This species is known for its fast-swimming abilities and has an oval, flat shape. It holds significant economic value due to its relatively high market price and potential as an export commodity. At sizes of 5–10 cm, the golden trevally is valued as a marine ornamental fish, referred to as yellow penal fish or yellow simba. In nature, these fish typically inhabit shallow coastal waters, coral reefs, and rocky areas, and are classified as benthopelagic species (Widodo & Burhanuddin, 2003). Golden trevally can reach juvenile stages at 30–35 days of age and grow relatively quickly, achieving lengths of 23.9–26.6 cm and weights of 282.2–383.9 g within 7–9.5 months of cultivation. The kuwe fish, which can grow up to 60 cm, typically resides at depths of 12 m and is commonly found in tropical and subtropical seas. Its meat is known for being high in protein and has a desirable taste. The golden trevally is an economically important fish, primarily marketed in Japan and other Asian countries such as China, Hong Kong, Singapore, and Malaysia. An analysis of global catch data indicates that populations of giant trevally, particularly the tiger and silver trevally species, have been harvested from the wild as seeds for aquaculture. Therefore, there is a pressing need to develop innovative approaches in aquaculture, starting with advancements in seed technology.

T. S. Dharma (✉) · J. H. Hutapea · K. Mahardika
Research Center for Fishery, National Research and Innovation Agency, Cibinong, Indonesia
e-mail: tony005@brin.go.id

A. Supii
Research Center for Marine and Land Bioindustry, National Research and Innovation Agency, Cibinong, Indonesia

E. Nugroho et al. (eds.), *Aquaculture Development in Archipelago*,
https://doi.org/10.1007/978-981-95-1084-9_24

24.2 Taxonomy and Morphology

The classification of the pompano is as follows:

Kingdom:	Animalia
Phylum:	Chordata
Class:	Actinopterygii
Order:	Perciformes
Family:	Carangidae
Genus:	*Gnathanodon*
Species:	*Gnathanodon speciosus*

The form of the pompano is oval and flat (Fig. 24.1). The body color varies, ranging from yellow with black crescent stripes from top to bottom, to silver or whitish on the underside. The body is covered with fine, cycloid-shaped scales, which are small and feature a branched lateral line. On the chest, the scales are reduced or absent. There are three spines, with the first two separated from the fixed fin. The caudal fin is wavy (Priyambodo et al, 2004). Pompanos are nocturnal feeders that primarily consume fish and crustaceans. They exhibit the following physical characteristics: nine dorsal fins with 19–22 soft dorsal fin rays, and three anal fins with 14–17 soft anal fin rays. The body is colorful, displaying light green on the back and silvery white on the underside, complemented by sharp, curved pectoral fins. Another name for the golden trevally is the "leopard fish," as well as "yellow

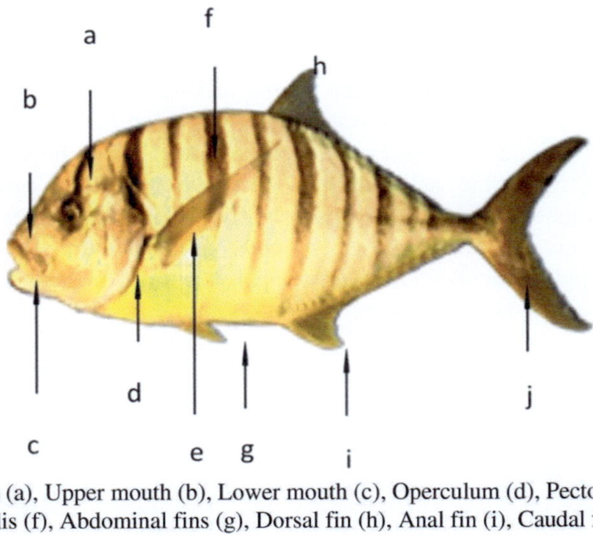

Fig. 24.1 Morphology of the pompano fish, *Gnathanodon speciosus*. Description: Eyes (**a**), upper mouth (**b**), lower mouth (**c**), operculum (**d**), pectoral fins (**e**), linea lateralis (**f**), abdominal fins (**g**), dorsal fin (**h**), anal fin (**i**), caudal fin (**j**)

criminal," "golden jack," or "pilot fish." The body of this fish features black transverse stripes on a golden-yellow base, resembling the coloration of a tiger, which gives rise to the name "kuwe tiger." The mouth is thick-lipped and toothless. Kuwe tigers inhabit reef environments and often surface to chase their prey. With their large and strong caudal fin, these fish can swim quickly through alcoves to capture food. Kuwe tigers are typically caught by fishermen using purse seine nets, fishing rods, and traps. As a visually appealing species, the kuwe tiger, or penal fish, is highly sought after as a marine ornamental fish. Its international trade names, "golden jack" or "pilot fish," contribute to its relatively high export value.

24.3 Distribution

The habitats of pompano vary greatly depending on the species, ranging from coastal waters to the open sea (oceanic) and from pelagic to demersal environments near the bottom. The genera *Caranx* and *Gnathanodon* are particularly distinctive as reef-associated inhabitants. Almost all kuwe fish exhibit schooling behavior and are carnivorous, with their main diet consisting of fish and other macrofauna. Pompano typically inhabits shallow waters and coral reefs, often forming small schools. They can reach a total length of approximately 30–75 cm. In the wild, their diet includes small fish and crustaceans. Fishing methods for pompano include fishing rods, traps, and gill nets, such as moroami gill nets. They are marketed in various forms, including fresh and dry salted, with the latter being relatively more expensive. Their distribution includes shallow coastal areas, coral reefs in Indonesia, the Bay of Bengal, the Gulf of Siam, and along the coast of the South China Sea and the Philippines. The feeding habits of these fish are influenced by the quality and quantity of available food, which in turn depends on their environment. The size of the fish population in a given area is determined by the availability of food. Factors related to food include the quantity and quality of resources available and the duration of time fish spend feeding. The food consumed by fish directly impacts the overall food supply in their habitat. This consumption affects growth, maturity, and survival rates for individual fish. Consequently, the population dynamics, growth, reproduction, and overall health of fish populations are closely linked to the availability of food in specific waters (Effendi, 2003).

24.4 Reproduction

Golden trevallies are primarily caught from the natural waters of East Java and Bali using fishing nets and rods. The broodstock are initially accommodated and maintained in floating net cages for 10 to 15 days until they are healthy and feeding well. Once the broodstock, which weigh between 1.0 and 2.5 kg, have adapted to the rearing environment, they are transferred to a tank with a volume of 30 m^3 that utilizes a running water system. During this period, the broodfish are fed a diet of fresh fish and squid, supplemented with a vitamin mix at a rate of 5% of their biomass per

Fig. 24.2 Types of prospective goldfish breeder (**a**) natural, and (**b**) culture. (Setiadharma et al, 2006). Tank of rearing broodfish. (Setiadharma et al, 2006)

day. After 4 to 5 months of maintenance, the broodstock typically spawn and produce eggs. The main tank serves as both a maintenance and spawning area for the broodstock. Its shape and size are tailored to accommodate the spawning habits of the fish. While spawning, the broodstock chase each other, necessitating a larger tank. The ideal tank size for the spawning process is between 30 and 100 m^3, with a depth of 2 m and a round shape. This design prevents the fish from gathering at a single point and facilitates the removal of leftover feed and waste by allowing for drainage at the center of the tank's bottom. The main tank is equipped with pipes for the inflow and outflow of water and for aeration. Each brood tank must also have an overflow pipe connected to an egg storage net (Fig. 24.2).

24.5 Domestication and Maintenance of Broodstock

The broodfish are reared in two round concrete tanks with a volume of 30–100 m^3. The broodstock used consists of fish caught with a fishing rod, ranging in size from 1200 to 3500 g. These fish are selected and acclimated in concrete tanks, where they are reared using a circulating system with a total water change of approximately 200–300% per day. The food provided is in the form of fresh fish and squid, supplemented with a vitamin mix at a rate of 5–10 g/kg. The amount of feed given is 5% of the biomass per day, distributed in the morning and evening. To maintain optimal dissolved oxygen levels in the water, each tank is equipped with four aeration channels. Measurements of growth in length and weight of the fish are conducted, along with observations of the maturity levels of the gonads of prospective broodstock every 2 months. The initial weight of the prospective broodfish caught from the wild ranges from 1.20 to 3.50 kg. These fish are initially acclimated in a fiber tank with a volume of 5 tons and treated with Elbazu every 2 days to prevent the growth of bacteria or fungi. After approximately 7–10 days, the fish are transferred to the main tanks. A total of about 20 individual fish are kept, and observations are made regarding the development of the gonads to determine male and female breeders. The gonads of prospective broodfish develop and spawn, producing fertilized eggs.

Natural spawning occurs monthly between 12:00 and 02:00, with hatching time taking around 14–18 h.

24.6 Seed Production

The initial step in larval rearing is the preparation of the tank. The tanks used should be round or square in shape, made of fiberglass or concrete, with a volume of 1–6 m^3. Before using, the tanks must be thoroughly cleaned to remove dirt and eliminate pests and diseases. This involves brushing the walls and bottom of the tank and rinsing with a chlorine solution to ensure even distribution. Once dry, fill the tank with seawater and filter it using a filter bag or sand filter until the water level reaches 75 cm. The larval rearing tank should be equipped with aeration, providing one aeration unit per square meter. Additionally, a running water piping system should be installed to facilitate water changes and allow for siphoning of the tank bottom. Natural food for rearing pompano larvae includes phytoplankton (e.g., *Chlorella* sp.), rotifers, Artemia nauplii (small shrimp), and artificial feed. Phytoplankton should be provided from days 1 to 20, while retirees are given from days 2 to 20. Artemia nauplii should be introduced from days 15 to 25, followed by artificial feed as an additional food source from days 15 to 40. Water changes should be conducted when the larvae are 10 days old, starting with a 20% change and increasing to 50–80% as needed. After 30 days, maintenance can be performed using running water or circulation system. Siphoning of the tank bottom should begin when the larvae are 12 days old and can be carried out every 2–3 days thereafter. By 35 days of age, the pompano larvae typically reach a size of 3–4 cm and are ready for harvesting (Fig. 24.3).

24.7 Nursery and Growth

The behavior of this fish like the spooling used can be circular or rectangular (Fig. 24.4). However, using a circular tank will cause the juveniles to be more responsive to the feed given and easy to clean. The tank spindle is equipped with a water intake and output system as well as aeration. The tank is easy to clean and drain. The walls of the tank should not be a good medium for the attachment of algae. Juveniles that have just been harvested from larvae tanks are still small with a total length (TL) ranging from 2.0 to 3.0 cm and are still not strong enough to be reared in floating nets cage, so further maintenance or nursery needs to be carried out for a long time in tanks so that reach a spindle size of 5–8 cm (Fig. 24.6). It takes between 2 and 4 weeks to reach the size of the logs. Juveniles of pompano that have just been transferred often cluster and move in circles. In the juvenile stage, nursery rearing tanks given a combination of artificial feed and small shrimp (Fig. 24.6). By giving combination feed to juveniles, the quality of the saplings produced is better and stable, producing a bright yellowish color because the artificial feed given has been adjusted to the nutritional needs of the fish.

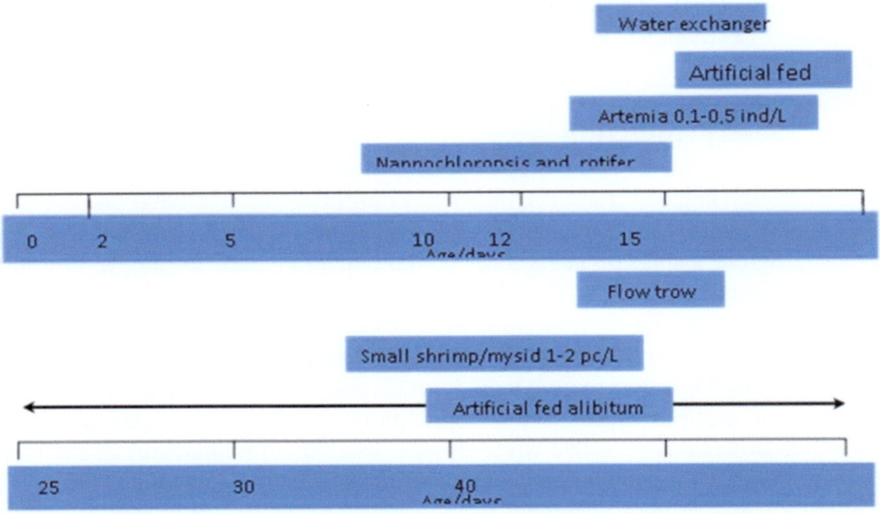

Fig. 24.3 Management of live food the fish larvae and juvenile. (Setiadharma et al, 2009)

(a) (b) (c) (d)

Fig. 24.4 (**a**) Seed size 3 cm, (**b**) 5 cm, and (**c**) 5–8 cm. At a size of 5–8 cm, it is ready to be kept in floating net cages. (Setiadharma et al., 2007; 2009)

Pompano fish hatcheries, with an average weight of 10.90 g, were reared in tanks measuring 3 × 2 × 1 m^3 at a density of 250 individuals. The tanks were equipped with a running water system. The fish were fed dry pellets three times a day. Maintenance included weighing the fish every 2 weeks at the beginning, followed by monthly weigh-ins. The development of fish weight can be seen in Fig. 24.6.

Kuwe fish, with an initial average weight of 10.90 g, can reach an average weight of 400–500 g after being reared in tanks for 6 months. Fish weighing between 350 and 500 g meet the requirements for local consumption and export.

24.8 The Development of Aquaculture Technology

The development of the pompano hatchery business includes broodstock management, larval rearing, and water and feed management until the larvae become juvenile. The technology for seeding pompano and eggs produced has been adopted and used by the community. It was observed in some small-scale hatchery. Hatcheries have been successful and able to adopt technology for seed production of pompano fish to produce seeds, but it needs to be developed in a sustainable manner in accordance with existing standard operating procedures (SOPs).

24.9 Prospects and Market

The pompano, or golden trevally fish, is a promising commodity that has not been widely cultivated. With successful mastery of domestication technology, natural gonadal maturation, and the use of artificial feed for seeding and batching, the cultivation business can be further developed. When they reach a size of 5–10 cm in total length, these fish have the potential to be marketed as marine ornamental fish. At a total length of 30–38 cm, with a weight of 350–500 g, they are suitable for consumption both domestically and internationally. Observations indicate that these fish possess significant economic value, presenting prospects and opportunities as alternative commodities for cultivation. Recently, several household-scale hatcheries (small-scale hatcheries) have begun to experiment with hatching these fish and have succeeded in producing fry.

References

Effendi, H. (2003). *Telaah kualitas air bagi pengelolaan sumber daya dan lingkungan Perairan [Water quality study for the management of aquatic resources and environment]* (p. 258). Penerbit Kanisius Yogyakarta.

Priyambodo, & Tri Wahyuningsih. (2004). *Budidaya pakan alami untuk ikan [Natural feed cultivation for fish]*. Penebar Swadaya.

Setiadharma, T., Prijono, A., Giri, I. N. A., & Hanafi, A. (2006). Pengamatan pola pemijahan alami induk ikan kuwe (*Gnathanodon speciosus*, Forsskal) pada pemeliharaan secara terkontrol [Observation of natural spawning patterns of mother fish (*Gnathanodon speciosus*, Forsskal)] in controlled maintenance. *Bulletin Teknisi Litkayasa Akuakultur, 8*(1), 61–65.

Setiadharma T., Syahidah D., Prijono A. (2007). *Konsumsi Oksigen Larva Ikan Golden Trevally (Gnathanodon speciosus, Forsskall). Prosiding Seminar Nasional Perikanan dan Kelautan.* ISBN : 978.979.704.595.1.Badan Penerbit Universitas Diponegoro Semarang. Hal 337–339.

Setiadharma T., Hanafi, A., Giri, I. N. A., Prijono, A., Asmanik, A. A., Kt Alit, Syahidah, D., & Suharyyah, S. (2009). Teknologi Pembenihan ikan kuwe macan *(Gnathanodon speciosus, Forsskal)* [Tiger fish *(Gnathanodon speciosus, Forsskal)* hatchery technology]. Balai Besar Riset Perikanan Budidaya Laut Gondol. Pusat Riset Perikanan Budidaya. Badan Riset Kelautan dan Perikanan, Kementerian Kelautan dan Perikanan. 28 p.

Widodo, J., & Burhanuddin. (2003). *Systematics of the small pelagic fish species. BIOSYNEX (Biology, Dynamics, Exploitation) of the small felagic fishes in the Java Sea* (2nd ed.). Marine and Fisheries Research Project. The Agency for Marine and Fisheries Research.

Breeding Technology and Domestication of Rabbitfish (*Siganus guttatus* Bloch, 1787)

25

Samuel Lante, Herlinah Herlinah, Rosmiati, and Andi Tenriulo

25.1 Introduction

Rabbitfish (*Siganus guttatus*) (Fig. 25.1) is an important species with high economic value. This native species of Southeast Asia, including Indonesia, is a significant commercial fish in the region (Hara et al., 1986; Mejos et al., 2023). The distribution of rabbitfish is primarily in tropical coastal waters (Darsiani et al., 2024). This fish has become a candidate for cultivation in marine aquaculture development (Laining et al., 2019) to compensate for the capture of wild populations, as continuous fishing can lead to declines in natural populations (Lante et al., 2012b; Syah et al., 2020). Rabbitfish have several advantages. They are highly preferred by local consumers due to their delicious taste and tender white meat (Cabanilla-Legaspi et al., 2021; Kamaruddin et al., 2015) and are commonly served in restaurants in South Sulawesi. Additionally, they command a relatively high price at the farmer level. These fish can thrive in high densities, respond well to artificial feed, and exhibit a high growth rate (Usman et al., 2021). They can be cultivated in floating net cages and ponds (Syah et al., 2020) and can spawn in a controlled manner using hormonal stimuli (Lante et al., 2007) or naturally.

S. Lante (✉) · Rosmiati · A. Tenriulo
Research Center for Fishery, National Research and Innovation Agency, Cibinong, Indonesia
e-mail: samu005@brin.go.id

H. Herlinah
Research Center for Fishery, National Research and Innovation Agency, Cibinong, Indonesia

Faculty of Vocational Studies, Hasanuddin University, Makassar, Indonesia

E. Nugroho et al. (eds.), *Aquaculture Development in Archipelago*,
https://doi.org/10.1007/978-981-95-1084-9_25

Fig. 25.1 Morphology of rabbitfish (*Siganus guttatus*). mouth, (2) eyes, (3) operculum, (4) pectoral fins, (5) pelvic fins, (6) dorsal fins, (7) fins, and (8) tail. (Lante et al., 2015)

25.2 Taxonomy and Morphology

The biological classification of the rabbitfish *Siganus guttatus* was elaborated by (Duray, 1998):

Phylum:	Chordata
Class:	Osteichthyes
Order:	Perciformes
Famili:	Siganidae
Genus:	*Siganus*
Species:	*Siganus guttatus*

The body of the rabbitfish is wide and flattened, covered with smooth, small cycloid scales. Its nose resembles that of a rabbit, and the tiny tip of its mouth has small teeth, while the lateral line is simple. The general color is brownish to greenish black. There are white, brown, gray, or golden spots on its back, while on the abdomen, the spots are blurred and appear as stripes. These spots can be black or may disappear on the upper gill covers. The fish has venomous spines found on 13 stiff spines of the dorsal fin, four rigid spines of the ventral fin, and seven hard spines of the anal fin. Based on their feeding habits, rabbitfish are classified as herbivorous fish (Lam, 1974). Rabbitfish are tropical herbivorous fish that primarily feed on seaweeds and other benthic algae (Parawansa et al., 2020). The color of the fish can change rapidly according to environmental conditions and to avoid predators. Rabbitfish living in the wild have a light or bright body color (Fig. 25.1), while those in ponds have a gray body color (Fig. 25.1).

25.3 Distribution

According to FAO-reported statistics, the major countries producing rabbitfish are illustrated in Fig. 25.2.

Fig. 25.2 Main produces countries of Siganus spp. from FAO Fisheries Statistic, 2018. (Source: FAO, 2023)

Since 2000, the leading rabbitfish-producing countries have been Indonesia and the Philippines (FAO, 2023). In Indonesia, as well as in other Asian countries such as the Philippines, Hong Kong, Singapore, Taiwan, China, and Malaysia, rabbitfish from the family Siganidae is highly popular and holds significant economic value. The genus *Siganus* consists of 12 species (FAO, 2023); however, only six species are found in Indonesian waters: *S. canaliculatus*, *S. fuscescens*, *S. guttatus*, *S. punctatus*, *S. spinus*, and *S. virgatus* (Jaikumar, 2012). These species inhabit almost all shallow marine environments. Among these, *Siganus guttatus* is notably subject to intense fishing pressure. In Indonesia, this rabbitfish is distributed throughout the country's waters, including the coastal areas of Sumatra, surrounding small islands, Java and the Thousand Islands, Kalimantan, Sulawesi, the Maluku Islands, and West Irian. Furthermore, *Siganus guttatus* is commonly found in various habitats, including vegetated coastal areas, river estuaries, and sandy mud. It is also occasionally located in mangrove forests and coastal regions rich in seagrass beds.

25.4 Domestication

Efforts to maintain sustainability, reduce dependence in nature, and support the development of domestication of rabbitfish begin with analyzing their genetic diversity before domestication seed is carried out in hatchery. Genetic diversity is the lowest hierarchy in the level of biodiversity and is an important key for a species to survive, maintain population sustainability, and increase its productivity of a species (Soewardi, 2007). To determine the genetic diversity of rabbitfish, fish sampling was carried out from populations in the waters of Makassar Strait including Takalar, Barru, and Polman, and in the waters of Bone Bay, represented by Malili, Sulawesi Island. Observations of genetic diversity of *random amplified polymorphic* DNA (RAPD) markers show that the rabbitfish of the Malili population is higher than the genetic diversity of the Makassar Strait rabbitfish population (Table 25.1).

Table 25.1 Genetic distance of RAPD markers in four populations of rabbitfish, *S. guttatus* from Barru, Polman, and Takalar, Makassar Strait waters, and Malili Bone Bay waters

Population	Polman	Barru	Takalar	Malili
Polman	–	0.1780	0.1863	0.3768
Barru		–	0.1967	0.4256
Takalar			–	0.3114
Malili				–

Source: Lante (2010) and Lante et al. (2011)

Table 25.2 Genetic distance of RAPD markers in three populations of rabbitfish, *S. guttatus* from Barru waters (Sulawesi Island), Lampung (Sumatra Island), and Sorong (Papua Island) in Indonesia

Populasi	Sorong	Barru	Lampung
Sorong	–	0.0426	0.2237
Barru		–	0.2202
Lampung			–

Source: Lante et al. (2012b)

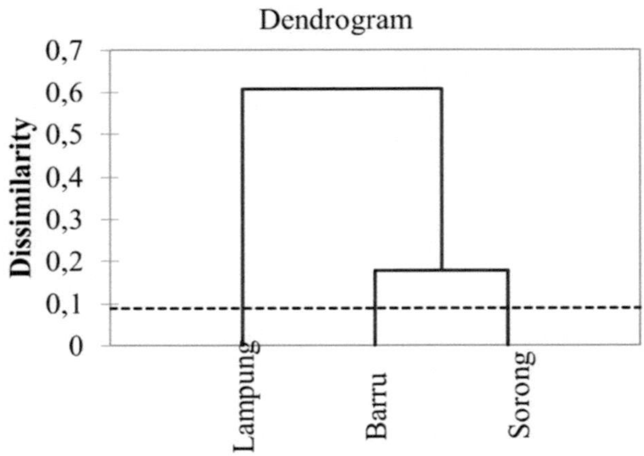

Fig. 25.3 Dendrogram that describes the level of morphological similarity among three rabbitfish populations in Barru (Sulawesi Island), Lampung (Sumatra Island), and Sorong (Papua Island) in Indonesia. (Source: Herlinah et al. (2020))

The analysis of genetic diversity and morphometric traits of rabbitfish (RAPD and morphometric truss) was expanded to include three populations from Indonesia: Barru waters in the Makassar Strait (Sulawesi Island), Lampung waters (Sumatra Island), and Sorong waters (Papua Island). Observations indicate that rabbitfish from Lampung waters exhibit higher genetic diversity and morphometric truss measurements compared to those from the Barru and Sorong populations (Table 25.2 and Fig. 25.3). Morphometric kinship also shows that rabbitfish of Barru are closer or more similar to the rabbitfish from Sorong. This diversity observation complements the development of rabbitfish reproduction technology in Indonesia.

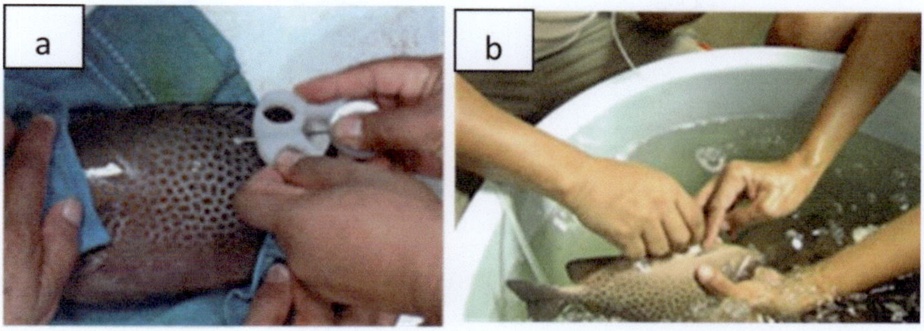

Fig. 25.4 Implant pellet hormone LHRH-a (**a**) and monitoring of gonadal maturity level (TKG) with cannulation (**b**). (Source: Lante et al. (2007))

25.5 Reproduction

Observations of fish reproduction using natural broodstock indicate that the broodstock is typically caught by fishermen in sand-vegetated coastal areas, with body weights ranging from 311 to 382 g per individual. To assess the reproductive performance of rabbitfish, the maturation of broodstock is conducted using a male-to-female ratio of 1:1 (Lante & Palinggi, 2009). Generally, broodstock require about 8 to 12 months to spawn naturally. To enhance the maturation of rabbitfish gonads, LHRH-a hormone pellet implants are administered at a dosage of 10 g per kg of fish weight (Fig. 25.4a). These pellets are applied monthly for 4 months. The determination of gonadal maturity level (TKG) is performed through cannulation (Fig. 25.4b).

To treat fish wounds caused by external parasites attached to the body or gills, and to prevent secondary damage from bacterial infections, broodstock fish are soaked in a disinfectant solution of sodium nifurstyrenate ($C_{13}H_8NO_5{\cdot}Na$) or Elbacin at a dosage of 10 to 20 ppm for 15 to 30 min (Fig. 25.5a). The feed for rabbitfish broodstock consists of pellets (Fig. 25.5b) with a protein content of 46%, fat content of 20%, and a total energy value of 5050 kcal/kg of feed. The feeding rate is set at 3% to 5% of body weight per day, with a feeding frequency of three times daily at 08:00, 12:00, and 16:00 (Lante et al., 2007). Additionally, fresh seaweed (*Gracillaria* sp.) is provided as supplementary feed three times a week, constituting 20% to 25% of the total body weight (Fig. 25.5c).

In the reproduction of barred rabbitfish, parasites that infect the broodstock typically attach to the body, gills, pectoral fins, ventral fins, and caudal fins. One common parasite affecting rabbitfish broodstock is *Pseudocaligus* sp. (Fig. 25.6). Broodstock with parasitic wounds are treated by applying Albaju to the affected areas. The seawater source for the broodstock tanks should be clean, with a salinity of 31 to 34 ppt and a temperature of 28 °C to 30 °C. These salinity and temperature ranges can stimulate the broodstock to spawn monthly. Water changes in the broodstock maintenance are conducted at 200% to 250% per day to prevent the emergence of parasites. Remaining inedible feed is removed by scooping or opening the water drain faucet. Before spawning, rabbitfish exhibit specific characteristics: the

Fig. 25.5 Sodium nifurstyrenate (**a**), feed pellets (**b**), and seaweed (**c**). (Source: Lante et al., 2007)

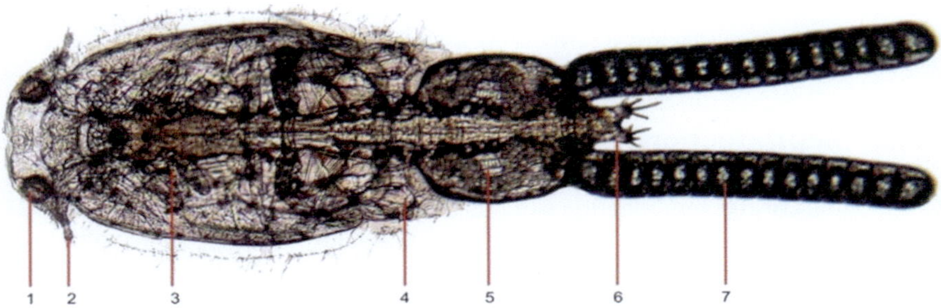

Fig. 25.6 *Pseudocaligus uniartus* parasites (1) eyes, (2) lanule, (3) chepalothora, (4) swimming legs, (5) abdomen, (6) tail, and (7) egg bag. (Source: Lante and Palinggi (2010))

belly of the fish becomes enlarged (distended), appetite decreases until the fish stops eating, and the fish move continuously in schools. Spawning eggs adhere to the walls of the tank or to collectors placed in the trough. The eggs are round and clear in color, while unfertilized eggs either float or sink to the bottom of the trough. One rabbitfish broodstock can produce between 245,000 and 500,000 eggs, with an egg diameter of 546 to 550 μm (Lante et al., 2007), and a hatchability rate ranging from 80% to 95%.

Rabbitfish eggs contain oil bubbles measuring between 70 and 150 μm (Lante & Palinggi, 2010). The developmental timeline of the eggs is as follows: they reach the two-cell stage in 20 min, the four-cell stage in 30 min, the eight-cell stage in 45 min, the 16-cell stage in 60 min, the 32-cell stage in 80 min, and the many-cell stage in 85 min. The transition to the morula phase occurs at 90 min, characterized by the aggregation of cells. The blastula phase forms 3 h after spawning, followed by the gastrula phase, which consists of three layers: ectoderm, mesoderm, and endoderm. This gastrula phase takes 3 h and 25 min to develop from the blastula phase and is the longest stage of embryonic development. During this time, the gastrula undergoes growth and the formation of various body organs. After the gastrula phase, head formation begins, taking 8 h and 50 min post-spawning. At this stage, differentiation from the middle layer occurs. The eye formation phase takes place 10 h and 10 min after spawning, with eyes developing in rudimentary kuffer bubbles. The heart organ begins to beat 12 h and 30 min after spawning, typically establishing a consistent rhythm after 14 h and 5 min, followed by the tail beginning to move.

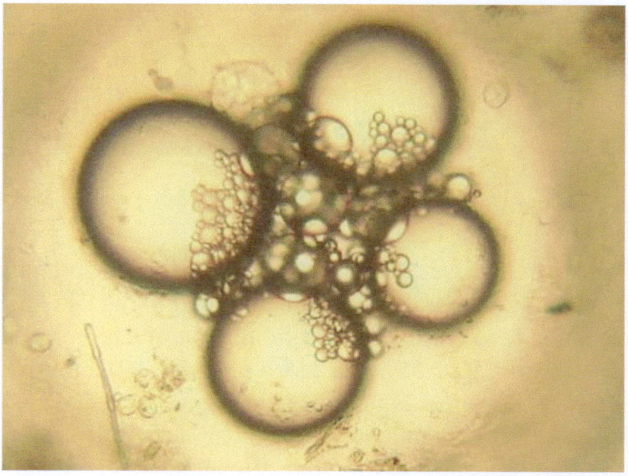

Fig. 25.7 Embryonic development of rabbitfish, *Siganus guttatus*. 1 cel, (**b**) 2 cells, (**c**) 4 cells, (**d**) 8 cells, (**e**) 6 cells, (**f**) 32 cells, (**g**) morula, (**h**) blastula, (**i**) gastrula, (**j**) head formed, (**k**) eye formed, (**l**) heartbeat, (**m**) moving tail, (**n**) hatch, and (**o**) larvae day 0. (Source: Lante et al. (2012b))

Within 16 h and 30 min after spawning, the embryo starts to break through the egg's membrane using its head. At this stage, the yolk sac is fully formed, and the larvae begin to hatch completely. The entire process from the formation of two cells to the larval stage (H0) takes 16 h and 45 min, as illustrated in Fig. 25.7.

25.6 Rabbitfish Seed Production

Seed production of rabbitfish can be conducted in tanks with a volume of 0.5 to 3 tons, utilizing grayish or yellow colors. These colors promote good survival and growth for barred fish larvae. The grayish color reduces predation, as the reflection of the tank's paint helps larvae locate their food more easily. Similarly, yellow provides a contrasting background, making it easier for fish larvae to find and consume the provided feed (Lante & Palinggi, 2009). Seawater is filtered using a membrane filter, and the density of stocked larvae ranges from 10 to 20 individuals per liter (Lante et al., 2016). Newly hatched larvae (H-0) measure approximately 1.7 mm in length and possess a clear yolk sac, which serves as their energy reserve until they begin to feed externally. At 1 day old, the larvae grow to 2.0 mm, and the yolk sac starts to diminish. At 2 days old, the larvae measure 2.1 mm, with the yolk almost gone, and their pectoral and tail fins become more distinct, while gills begin to appear, and the digestive tract starts to form. By 3 days old, the larvae reach a length of 2.2 mm and begin to utilize external food, as indicated by the presence of natural feed in their digestive tract (Lante et al., 2016).

The initial feed provided to larvae aged 1 to 7 days consists of a combination of Chlorella, oyster trochophores, and SS-type rotifers. Administering oyster eggs is crucial for enhancing the survival of larvae during this critical period (Lante et al.,

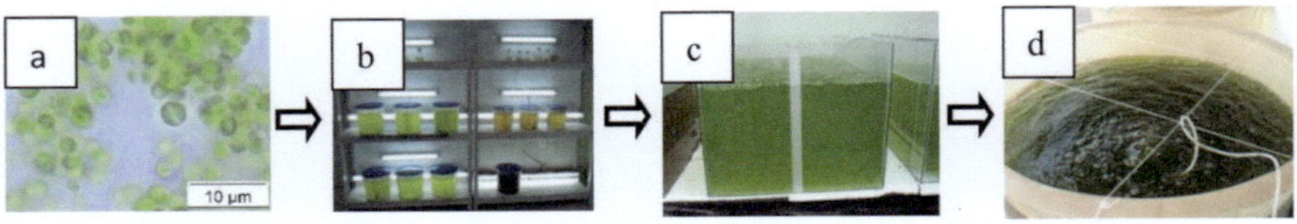

Fig. 25.8 Chlorella culture (*Nannochloropsis occulata*). (Source: Lante et al. (2016))

2007). Therefore, sourcing live oysters for oyster eggs or trochophores is essential. Additionally, oyster trochophore filtrates are provided to the fish larvae. For larvae aged 8 to 24 days, the feed consists of a combination of Chlorella and S-type rotifers, while larvae aged 25 to 35 days receive a mix of Chlorella, rotifers, and Artemia. The density of rotifers is maintained at approximately 20 individuals per milliliter, while nauplii of Artemia are provided at a rate of 2 to 3 individuals per larva. Artemia culture is conducted in conical containers equipped with aeration, with hatching occurring at a salinity of 25 to 35 ppt. Chlorella is administered in the morning and evening to ensure the quality of rotifers in the larval rearing troughs. This addition of Chlorella also enhances the nutritional content of the rotifers, thereby improving larval performance. During the hatchery phase, the primary parasite infecting rabbitfish fry is *Pseudocaligus* sp. This parasite attaches to the entire surface of the fry's body during the copepodid phase (Fig. 25.8). Caligidae parasites can affect various organs, including fins, heads, scales, tails, and gills. *Pseudocaligus* sp. is large enough to be observed without a microscope. Treatment for larvae affected by parasites is straightforward; soaking the fish in freshwater for 15 min is effective for removing parasites from the larvae's body (Lante et al., 2016).

Rotifer culture is highly dependent on the provision of Chlorella as the main feed for rotifers, which supports the maintenance of fish larvae. Rotifer culture can be conducted in a 4 ton tank or as needed. Once the density reaches 100 to 150 individuals per milliliter, the rotifers are harvested for feeding fish larvae. SS-type rotifers, which measure 100 μm, are suitable for maintaining small fish larvae and should be used from the time the larvae hatch until they are 7 days old. After this period, rotifers measuring 120 μm can be introduced. The ratio of the volume of Chlorella, rotifers, and larvae should be 5:3:1. The provision of Artemia can be implemented when the stock of rotifers is limited. The number of hatched Artemia depends on the number of larvae being reared. Hatching is performed in a conical container made of fiberglass or plastic (see Fig. 9a). The oxygen requirement for Artemia hatching is at least 3 ppm, with temperatures ranging from 28 to 30 °C, a pH of 8.0 to 8.5, and an Artemia density of 10 g/L. Harvesting involves closing the aeration faucet, wrapping the Artemia culture tub with black plastic, and placing a light source (such as a flashlight) at one corner or gap of the tub. Nauplii of Artemia will gather in the brighter area created by the light. By slightly opening the faucet of the Artemia culture, the nauplii can be collected slowly and are then ready to be used as live feed for the larvae of rabbitfish. The Artemia hatching process lasts from the morning until the following morning, when the Artemia are fully hatched (Fig. 25.9).

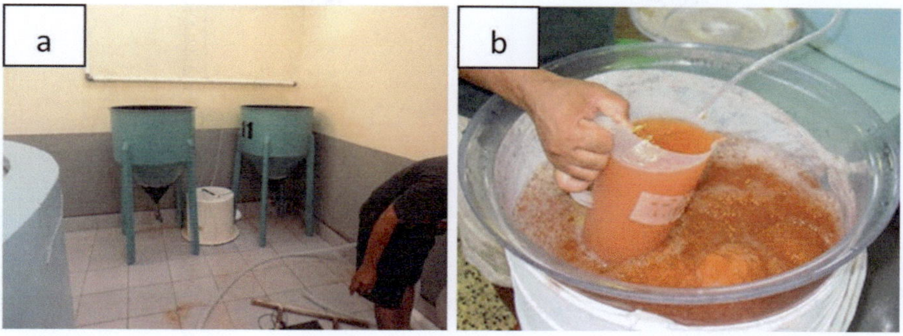

Fig. 25.9 Artemia cultures in controlled baths (**a**) and artemia larvae (**b**). (Source: Lante et al. (2016))

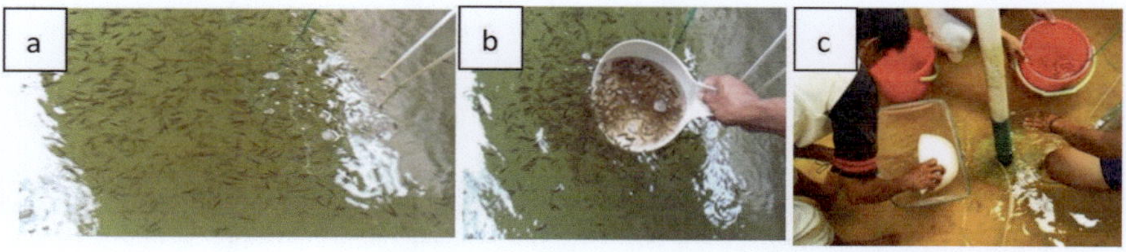

Fig. 25.10 Siphon larval fecal (**a**), larvae aged 45 days (**b**), and harvest larvae aged 50 days (**c**). (Source: Lante et al. (2016))

A 10% water change in the larval rearing trough is initiated when the larvae reach 10 days old. Subsequently, water is added to restore the original water level using a filter bag to reduce stress in the fish larvae. After the larvae reach 15 to 20 days of age, water changes are increased to 15% to 20% per day, and further increased to 20% to 30% per day for larvae aged 21 to 35 days. Additionally, the removal of feces and leftover feed from the larval maintenance tank is conducted every three days after the larvae are 15 days old. From 21 to 35 days of age, this cleaning process occurs daily until the larvae reach 50 days old (Fig. 25.10).

The results of maintaining rabbitfish larvae at a density of 20,000 to 40,000 individuals per tank for 50 days indicate a production of 2000 to 4000 larvae per tank, resulting in a survival rate of 10% (Lante et al., 2016). During the larval rearing period, water quality parameters are maintained within the following ranges: salinity of 31 to 33 ppt, temperature of 28 °C to 30 °C, pH of 7.5 to 8.5, and dissolved oxygen levels above 5 ppm. The collected seeds are then counted and transferred to tanks for cultivation, utilizing a volume of 1 to 2 tons or floating net cages in the sea. The development of rabbitfish fry occurs in these floating net sea cages (Fig. 25.11).

The maintenance of rabbitfish fry aged 50 days in floating net cages involves feeding them commercial pellet feed measuring 100 to 150 μm, with a crude protein content of 28%, crude fat of 18%, crude fiber of 1%, and ash content of 7%. A significant challenge in fish farming technology is the difficulty in producing stable

Fig. 25.11 Development of rabbitfish fry (*a* = 45 days), (*b* = 50 days), and (*c* = 115 days). (Source: Lante et al. (2007))

SS-type rotifers that correspond to the larval mouth opening (less than 100 μm). Continuous provision of SS-type rotifer feed can enhance the production of juvenile rabbitfish in the hatchery, positively affecting the future development of rabbitfish farming in floating nets in the sea in Indonesia.

References

Cabanilla-Legaspi, M. I. C., Traifalgar, R. F. M., de Jesus-Ayson, E. G. T., Andrino-Felarca, K. G. S., & Mamauag, R. E. P. (2021). Growth, metamorphosis and survival of orange-spotted rabbitfish (*Siganus guttatus* larvae fed sodium iodide-supplemented brine shrimp (*Artemia* sp.)). *Aquaculture, 536*, 1–17. https://doi.org/10.1016/j.aquaculture.2021.736443

Darsiani, Setiawati, M., Jusadi, D., Suprayudi, M. A., & Laining, A. (2024). Influence of rotifer enrichment with taurine on larval eye development and growth performance of golden rabbitfish (*Siganus guttatus*). *Biotropia, 31*(2), 217–227. https://doi.org/10.11598/btb.2024.31.2.2057

Duray, M. N. (1998). *Biology and culture of siganids*. Aquaculture Department Southeast Asian Fisheries Development.

FAO. (2023). Keragaman fenotipe truss morfometrik populasi ikan beronang *Siganus guttatus* (bloch, 1787) asal perairan Barru, Lampung, dan Sorong [Phenotypic diversity of morphometric truss populations of *Siganus guttatus* (Bloch, 1787) from the waters of Barru, Lampung, and Sorong].

Hara, S., Kohno, H., & Taki, Y. (1986). Spawning behavior and early life history of the rabbitfish, *Siganus guttatus*, in the laboratory. *Aquaculture, 59*(3–4), 273–285. https://doi.org/10.1016/0044-8486(86)90009-8

Herlinah, H., Lante, S., Tenriulo, A., & Nawang, A. (2020). Keragaman fenotipe truss morfometrik populasi ikan beronang *Siganus guttatus* (bloch, 1787) asal perairan Barru, Lampung, dan Sorong [Phenotypic diversity of morphometric truss populations of *Siganus guttatus* (Bloch, 1787) from the waters of Barru, Lampung, and Sorong]. *Media Akuakultur, 15*(129), 61–70.

Jaikumar, M. (2012). A review on biology and aquaculture potential of a rabbit fish in Tamilnadu (*Siganus canaliculatus*). *International Journal of Plant, Animal and Environmental Sciences, 2*(2), 57–64.

Kamaruddin, Usman, & Lante, S. (2015). *Digestive enzyme activity of protease, a-amylase, and lipase of rabbitfish (Siganus guttatus) larvae* (pp. 401–407). Forum Inovasi Teknologi Akuakultur.

Laining, A., Trismawanti, I., Usman, U., & Masruri, M. H. (2019). Influence of squid liver meal in maturation diet on gonadosomatic index and gonadal amino acid content of golden spotted rabbitfish, Siganus guttatus. *Indonesian Aquaculture Journal, 14*(1), 31–38. https://doi.org/10.15578/iaj.14.1.2019.31-38

Lam, T. J. (1974). Siganids: Their biology and mariculture potential. *Aquaculture, 3*(4), 325–354. https://doi.org/10.1016/0044-8486(74)90001-5

Lante, S. (2010). *Analisis keragaman genetik populasi ikan beronang (Siganus guttatus) di Selat Makassar dan Teluk Bone [Analysis of genetic diversity of rabbitfish (Siganus guttatus) populations in the Makassar Strait and Bone Bay]*. Universitas Hasanuddin.

Lante, S., & Palinggi, N. N. (2009). Pemeliharaan larva ikan beronang (*Siganus guttatus*) dengan nuansa warna wadah yang berbeda [Maintenance of rabbitfish larvae (*Siganus guttatus*) with different container color shades]. *Prosiding Forum Inovasi Teknologi Akuakultur, 2009*, 203–208.

Lante, S., & Palinggi, N. N. (2010). Pematangan gonad dan pemijahan induk beronang (*Siganus guttatus*) dengan rasio jantan dan betina yang berbeda [Gonad maturation and spawning of rabbitfish broodstock (*Siganus guttatus*) with different male-to-female ratios]. *Prosiding Forum Inovasi Teknologi Akuakultur, 2000*, 205–210.

Lante, S., Usman, & Rachmansyah. (2007). Pemijahan dan pemeliharaan larva ikan baronang (*Siganus guttatus*) [Spawning and maintenance of rabbitfish larvae (*Siganus guttatus*)]. *Media Akuakultur, 2*(2), 57–61.

Lante, S., Tenriulo, A., Parenrengi, A., Rachmansyah, R., & Malina, A. C. (2011). Keragaman genetik populasi ikan beronang (*Siganus guttatus*) di Selat Makassar dan Teluk bone menggunakan metode random amplified polymorphic DNA (RAPD) [Genetic diversity of the population of rabbitfish (*Siganus guttatus*) in the Makassar Strait and Bone Bay using the random amplified polymorphic DNA (RAPD) method]. *Jurnal Riset Akuakultur, 6*(2), 211–224. https://doi.org/10.15578/jra.6.2.2011.211-224

Lante, S., Tenriulo, A., & Palinggi, N. N. (2012b). Variasi genetik ikan beronang (*Siganus guttatus*) asal perairan Barru, Lampung, dan Sorong menggunakan penanda RAPD (random amplified Polymorfism DNA) [Genetic variation of rabbitfish (*Siganus guttatus*) from the waters of Barru, Lampung, and Sorong using the marker RAPD (random amplified polymorphism DNA)]. *Jurnal Riset Akuakultur, 7*(2), 195–204. https://doi.org/10.15578/jra.7.2.2012.195-204

Lante, S., Usman, U., Palinggi, N. N., & Santiajinata, W. (2015). Petunjuk teknis pembenihan ikan beronang *Siganus guttatus* [Technical instructions for hatchery of rabbitfish (*Siganus guttatus*)].

Lante, S., Usman, Palinggi, N. N., & Santiadjinata, W. (2016). Pembenihan ikan baronang *Siganus guttatus* [Spawning of rabbitfish (*Siganus guttatus*)]. In *Petunjuk Teknis* (p. 36). Badan Penelitian dan Pengembangan Kelautan dan Perikanan.

Mejos, J. S., Cotamora, G. D., Vidal, P. M., Baguio-Gonzales, A., Tito, O. D., & Alejos, M. S. (2023). Determination of microplastic accumulation and its effect on rabbitfish (*Siganus guttatus*) and Parrotfish (*Scarus ghobban*): A case study from the local market of Zamboanga City, Philippines. *Advanced Research in Biological Science, 5*(59), 86–95. https://doi.org/10.9734/bpi/arbs/v5/1911G

Parawansa, B. S., Ali, S. A., Nessa, N., Rappe, R. A., & Indar, Y. N. (2020). Biological analysis of adult rabbitfish (*Siganus guttatus* bloch, 1787) in seagrass and coral reef ecosystems at Laikang bay, Takalar regency. *IOP Conference Series: Earth and Environmental Science, 473*(1), 1–17. https://doi.org/10.1088/1755-1315/473/1/012006

Soewardi, K. (2007). *Pengelolaan keragaman genetik sumberdaya perikanan dan kelautan.* Management of genetic diversity of fishery and marine resources.

Syah, R., Makmur, Tampangallo, B. R., Undu, M. C., Asaad, A. I. J., & Laining, A. (2020). Rabbitfish (*Siganus guttatus*) culture in floating net cage with different stocking densities. *IOP Conference Series: Earth and Environmental Science, 564*(1), 1–11. https://doi.org/10.1088/1755-1315/564/1/012022

Usman, U., Kamaruddin, K., Laining, A., & Lante, S. (2021). Performansi reproduksi induk ikan baronang, *Siganus guttatus* yang diberi pakan mengandung rumput laut [Reproductive performance of the rabbitfish broodstock, *Siganus guttatus* fed with seaweed]. *Jurnal Riset Akuakultur, 16*(1), 19–30. https://doi.org/10.15578/jra.16.1.2021.19-30

Domestication and Recent Status of Indonesian Abalone

26

I. Gusti Ngurah Permana, Ngurah Sedana Yasa, and Ibnu Rusdi

26.1 Introduction

Abalone is a shellfish of high economic value, selling for US$10–12 per kg when alive. Additionally, abalone meat is highly nutritious, containing 71.99% protein, 3.20% fat, 5.6% fiber, and 11.11% ash. The abalone shells are also aesthetic, making them suitable for jewelry, button making, and other handicrafts (Romimohtarto & Juawana, 2001). Seven species of abalone are found in Indonesia: *Haliotis asinina*, *H. varia*, *H. squamata*, *H. ovina*, *H. glabra*, *H. planata*, and *H. crebrisculpta* (Dharma, 1988; Setyono, 2004). In this domestication effort, the species developed is *H. squamata*, commonly known as the Tokobushi or yellow-footed abalone, marketed under the trade name "abalone" (Setyono, 2004). Images of *H. squamata* abalone are presented in Fig. 26.1.

The potential for abalone is huge in several Indonesian waters, especially in rocky areas, such as NTB, East Java, Bali, and Bajo, South Sulawesi (Priyambodo et al., 2005). Abalones are endemic; in the larval stage, they eat benthic diatoms, while adults eat seaweed. Abalone production in Indonesia is mainly obtained from catches, giving rise to fears of uncontrolled exploitation. This abalone has a prolonged growth, so it will cause scarcity, which will result in extinction. Therefore, it is necessary to make a domestic effort to raise abalones in a controlled artificial environment and turn them into a commercial commodity that can be commercialized into a profitable business.

I. G. N. Permana (✉) · N. S. Yasa
Research Center for Fishery, National Research and Innovation Agency, Cibinong, Indonesia
e-mail: igus022@brin.go.id

I. Rusdi
Research Center for Marine and Land Bioindustry, National Research and Innovation Agency, Jakarta, Indonesia

E. Nugroho et al. (eds.), *Aquaculture Development in Archipelago*,
https://doi.org/10.1007/978-981-95-1084-9_26

Fig. 26.1 Abalone, *Haliotis squamata*. (Source: Personal documentation)

Domestication of abalone in Indonesia has been carried out at the Gondol Research Institute for Mariculture starting in 2007 by collecting broodstock from wild catches by fishermen/collectors in Pekutatan Village, Jembrana Regency, Bali and Lebak Regency, Banten Province. Prospective abalone broodstock from Bali and Banten were transported to the BBPPBL hatchery, Gondol, using the dry method (Rusdi et al., 2009; Susanto et al., 2008). In addition, abalone seeds have been mass-produced in hatcheries, with 2-month-old seed survival fluctuations ranging from 3.35% to 18.4% (Susanto et al., 2009). It is hoped that this paper on Domestication and the Recent Status of Indonesian Abalone can provide information about the resources of Indonesian mussels and increase their numbers. The potential for abalone farming is significant in several Indonesian waters, particularly in rocky areas such as NTB, East Java, Bali, and Bajo in South Sulawesi (Priyambodo et al., 2005). Abalones are endemic species; they feed on benthic diatoms during their larval stage, while adults primarily consume seaweed. Most abalone production in Indonesia is derived from wild catches, raising concerns about uncontrolled exploitation. Their slow growth rate can lead to scarcity and potential extinction. Therefore, domestication efforts are essential for cultivating abalones in a controlled artificial environment, transforming them into a commercial commodity that can be developed into a profitable business.

Domestication of abalone in Indonesia has been conducted at the Gondol Research Institute for Mariculture since 2007. This initiative began with broodstock collection from wild catches by fishermen and collectors in Pekutatan Village, Jembrana Regency, Bali, and Lebak Regency, Banten Province. Prospective abalone broodstock from Bali and Banten were transported to the BBPPBL hatchery in Gondol using a dry method (Rusdi et al., 2009; Susanto et al., 2008). Additionally, abalone seeds have been successfully mass-produced in hatcheries, with survival rates for 2-month-old seeds fluctuating between 3.35% and 18.4% (Rusdi et al., 2009). It is hoped that this paper on the domestication and recent status of Indonesian abalone will provide valuable information about the resources of Indonesian mussels and enhance the potential for cultivating abalones, utilizing the domestication technology that has been developed.

26.2 Taxonomy and Morphology

The following is an abalone taxonomy (Setyono, 2009; Stang & David, 2012):

Kingdom:	Animalia
Phylum:	Mollusks
Class:	Gastropods
Sub Class:	Prosobranchia
Ordo:	Archeogastropoda
Family:	Haloimide
Genus:	*Haliotis*
Spesies:	*Haliotis squamata, Haliotis asinina*

Abalone has a robust and single shell and has adapted well to harsh environmental conditions. The shells are almost rounded in shape and feature a series of holes, typically numbering four to seven. These holes are neatly arranged from the shell's left front to the left rear. The number of holes correlates with the abalone size; larger abalones tend to have more holes. Abalone possesses eyes, a mouth, and a sizable exhaling tentacle. Within their mouth is a radula, a grated tongue that scrapes algae into digestible sizes. Abalone breathes by absorbing water beneath their hollow shell, allowing oxygenated water to pass through the gills before being expelled through the holes. Although abalone do not have a well-defined brain structure, they are considered primitive animals. They also possess a liver located on the upper side. Blood circulates through main arteries, vessels, and sinuses (Baldwin et al., 2007). The anatomy and morphology of the abalone can be seen in Fig. 26.2.

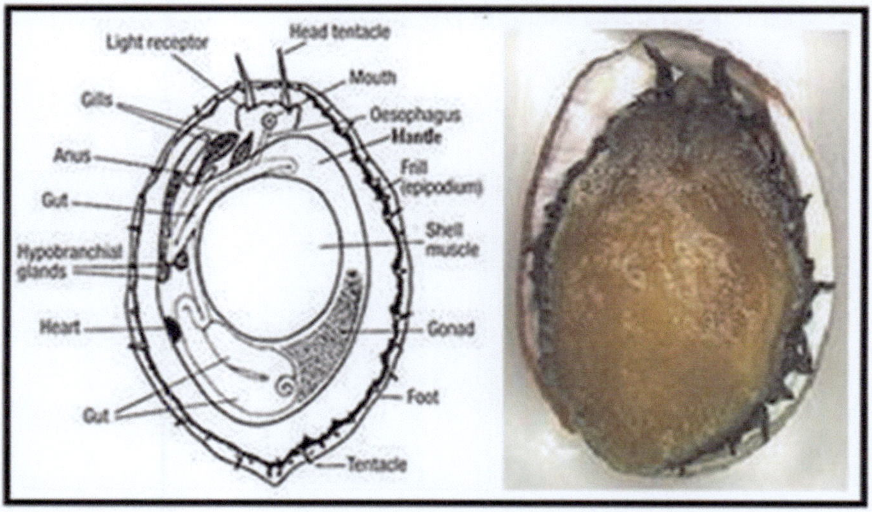

Fig. 26.2 Anatomy of abalone (*H. squamata*). (Source: Personal documentation)

26.3 Abalone Distribution

Abalone is found in the craggy coastal waters of the open sea, ranging from the edge of shallow coastal waters to a depth of 2 m. They are typically located behind corals or rocks with rough, dark surfaces. Female abalones are more accessible to catch when resting, allowing fishermen to use flat-ended iron tools for removal. Seven species of abalone are found in Indonesia: *Haliotis asinina, H. varia, H. squamosa, H. ovina, H. glabra, H. planata,* and *H. crebrisculpta.* The largest species, *H. asinina*, can reach up to 12 cm and is distributed throughout the Indo-Pacific, including Indonesian waters such as Lombok, Sumbawa, Sulawesi, Maluku, and Papua (Romimohtarto & Juawana, 2001).

26.4 Reproduction

To distinguish between male and female abalones, one can observe the color of the gonads located under the shell. Male gonads are yellowish, while female gonads are grayish blue. The maturation tank uses a circulation system with 200 × 50 × 50 cm dimensions and a water discharge of 5–6 L/min (see Fig. 26.3). The density of abalone broodstock is maintained at 15 individuals per basket, with males and females kept separately. For gonad maturation, the feed consists of the seaweed *Gracilaria* sp. and *Ulva* sp., provided at a rate of 10–15% of biomass per day, every 2 days.

The development of *H. squamata* gonads from TKG-1 to TKG-3 requires at least 60 days of maintenance. The percentage of abalone broodstock experiencing gonadal development is higher when fed a combination of *Gracilaria* sp., *Ulva* sp., and *Sargassum* sp. (in a 1:1:1 ratio) compared to single species feeding (Rusdi et al., 2009). The development of abalone gonads was carried out by following the methods as follows (Singhagraiwan & Doi, 1992a):

Level 0 = Undeveloped gonads.
Level 1 = The gonads slightly cover the hepatopancreas section.

Fig. 26.3 Abalone *H. squamata* broodstock maintenance container. (Source: Personal documentation)

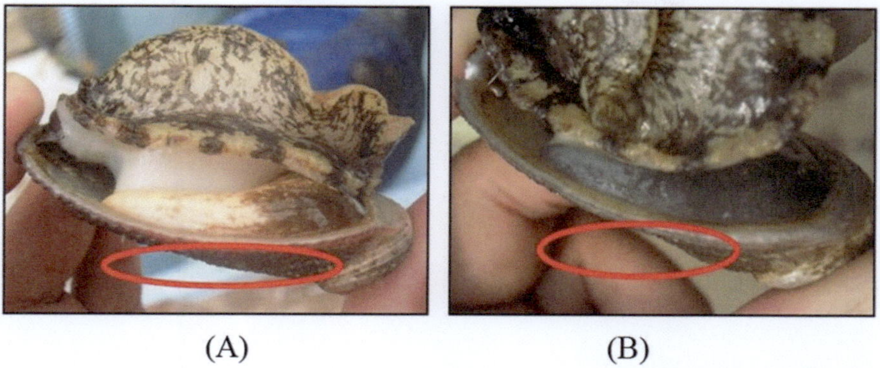

Fig. 26.4 Abalone broods of males (a) and females (**b**) mature gonads. (Source: Personal documentation)

Level 2 = Gonads, about 25% cover part of the hepatopancreas.
Level 3 = The gonads, about 50%, cover part of the hepatopancreas.

The maturity of the gonads on abalone will look bulging and blunt at the ends of the hepatopancreas (Fig. 26.4).

Abalone spawning is conducted using the dry-up method. Spawning occurs with the addition of pure oxygen through aeration for 3 to 5 h in a spawning tank (Fig. 26.5). After this, the eggs are fertilized, after which the eggs are siphoned and thoroughly washed before being placed in a hatching trough. The duration for egg hatching is approximately 6 ± 1 h (Table 26.1).

26.5 Domestication

Broodstock collection is conducted during the full moon and new moon phases, as these times coincide with the lowest tides, facilitating easier capture by fishermen. An excellent female abalone exhibits an intact and non-cracked shell, is free from scars, displays agile movements, and has a body length ranging from 3.5 to 6 cm, along with mature gonads (Singhagraiwan & Doi, 1992b).

26.6 Morphometric

The abalone has a single shell, or mono valve, that covers almost its entire body. The shells are oval-shaped, with an axis extending from the anterior (front) to the posterior (back). Although abalone shells are spiral, they are flattened and do not form cones. The head is positioned at the anterior end, while the apex of the spiral is located at the posterior end on the right side. The shell's outer surface is somewhat rough, while the interior is generally smooth and resembles a layer of pearls, with

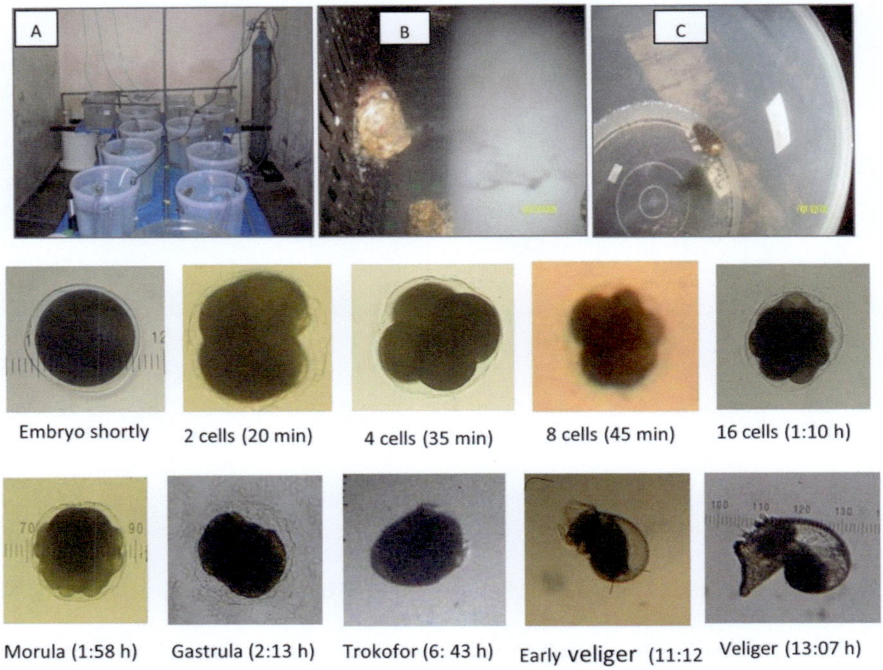

Fig. 26.5 Spawning stimuli of abalone in containers placed in dark conditions (**a**); male abalone release sperm (**b**) and female abalone release eggs (**c**). (Source: Personal documentation)

Table 26.1 Observation of reproductive aspects in each generation

Parameter	G0	G1
Fecundity (grain)	1.025.646 ± 300.231	1.326.300 ± 336.432
Fertilization rate/FR (%)	75,42 ± 9,65	84,66 ± 15,62
Hatching rate/HR (%)	42,30 ± 21,95	61,43 ± 7,86
Egg diameter (μm)	181,25 ± 0,96	183,67 ± 2,12
Spawning season	Depending on the season	Season-independent
Reforestation (days)	Unknown age	90
Productive period (years)	Unknown age	2

some species displaying vibrant colors. The results of morphological observations are presented in Table 26.2.

26.7 Growth

Abalones grow from larvae to adults (Table 26.3). Egg development begins with the division of the embryo into two cells 20 to 30 min after fertilization, followed by the development of larvae (trochophores) that swim at the surface 5 to 6 h later. The

Table 26.2 Morphological characters of abalone parent from Bali collection ($n = 30$)

Morphological characters	Description
Weight (g)	53,32 ± 6,12
Shell length (mm)	67,46 ± 1,58
Shell width (mm)	41,93 ± 1,47
Shell color	Reddish brown
Shell brightness	Dark
Shell texture	Rugged rough
Foot/body muscles	Shell covered
The color of the tentacles	Black
The appearance of cilia	Not visible
Muscle/flesh color	Yellowish black

Table 26.3 Description of the growth of larvae, seeds, and mature brood abalone gonads of Balinese origin

No.	Abalone	Age (day after hatching)	Description		Information
			Length (mm)	Weight (g)	
1	Spat	3–4	0,30 ± 0,05	–	Start sticking
2	Spat	10	0,89 ± 0,01	–	Survival 24.52%
3	Spat	17–19	1,60 ± 0,10	–	The first respiration hole
4	Spat	20	1,71 ± 0,29	–	Respiration hole 1–2
5	Spat	30	3,39 ± 0,64	–	Respiration pit 2–4 Survival 6.74%
6	Spat	40	3,91 ± 0,72	–	Respiration holes 3–5 able to optimally digest diatoms in the plate
7	Spat	50	4,12 ± 1,05	–	Respiration pit 4–7
8	Seed	60	7,18 ± 0,16	–	Respiration pit 4–7 Macroalgae feed
9	Male mature gonad	760	66,73 ± 1,67	50,94 ± 6,26	The first mature gonad
10	Female mature gonad	760	68,19 ± 1,10	55,69 ± 5,14	The first mature gonad

initial stocking of trochophores measured 196.39 ± 14.93 μm. Trochophores possess planktonic characteristics that allow them to float on the water's surface; therefore, natural floating feeds such as *Chaetoceros* sp. and *Isochrysis* sp. are used to support their growth and survival (Susanto et al., 2013).

Abalone larvae typically begin to adhere to the rearing plate between days 3 and 4. During this phase, the larvae transition from a planktonic to a benthic lifestyle, attaching themselves to the plate's surface or the maintenance basin's walls. The success of this adhesion phase largely depends on the larvae's ability to identify a suitable substrate that supports their survival and growth. The formation of the first respiration hole occurs between 17 and 19 days of age, with an average shell length of 1.60 ± 0.10 mm.

Abalone has a lifespan of 45 to 60 days during maintenance. Seed harvesting involves abalone measuring 7.18 ± 0.16 mm, which is conducted in stages, yielding a seed survival rate of 3.35% to 18.4% (Susanto et al., 2013). The abalone seeds from the hatchery are then raised in nurseries using rectangular perforated plastic baskets measuring 35 × 25 × 8 cm^3, with a density of 250 individuals per container. Growth rates in Abalone are relatively slow. For seeds ranging in size from 10.2 ± 0.10 mm to 22.6 ± 0.20 mm, the growth period is approximately 105 days, with a daily shell length increase of 1.84 ± 0.13 μm/day and a daily body weight growth of 17.51 ± 0.13 μg/day, resulting in a survival rate of 98.7% (Rimmer et al., 2013). Juveniles with a shell length of 28.17 ± 1.92 mm and a body weight of 3.79 ± 0.68 g require five months of maintenance to reach a shell length of 45.23 ± 0.96 mm and a body weight of 15.98 ± 0.91 g. During this period, the daily shell length growth is 115.67 ± 0.79 μm/day, and the daily body weight growth is 81.3 ± 0.13 μg/day, achieving a survival rate of 89.65% [13]. Mature abalone reaches a ripe size with a gonadal shell length of 67.46 ± 1.58 mm and a weight of 53.32 ± 6.12 g after 25 months of maintenance (Permana et al., 2012).

26.8 Types of Feed and Eating Habits

26.8.1 Types of Feed

Macroalgae, which serve as feed for abalone, are distributed along the coast of Bali, from Pekutatan in Jembrana Regency to Canggu Beach in Tabanan Regency. Abalone typically inhabits rocky areas along the coastline, mainly where macroalgae are abundant. The types of macroalgae found in these waters include *Ulva* sp., *Gracilaria* sp., and *Sargassum* sp.

26.8.2 Eating Habits

Abalone are herbivorous, plant-eating animals. They are primarily nocturnal, feeding more actively in dark conditions (Setyono, 2004). In their larval stadium, abalone in nature feeds on benthic diatoms, while adult abalone consumes macroalgae, which can be classified into three groups based on color: red algae (*Gracilaria* sp.), brown algae (*Sargassum* sp.), and green algae (*Ulva* sp.).

26.9 Disease Resistance

Challenge tests were conducted to determine the lethal dose (LD50) of *Vibrio harveyi* on 20 abalone samples, with infection treatments applied at densities of 10^5, 10^6, 10^7, and 10^8 cfu/mL over 96 h. An infection density of 1 × 10^7 cfu/mL resulted in 50% mortality among the abalone samples, establishing this density as the LD50. Further challenge tests (G1 and G2) were performed to evaluate the resistance of G1 and G2 groups to the pathogenic bacteria *V. harveyi*.

26.10 Current Status

26.10.1 Seeding

A total of 81,500 veliger larvae are maintained in a larval rearing basin measuring 2 × 3 × 1 m^3. Harvesting abalone spats, measuring 7.18 ± 0.16 mm in shell length, resulted in a survival rate of 18.40% after 60 days of maintenance (Susanto et al., 2009). The harvested seeds are reared in perforated rectangular plastic baskets, each measuring 35 × 25 × 8 cm^3, mounted opposite each other and secured with pieces of PVC pipe. All baskets are then placed in a rectangular tub with a capacity of 5 m^3, designed with a flowing water system providing 1.0–1.2 L/min. Each basket contains abalone seeds with a shell length of 10.14 ± 0.27 mm and a weight of 0.16 ± 0.01 g, at a density of 250 individuals per basket, achieving a survival rate of 98.3%. Thinning is performed, reducing the total density of abalone seeds by 50% on the 30th day of maintenance in the baskets (Soelistyowati et al., 2014).

26.10.2 Enlargement

Abalone enlargement is carried out until they reach prospective parent stock, which can then be re-spawned to obtain seeds. Hatchery seeds have been successfully raised in the sea using various methods and types of feed until they achieve the size of prospective broodstock. The grow-out of abalone in floating cages utilizes hanging basket methods at different depths (Giri et al., 2014; Sutarmat et al., 2009). Floating cages, basket containers, or hanging basins can be used for abalone maintenance. The maintenance containers consist of plastic baskets with holes that measure 47.5 cm in length, 36.5 cm in width, and 12.4 cm in height or hollow plastic basins adapted to field conditions. The abalone maintenance system typically consists of two vertically tied baskets, or two hollow basins hung on floating cages. Each unit (one basket) is placed 2–3 m from sea level. Abalone biomass yields range from 2.5 kg to 3.04 kg per hanging basket container.

References

Baldwin, J., Elias, J. P., Wells, R. M. G., & Donovan, D. A. (2007). Energy metabolism in the tropical abalone, *Haliotis asinina* Linné: Comparisons with temperate abalone species. *Journal of Experimental Marine Biology and Ecology, 342*(2), 213–225. https://doi.org/10.1016/j.jembe.2006.09.005

Dharma, B. (1988). Indonesian Shell I. PT. Graha Facilities.

Giri, I. N. A., Sutarmat, T., Yudha, H. T., Rusdi, I., & Susanto, B. (2014). Grow-out of abalone *Haliotis squamata* in floating cages fed different proportions of seaweed. *Indonesian Aquaculture Journal, 9*(1), 15–21. https://doi.org/10.15578/iaj.9.1.2014.15-21

Permana, G. N., Rusdi, I., Susanto, B., & Khotimah, F. H. (2012). *Haliotis squamata* F2 seed production and hybridization (*H. squamata* and *H. asinina*).

Priyambodo, B.Y., Sofyan, & Jaya I.S. (2005). Abalone mussel seed production (*Haliotis asinina*) in Lombok marine aquaculture workshop. *Annual National Seminar on Fisheries and Marine Research Results* (pp. 144–148).

Rimmer, M. A., Sugama, K., Rakhmawati, D., Rofiq, R., & Habgood, R. H. (2013). A review and SWOT analysis of aquaculture development in Indonesia. *Reviews in Aquaculture, 5*(4), 255–279. https://doi.org/10.1111/raq.12017

Romimohtarto, K., & Juawana, S. (2001). *Marine biology. Science of marine life*. Djambaran.

Rusdi, I., Susanto, B., Rahmawati, R., Ismi, S., Yudha, H. T., & Septory, R. (2009). Improvement of the quality of abalone brood and larvae-juveniles (*Haliotis squamata*) through environmental management.

Setyono, D. E. (2004). Conditioning broodstock of tropical abalone (*Haliotis asinina*) under different photoperiod. *Oseanologi dan Limnologi di Indonesia, 36*, 1–13.

Setyono, D. E. (2009). *Abalon: Biologi dan reproduksi [abalone: Biology and reproduction]*. LIPI Press Shepherd.

Singhagraiwan, T., & Doi, M. (1992a). *Seed production and culture of a tropical abalone Haliotis asinina* (p. 10900). Linne.

Singhagraiwan, T., & Doi, M. (1992b). Spawning pattern and fecundity of the donkey's ear abalone, *Haliotis asinina* Linne, observed in captivity. *Thai Mar Fish Res Bull, 3*, 61–69.

Soelistyowati, D. T., Kusumawardhani, A., & Junior, M. Z. (2014). Phenotype characterization of interspecific hybrid abalone *Haliotis asinina* and *Haliotis squamata* seed. *Jurnal Akuakultur Indonesia, 12*(1), 25–30. https://doi.org/10.19027/jai.12.25-30

Stang, & David. (2012). *Haliotis squamata*. http://www.zipcodezoo.com/Animals/H/Haliotis_squamata

Susanto, B., Rusdi, I., & Buda, M. (2008). Trial of abalone transport (*Haliotis squamata*) living conditions by dry method. *Proceedings of the National Seminar on Biodiversity, II*, 235–238.

Susanto, B., Rusdi, I., Ismi, S., & Rahmawati, R. (2009). Abalone seeding and enlargement (*Haliotis squamata*) at the Center for Marine Aquaculture Research, Gondol Bali.In *Proceedings of the National Seminar on Mollusks 2. Faculty of Fisheries and Marine Affairs* (pp. 149–161).

Susanto, B., Rusdi, I., & Khotimah, F. H. (2013). Test of the technique of nursery and abalone enlargement (*Haliotis squamata*) in society. In *Proceedings of the aquaculture technology innovation forum (FITA) 2013. Field of Sustainable Aquaculture Technology* (pp. 239–247).

Sutarmat, T., Yudha, H. T., & Giri, I. N. A. (2009). *Substrate optimization on abalone enlargement (Haliotis squamata) in floating cages*. Center for Marine Aquaculture Research, Gondol, Bali. Thing.

The Status of Mud Crab *Scylla* spp. Domestication in Indonesia

27

Herlinah Herlinah, Gunarto Gunarto, and Sulaeman Sulaeman

27.1 Introduction

Mud crabs of the genus *Scylla* (*Scylla* spp.) are a highly valuable coastal fishery commodity in Asia-Pacific countries, including Indonesia, Malaysia, Singapore, Thailand, Vietnam, India, China, Japan, and Australia. Based on electrophoresis allozyme fingerprints and mitochondrial DNA (mt-DNA) gene sequencing, *Scylla* consists of four species: the red mangrove crab (*Scylla olivacea*), also known as the "red/orange mud crab"; the green mangrove crab (*Scylla serata*), referred to as the "giant mud crab" due to its potential size of 2–3 kg per individual; the purple mangrove crab (*Scylla tranquebarica*), which can reach sizes of 0.6–1 kg per individual; and the white mud crab (*Scylla paramamosain*) (Keenan & Blackshaw, 1999).

Despite recent advancements, Indonesia's mud crab hatchery technology has seen limited success (Gunarto et al., 2021, 2023). Increasing crab seed production from hatcheries is crucial to address the challenges of limited seed supply for grow-out in brackish water ponds. The market demand for mangrove crab commodities has steadily increased each year, leading to intensified crabbing activities in the wild. This has resulted in a decline in the mud crab population in certain Indonesian waters due to overfishing. Therefore, the development of mud crab aquaculture, particularly through the use of hatchery-produced seeds, is essential. A significant

H. Herlinah (✉)
Faculty of Vocational Studies, Hasanuddin University,
Makassar, Indonesia

Research Center for Fishery, National Research and Innovation Agency, Cibinong, Indonesia
e-mail: hjompa@unhas.ac.id

G. Gunarto · S. Sulaeman
Faculty of Vocational Studies, Hasanuddin University, Makassar, Indonesia,
Makassar, Indonesia

E. Nugroho et al. (eds.), *Aquaculture Development in Archipelago*,
https://doi.org/10.1007/978-981-95-1084-9_27

aspect of advancing mud crab culture is the effective utilization of brackish water ponds.

Domestication can be defined as the process of adapting an organism to a cultivation environment for generations. Seed production technology encompasses various stages, including gonad maturation, spawning, egg incubation, larval maintenance, crablet rearing, and disease control. Successful domestication supports the availability of farmed crab production, helping to mitigate scarcity and ensure the sustainability of wild populations.

From 2011 to 2021, the Research Institute for Brackish Water Aquaculture and Fisheries Extension conducted research on the seed production of several mud crab species in hatchery. The first species to undergo domestication was *Scylla olivacea*, sourced from the estuary waters of the Cenranae River in Bone Regency, South Sulawesi. Subsequently, *Scylla paramamosain* was sourced from Bangil, Pasuruan waters, on the north coast of East Java. Since 2015, mud crab seed production has focused on *Scylla tranquebarica*, obtained from Malili, East Luwu Regency, South Sulawesi.

In the domestication process, *S. tranquebarica* has shown it to be easier to seed, grow more quickly, and attain larger sizes compared to *S. olivacea*. Between 2015 and 2017, domesticated mud crab seeds were stocked in ponds across various regencies, including Maros, Pangkep, Takalar, Barru, Bone, Polewali Mandar, and East Luwu. As of 2018, the fourth-generation (F4) descendants of *Scylla tranquebarica* have been monitored for growth performance. The objectives of mud crab domestication include producing *S. tranquebarica* strains that are both adaptive and suitable for cultivation in brackish water ponds, as well as developing aquaculture technologies that are readily adopted and applied by local communities.

27.2 Taxonomy and Morphology

Mud crab of the genus *Scylla* comprise four species found in Indonesia namely: Red mud crab (*Scylla olivacea*) internationally known as "red orange mud crab," green mud crab (*Scylla serrata*) commonly referred to as the "giant mud crab," purple mud crab (*Scylla tranguebarica*), and white mud crab (*Scylla paramamosain*) (Keenan et al., 1998).

Classification:

Phylum:	Arthropoda
Class:	Crustacea
Ordo:	Decapoda
Familia:	Portunidae
Genus:	*Scylla* (De Haan, 1833
Species:	*Scylla serrata* (Forsskal, 1755), *Scylla tranquebarica* (Fabricius, 1798), *Scylla olivacea* (Herbs, 1796), *Scylla paramamosain* (Estampador, 1949)

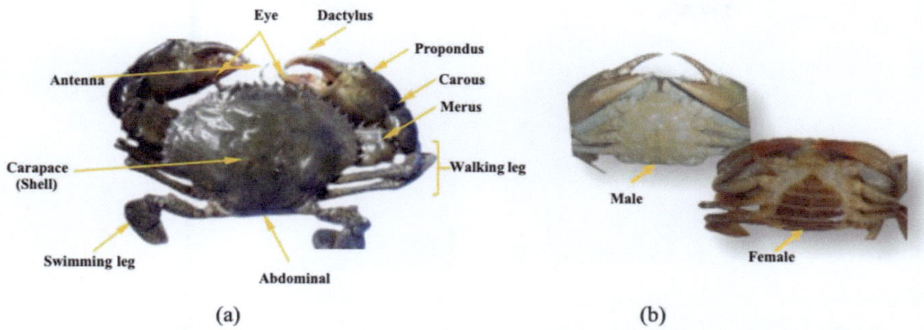

Fig. 27.1 Morphology of mud crabs (**a**) and the abdominal shape of male and females (**b**). (Source: Personal documentation)

27.3 Morphological Characteristics

Mud crabs possess fused heads and bodies, protected by a robust carapace. They have five legs, which include a pair of claws, three pairs of walking legs, and one pair of swimming legs. Morphologically, mud crabs display a round carapace shape that is slightly oval and thick, resembling a shield composed of lime (horn substance) to safeguard their internal structures. The crabs feature thorns located at the front of their bodies, with each side having one thorn, alongside four thorns situated between their two eye stalks. The first pair of legs is notably larger and claw-shaped, serving as a means of grasping and defense. The last pair of legs has undergone paddle-shaped modifications, which assist in swimming and maneuvering in the water (see Fig. 27.1a). Sexual dimorphism is evident in mud crabs, particularly in the shape of the abdomen. Females exhibit a rounded abdominal shape, while males have a slimmer, tapered abdomen. This distinction begins to manifest when the carapace width reaches 20–31 mm. Below this size, the structures are similar, making it challenging to differentiate between the sexes (see Fig. 27.1b).

The color of the mud crab carapace is influenced by environmental conditions. Crabs with a blackish-green base are typically found in open water, while those with a brownish-red coloration are usually caught in burrows within mangrove areas (Fig. 27.2).

27.4 Distribution

Mud crabs of the genus *Scylla* exhibit a wide distribution along the Indo-Western Pacific, with the distribution of each species defined according to Keenan et al. (1998) (Table 27.1). *Scylla serrata* (synonym: *S. oceanica* Estampador) has the most extensive distribution, spanning from the Western Indian Ocean to the South Pacific Islands. *Scylla tranquebarica* (formerly *S. serrata* var. *paramamosain* Estampador) and *Scylla olivacea* (formerly *S. serrata* Estampador) have a more focused distribution, primarily in the South China Sea, extending into the Indian

Fig. 27.2 Differences color of claws in some species of mud crabs. (Source: Personal documentation)

Table 27.1 Distribution areas of *Scylla* spp. species worldwide

Species	Distribution area	Habitat
S. serrata	Indo-West Pacific: South Africa, Red Sea, Australia, the Philippines, Indonesia. Pacific Islands, Taiwan, Japan	Associated with mangrove forests that are submerged by seawater throughout the year. Can tolerate low salinity conditions
S. paramamosain	South China Sea: Cambodia, Vietnam, Singapore, China, Taiwan, Hong Kong. Java Sea: Kalimantan, Central Java	Found in various habitats, including shallow coral rubble areas, shallow subtidal plains, tidal inundation zones around estuaries, and mangrove forests
S. olivacea	Indian Ocean: From Pakistan to Western Australia. South China Sea: Thailand, Indonesia, Singapore, Vietnam, Sarawak to southern China. Pacific Ocean: the Philippines, East Timor, Gulf of Carpentaria	Commonly found in mangrove forests and coastal areas influenced by brackish water with low salinity during the rainy season
S. tranquebarica	Indian Ocean: From Pakistan to Malaysia, Indonesia. South China Sea: Sarawak, Singapore. Pacific Ocean: the Philippines	Associated with mangrove forests and coastal areas that are submerged in low-salinity brackish water for most of the year

Source: Self Modification from Le Vay (2001)

Ocean and Western Pacific Ocean. In contrast, *Scylla paramamosain* (formerly *S. tranquebarica* Estampador) has a limited distribution, predominantly found in the South China Sea and Java Sea (Le Vay, 2001).

Mud crabs (*Scylla* spp.) are closely associated with mangrove forests and have a widespread distribution throughout Indonesia. According to Sulistiono et al. (1994, 2021), mud crabs inhabit brackish waters and are primarily caught in the coastal areas of Indonesian waters, including Sumatra, Java, Kalimantan, Sulawesi, Maluku, and Irian Jaya. Notable concentrations of mud crab farming areas are linked to

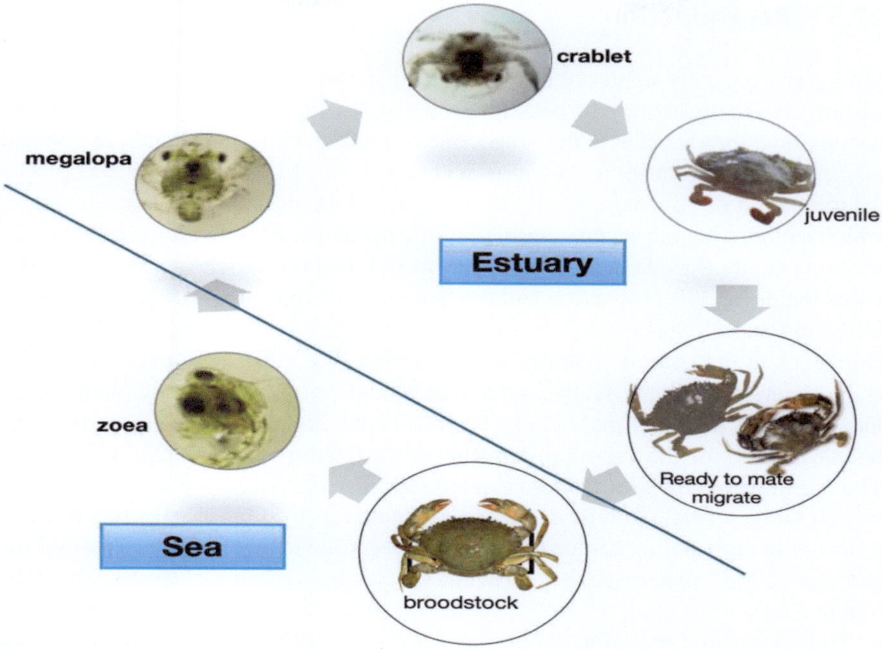

Fig. 27.3 Life cycle of mud crab. (Source: Personal documentation)

favorable habitats, such as southern Java (Cilacap), northern Java (Tanjung Pasir, Pamanukan), western Sumatra (Bengkulu, Riau), eastern Kalimantan (Kota Baru, Pasir, Balikpapan), Sulawesi (Bone Bay, Kolono Bay, Kendari), West Nusa Tenggara (Waworada Bay, Bima), and Irian Jaya (Bintuni Bay, Biak Nimfor). Furthermore, Sulaeman and Nevdal (2000) and Sulaeman and Watanabe (2002) identified *Scylla olivacea*, *Scylla serrata*, and *Scylla tranquebarica* in the Makassar Strait and Bone Bay waters of Indonesia.

Mud crabs exhibit migratory behavior, moving from coastal waters to the sea, where broodstock and their offspring return to coastal areas, estuaries, or mangrove forests for shelter and foraging. Crabs that are ready to mate migrate from mangrove waters to the shore and then to deeper waters for spawning. Male crabs that have mated typically return to mangrove forest waters or coastal muddy areas with abundant food sources (Fig. 27.3). Females migrate offshore to lay eggs, which have a diameter of approximately 0.3 mm. Each female can release between two to five million fertilized eggs, which will hatch in 2–4 weeks. During hatching, the female crab stands on her toes and moves her abdominal folds to assist in freeing the zoea. The life cycle then begins anew, with fertilized eggs developing into zoea (Z I, II, III, IV, V), megalopa, and eventually juvenile crabs, which return to beaches or mangrove areas in search of food and safe shelter (Bir, 2016).

27.5 Reproduction

Mud crabs typically mature their gonads and spawn in brackish waters. Gonadal maturity is usually achieved when mud crabs reach an age of 4 to 5 months, although copulation does not occur at this stage. The carapace width of mud crabs at gonadal maturity ranges from 99 to 114 mm, with a weight of approximately 200 g. The determination of gonadal maturity levels is based on morphological development, which includes factors such as color, size, and the extent of gonad fulfillment within the body cavity (Sagala et al., 2020). The size of the broodstock significantly influences the degree of fertilization and the frequency of spawning. Fecundity, defined as the number of eggs produced, correlates with the final body weight of the mud crab stock at the ovarian maturity stage (Tiurlan et al., 2019). Copulation in mud crabs occurs when water temperatures rise. Female mud crabs release pheromones into the water to attract male crabs. Prior to copulation, when the carapace of the female crab is soft, the male crab deposits spermatophores in the female's thelycum. The sperm within the spermatheca can later fertilize the female's eggs upon spawning. Notably, a single mating event can facilitate fertilization of eggs two to three times. Mud crab mating can be conducted in controlled environments, allowing for pairings of the same species or different species between male and female crabs (Table 27.2).

To enhance the reproductive performance of potential mangrove crab (*Scylla olivacea*) broodstock intended for hatchery use, squid is provided as feed during the maintenance phase. According to Herlinah and Septiningsih (2015), the inclusion of fresh squid in the diet of *S. olivacea* broodstock, particularly when reared within a recirculating aquaculture system, significantly improves reproductive outcomes, including egg production and the number of zoeae (larvae) generated. Notably, the

Table 27.2 Gonadal maturity stages of female mud crabs (*Scylla* spp.)

Gonadal maturity stage	Ovarian morphological characteristics
I (proliferation)	The ovary is transparent, covered by a thin peritoneal layer, and has a ribbonlike appearance
II (pre-vitellogenesis)	Ovaries range from pale white to creamy white, with egg granules not distinctly separated. They are about 2–3 mm thick and occupy approximately 1–2% of the body cavity
III (primary vitellogenesis)	The ovaries transition to a pale yellow or light yellow color. Egg granules become visible but remain partially covered by oil glands. Thickness ranges from 3 to 7 mm, occupying about 10–20% of the cavity
IV (secondary vitellogenesis)	Ovaries turn yellow to orange, with a thickness of 7–12 mm, occupying 20–75% of the cavity. Layers become distinguishable, and oil glands are reduced
V (tertiary vitellogenesis)	Individual egg granules are clearly visible. The ovaries range in color from orange-yellow to orange-red, with a thickness of 10–20 mm, occupying more than 75% of the cavity. Some extend beyond the cavity and can be seen from the posterior abdomen

Source: Modification from Herlinah and Septiningsih (2015); Islam et al. (2020)

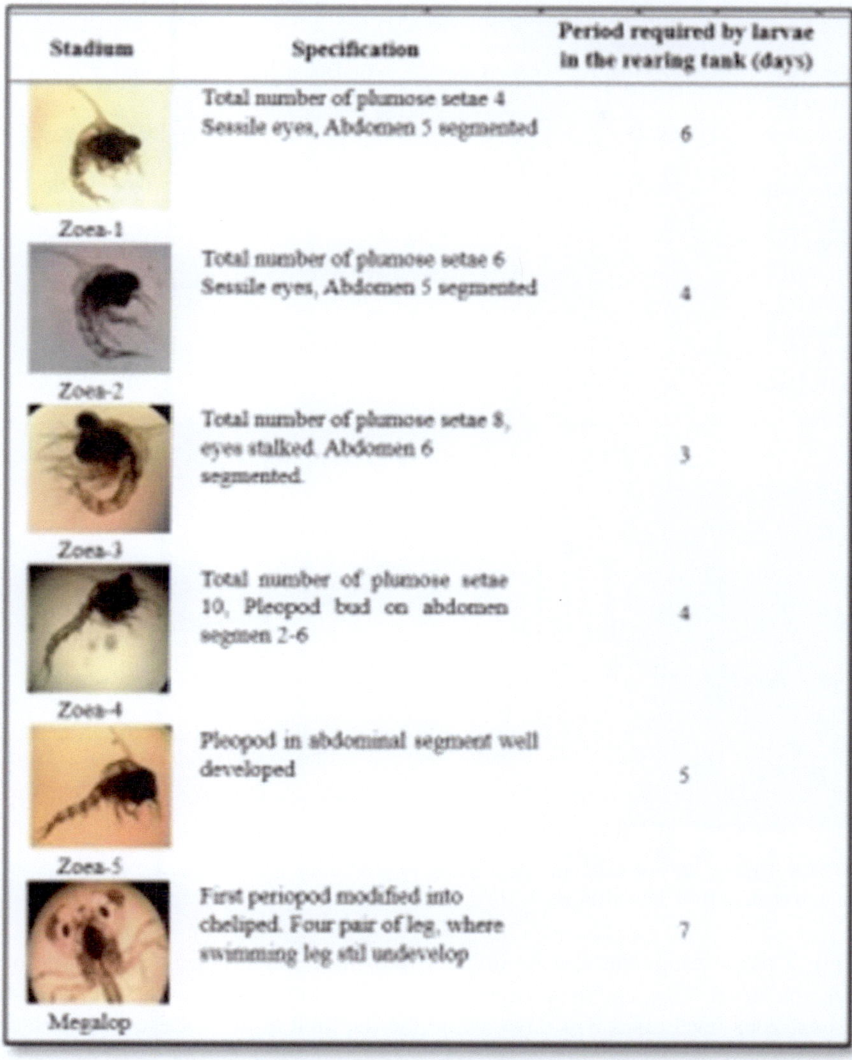

Stadium	Specification	Period required by larvae in the rearing tank (days)
Zoea-1	Total number of plumose setae 4 Sessile eyes, Abdomen 5 segmented	6
Zoea-2	Total number of plumose setae 6 Sessile eyes, Abdomen 5 segmented	4
Zoea-3	Total number of plumose setae 8, eyes stalked. Abdomen 6 segmented.	3
Zoea-4	Total number of plumose setae 10, Pleopod bud on abdomen segmen 2-6	4
Zoea-5	Pleopod in abdominal segment well developed	5
Megalop	First periopod modified into cheliped. Four pair of leg, where swimming leg stil undevelop	7

Fig. 27.4 Larval development of mud crab from zoea-1 stage to megalopa stage. (Gunarto et al., 2016a)

lipid content of squid is substantially lower—by factors of eight and five—compared to trash fish and golden snails, respectively. However, squid exhibits the highest protein content, which is considered a primary factor contributing to the enhanced reproductive performance observed in mud crab broodstock.

The life cycle of mud crabs progresses from egg to adult crab through several developmental stages, including the zoea, megalopa, juvenile crab, and adult crab stages (Fig. 27.4) (Khan, 2017; Ong, 1964; Thirunavukkarasu et al., 2014; Yi et al.,

Table 27.3 Growth rate and production of larvae up to cablet (Gunarto et al., 2014)

Treatment	Larval period Z1–Z5 (day)	Megalopa period (day)	Crablet production (ind./container)	Crablet size (g/ind)
Crablet D-7				
Rotifer with HUFA	18–25	7–15	1213	0,033 ± 0,039
Rotifer with Vit C	18–21	7–12	1120	0,067 ± 0,055
Crablet D-7				Survival (%)
Added HUFA			67,5	13,6
Added probiotic RICA-1			78,0	15,8
No HUFA No RICA			54	10,8

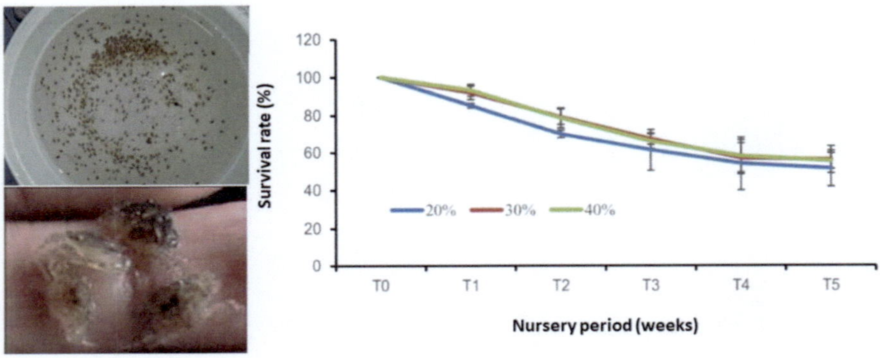

Fig. 27.5 Crablet D-7 of mud crab from hatchery production and the survival in the nursery. (Gunarto et al., 2016b; Usman et al., 2016)

2009). This development can be divided into three main stages: the embryonic stage, the larval stage, and the postlarval stage. The transition from the zoea level to the next zoea level takes 3–4 days, while the duration at the zoea level is at least 18 days.

During the development from the zoea level to the megalopa level, five molts occur. The body length at each zoea level after molting is as follows: 1.15 mm (zoea level 1), 1.51 mm (zoea level 2), 1.93 mm (zoea level 3), 2.40 mm (zoea level 4), and 3.45 mm (zoea level 5). Furthermore, the transition from the megalopa level to the young crab level (instar 1) takes 11–12 days at a salinity of 31–32 ppt; if conducted at a salinity of 21–27 ppt, the duration is reduced to 7–8 days. Mangrove crabs exhibit a single growth phase at the megalopa level, with a carapace length of 2.18 mm and a carapace width of 1.52 mm (Table 27.3 and Fig. 27.5).

27.6 Domestication

The effort to domesticate the mud crab, *Scylla* spp., encompasses various aspects, including the production of mud crab spawn, stimulation of spawning, larval rearing, crablet rearing, and grow-out in brackish water ponds. Despite these efforts, the mass production of mud crab seeds remains a significant challenge. Larval mortality is primarily attributed to infections caused by harmful bacteria, such as *Vibrio* spp. (Jithendran et al., 2010; Poornima et al., 2014).

The Research Institute for Brackishwater Aquaculture and Fisheries conducted studies on the efficacy of Elbayou, Rica-1 probiotic, and erythromycin in increasing the survival and development of *S. paramamosain* mud crab larvae. The results indicated that the production of crablets on Day 15 (D15) in the erythromycin treatment was 141.2 ± 34.1 individuals per tank, which was significantly different ($p < 0.05$) from the production of 70.0 individuals per tank in the Elbayou treatment. Consequently, a concentration of 15 mg/L of erythromycin produced the highest crablet yield during the larval rearing phase (Gunarto et al., 2021).

Research on the application of probiotics in mud crab larvae rearing aims to suppress pathogenic bacteria and improve water quality. Probiotic bacteria containing *Bacillus subtilis* and *B. licheniformis* were applied to enhance crablet production. The study revealed that crablet production using the probiotic *B. licheniformis* (280.5 ± 47.50 individuals per tank) was significantly higher ($p < 0.05$) compared to treatments combining *B. licheniformis* and *B. subtilis* (196.5 ± 29.50 individuals per tank) and treatments with only *B. subtilis* (161.0 ± 21.00 individuals per tank) (Gunarto et al., 2023).

Various salinity levels were tested for megalopa rearing to determine the optimal conditions for crablet production of *S. tranquebarica*. The salinity levels ranged from 26 to 30 ppt. Findings indicated that the total bacterial population was significantly lower ($p < 0.05$) at salinity levels of 29–30 ppt compared to 27–28 ppt and 25–26 ppt. Conversely, the population of *Vibrio* spp. was significantly higher at salinity levels of 29–30 ppt compared to the lower salinities. Moreover, megalopa reared at salinities of 27–28 ppt exhibited more total hemocytes ($p < 0.05$) than those at 25–26 ppt and 29–30 ppt, resulting in the highest crablet production at 27–28 ppt (Gunarto et al., 2023, unpublished).

Understanding nursery culture is crucial for effective commercial seed production of *Scylla* spp. When rearing mud crabs, whether indoors or outdoors, it is essential to provide adequate shelter and food for growth and survival. While live feed supplemented with artificial diets, such as micro-bound formulations, can enhance survival rates, Artemia nauplii remain vital, especially during the megalopa stage. The molting interval between the megalopa and first instar (C1) stages is approximately 6–7 days, but this interval increases steadily with subsequent crablet stages (Syafaat et al., 2021).

Crablet rearing also focuses on enhancing survival rates in nurseries. The application of seaweed as shelter has been shown to improve crablet survival rates. An individual system for crablet nurseries, utilizing floating plastic glasses in a raft, resulted in a significantly higher survival rate (95–100%) over one month. However, managing feeding and water quality in mass crablet rearing using this technique requires additional skills.

The grow-out phase of domesticated mud crabs aims to assess growth, survival, and production. Three pond locations Maros, Pangkep in South Sulawesi, and Polewali Mandar (Polman) in West Sulawesi were tested for mud crab growth in brackish water ponds. Crablets of *S. tranquebarica*, weighing between 0.05 and0.1 g each, were stocked at densities of 0.24 ind./m^2 in Pangkep, 0.27 ind./m^2 in Polman, and 0.53 ind./m^2 in Maros. Results indicated an average final weight ranging from 131.05 to 199.50 g per individual after three months (163.17 ± 34.42), with survival rates between 22% and 36.94% (30.41 ± 7.65) and production levels of 79.25 to 272.12 kg/ha (Syafaat & Gunarto, 2018).

A 2019 study also evaluated the growth rates of crablet progeny up to the broodstock stage. A total of 700 fast-growing crablets, with an average carapace width of 15.2 ± 1.29 mm and an average body weight of 0.55 ± 0.15 g, were compared against 700 slow-growing crablets, which had average body weights of 0.14 ± 0.08 g and carapace widths of 8.88 ± 3.11 mm. These crablets were placed separately in four 500 m^2 ponds, with a density of 350 individuals per pond. After three months, the fast-growing crablets reached an average weight of 185.7 ± 38.7 g, which was significantly different ($p < 0.05$) from the slow-growing crablets, which averaged 156.94 ± 15.30 g. During the fourth month of culture, female crabs with mature gonads were obtained from the fast-growing crablets (Gunarto et al., 2019).

References

Bir, J. (2016). Mud crab' s (*Scylla serrata*) reproduction and breeding technique course. *Basque*. https://doi.org/10.13140/RG.2.2.29718.70723

Gunarto, G., Parenrengi, A., & Septiningsih, E. (2016a). Crablet of mud crab *Scylla olivacea* production from the different stages of larvae fed Artemianauplii enriched using *Nannochloropsis* sp. *International Journal of Agriculture System, 4*(2), 132. https://doi.org/10.20956/ijas.v4i2.687

Gunarto, G., Syafaat, M. N., Herlinah, H., Andi, P., & Mustafa, A. (2016b). *Aspek biologi dan teknik produksi benih kepiting bakau Scylla spp. [Biological aspect and seeds production techniques of Scylla spp.]*. AMaFRaD PRESS Badan Penelitian dan Pengembangan Kelautan dan Perikanan KKP..

Gunarto, G., Syafaat, M. N., & Herlinah, H. (2019). Larva development and growth of mangrove crab, *Scylla tranquebarica* crablet from individual selection for the broodstock candidate. *Aquacultura Indonesiana, 20*(1), 24. https://doi.org/10.21534/ai.v20i1.132

Gunarto, G., Tampangallo, B. R., & Muliani, M. (2021). The application of erythromycin, elbayou, and rica-1 probiotic in the rearing of *Scylla paramamosain* mud crab larvae development into the crablet stage. *Thalassa, 37*, 465–475. https://doi.org/10.1007/s41208-021-00327-y

Gunarto, H., Sulaeman, M., & Tampangalo, B. R. (2023). The effectivity of probiotic containing *Bacillus* spp. applied in the rearing of mud crab *Scylla tranquebarica* larvae to produce crablet. *IOP Conference Series: Earth and Environmental Science, 1137*(1), 1–9. https://doi.org/10.1088/1755-1315/1137/1/012045

Herlinah, H., & Septiningsih, E. (2015). Reproduction performances of mud crab (*Scylla olivacea*) broodstocks with different feeds. *International Journal of Agriculture System (IJAS), 3*(2), 157–168.

Islam, M. J., Slater, M. J., & Kunzmann, A. (2020). What metabolic, osmotic and molecular stress responses tell us about extreme ambient heatwave impacts in fish at low salinities: The case of European seabass, *Dicentrarchus labrax. Science of the Total Environment, 749*, 141458. https://doi.org/10.1016/j.scitotenv.2020.141458

Jithendran, K. P., Poornima, M., Balasubramanian, C. P., & Kulasekarapandian, S. (2010). Diseases of mud crabs (*Scylla* spp.): An overview. *Indian Journal Of Fisheries, 57*(3), 55–63.

Keenan, C., & Blackshaw, A. (1999). Mud crab aquaculture and biology. In A. B. C.P, Keenan (Ed.), *ACIAR Proceedings*.

Keenan, C. P., Davie, P. J. F., & Mann, D. L. (1998). A revision of the genus Scylla de Haan, 1833 (Crustacea: Decapoda: Brachyura: Portunidae). *Raffles Bulletin of Zoology, 46*(1), 217–245.

Khan, M. (2017). *Hatchery-based mud crab production at BFRI Paikgacha Station*, Bangladesh.

Le Vay, L. (2001). Ecology and management of mud crab *Scylla* spp. *Asian Fisheries Science, 14*(2), 101–111. https://doi.org/10.33997/j.afs.2001.14.2.001

Ong, K. S. (1964). The early developmental stages of *Scylla serrata* Forsskal (Crustacea: Portunidae), reared in the laboratory. *Proceedings/Indo-Pacific Fisheries Council, 11*(2), 135–146.

Poornima, M., Singaravel, R., Rajan, S. J., Ramakrishnan, S., Alavandi, S., & Kalaimani, N. (2014). *Vibrio harveyi* infection in mud crabs (*Scylla tranquebarica*) infected with white spot syndrome virus. *International Journal of Research in Biological Sciences Universal Research Publications, 1*(3), 81–93.

Sagala, L. S. S., Idris, M., & Ibrahim, N. (2020). Perbandingan pertumbuhan kepiting bakau (*Scylla serrata*) jantan dan betina pada metode kurungan dasar [Growth comparison of male and female *Scylla serrata* through bottom confine]. *Jurnal Mina Laut Indonesia, 03*(12), 46–54.

Sulaeman, S., & Nevdal, G. (2000). Genetik differentiation of three morphologically distinguishable groups of mangrove crab (Genus: Scylla, Portunidea) in Makassar Strait and Bone Bay, South Sulawesi, Indonesia. *Indonesian Aquaculture Journal, 6*(1), 14–25.

Sulaeman, S., & Watanabe, S. (2002). Genetic differentiation of the mangrove crab, *Scylla tranquebarica*, in Makassar Strait and Bone Bay, Indonesia. *Fisheries Science, 68*, 337–340. https://doi.org/10.2331/fishsci.68.sup1_337

Sulistiono, Watanabe, S., & Tsuchida, S. (1994). Biology and fisheries of crabs in Segara Anakan Lagoon. *Ecological assessment for management planning segara anakan lagoon, Cilacap, Central Java*. NODAI Center for International Program. Tokyo University of Agriculture. Tokyo (JP): JSPS-DGHE Program, pp. 65–82.

Sulistiono, S., Yahya, N. M., & Riani, E. (2021). Distribusi *Scylla* spp. di perairan estuari Sungai Donan Segara Anakan Bagian Timur Cilacap [Distribution of Scylla spp. In the estuary waters of the Donan Segara River in Eastern Anakan, Cilacap]. *Habitus Aquatica, 2*(1), 1–11. https://doi.org/10.29244/haj.2.1.1

Syafaat, M. N., & Gunarto, G. (2018). Budidaya pembesaran kepiting bakau Scylla tranquebarica (Fabricius, 1798) hasil pembenihan pada lokasi tambak yang berbeda [Cultivation of mud crab *Scylla tranquebarica* (Fabricius, 1798) from hatchery in different pond locations]. *Media Akuakultur, 13*, 21–30.

Syafaat, M. N., Azra, M. N., Waiho, K., Fazhan, H., Abol-Munafi, A. B., Ishak, S. D., Syahnon, M., Ghazali, A., Ma, H., & Ikhwanuddin, M. (2021). A review of the nursery culture of mud crabs, genus Scylla: Current progress and future directions. *Animals, 11*(7), 2034. https://doi.org/10.3390/ani11072034

Thirunavukkarasu, N., Nesakumari, S. A., & Shanmugam, A. (2014). Larva rearing and seed production of mud crab Scylla tranquebarica (Fabricius, 1798). *International Journal of Fisheries and Aquatic Studies, 2*(2), 19–25.

Tiurlan, E., Djunaedi, A., & Supriyantini, E. (2019). Analisis Aspek Reproduksi Kepiting Bakau (Scylla sp.) Di Perairan Kendal, Jawa Tengah [Analysis of reproductive aspects of mud crab (Scylla sp.) in the waters of Kendal, Central Java]. *Journal of Tropical Marine Science, 2*(1), 29–36. https://www.researchgate.net/publication/334970232

Usman, Kamarudiin, Palinggi, N. N., & Laining, A. (2016). Performa pertumbuhan krablet kepiting bakau *Scylla olivacea*, yang diberi pakan dengan dosis berbeda selama periose pendederan [Growth performance of mud crab Scylla olivacea, which is fed at different doses during nursery periods]. *Media Akuakultur, 11*(1), 19–26. http://ejournal-balitbang.kkp.go.id/index.php/ma/article/view/2027

Yi, S. K., Lee, S. G., & Lee, J. M. (2009). Preliminary study of seed production of the Micronesian mud crab Scylla serrata (Crustacea: Portunidae) in Korea. *Ocean and Polar Research, 31*(3), 257–264. https://doi.org/10.4217/OPR.2009.31.3.257

Present Status of Indonesian Aquaculture with an Emphasis on Domestication and Species Diversification

28

Imron Imron and Estu Nugroho

28.1 Introduction

Indonesia's population is the fourth biggest after China, the United States, and India. While a large population may positively serve as a resource to drive country development, it demands sufficient support to ensure adequate food and nutrition security. Production of marine capture fisheries, which has long been served as a reliance for food, especially for protein security, during the last 5 years (2015–2020) has been stable at around 6 million metric tons (MMT) and increased only slightly (0.98%) (MMAF, 2022a). With the stagnation in capture fisheries production, fish production from capture fisheries alone will not be able to meet the human population's need for animal protein sources. Aquaculture production is, therefore, expected to play a crucial role in filling this supply-demand gap, as is the case with many other countries. This underscores the vital importance of aquaculture in addressing nutritional needs and the potential for the industry's growth (FAO, 2022).

At the global level, aquaculture production has increased dramatically and contributes about half of food fish production (FAO, 2022). In Indonesia, like in many other countries, aquaculture has also been a significant sector contributing toward the supply of fish food. By volume, out of 21.8 million metric tons (MMT) of Indonesian fish production in 2020, the aquaculture sector contributed around 14.8 MMT (66%). By value, Indonesian fisheries production is worth nearly IDR 400 billion, which was comparably contributed by capture and aquaculture fisheries. The fish production figures mentioned above show the comparatively predominant aquaculture over capture fisheries, following Phillips's projection made in 2015 that aquaculture production would surpass that of the capture fisheries by 2030 (Phillips

I. Imron (✉) · E. Nugroho
Research Center for Fishery, National Research and Innovation Agency, Cibinong, Indonesia
e-mail: imro005@brin.go.id

E. Nugroho et al. (eds.), *Aquaculture Development in Archipelago*,
https://doi.org/10.1007/978-981-95-1084-9_28

et al., 2015). While Indonesia is the fourth largest aquaculture-producing country globally, the sector must grow to meet future fish demand.

Considering the pivotal roles the aquaculture sector can play in supporting food security and foreign income, ensuring that the industry can continue to grow sustainably is crucial. One of the concepts of sustainable aquaculture approaching aquaculture from three perspectives, namely, economic, environmental, and sociological (Wurts, 2000), has been a paradigm adopted in most countries relying on aquaculture as a growth resource sector. At the operational level, this concept can be implemented in various ways. Supply of water in sufficient quantity and good quality, feeding with the nutrients needed by fish and using quality and superior seeds are determinant factors of aquaculture success (Nugroho, 2010; Nugroho & Haryadi, 2017). Using mechanical aeration, water circulation, and better nutritious feed in intensive aquaculture systems has improved economic sustainability. Improvements in feed ingredients, processing technologies, and the use of feed additives, such as pre- and probiotics and enzymes, have positively impacted financial and environmental sustainability. Additionally, domestication, which deals with providing new species or populations for aquaculture candidates and improving those already domesticated ones, may contribute to sustainable aquaculture (Boyd et al., 2020).

Referring to the success of China's aquaculture, an integrated approach involving species diversification, government support in policy and regulatory aspects, and the application of diverse farming technologies have been the major factors contributing to the phenomenal production increase (Yue et al., 2024). This chapter briefly overviews Indonesian aquaculture and discusses the prospects for further development. Emphasis is particularly given to the production performance of well-established commodities, particularly introduced species and the potential of native species of aquaculture that may diversify and enrich species choice for aquaculture. Throughout this chapter, we use the term "introduced species" to define those coming from outside the country (Barros, 2016), and native to describe those coming from the Government (Schwartz, 1997).

28.2 Present Status of Indonesian Aquaculture

28.2.1 Aquaculture Area and Production Systems

There is a great variety of production systems in aquaculture, some of which may result from differences in ecology, farming system, level of technology, or combination of these factors. Looking at the ecosystem and farming system, the aquaculture production figures were derived from a variety and combination of both ecosystem and farming systems. Based on the ecosystem, Indonesian aquaculture occurs in marine, brackish, and freshwater environments.

In the last 5-year period (2018–2022), the total aquaculture area in Indonesia, which is a combination of the three ecosystems, is around 1.2 million hectares (ha). The area peaked in 2019, reaching 1.5 million hectares (Mha). After that, the area

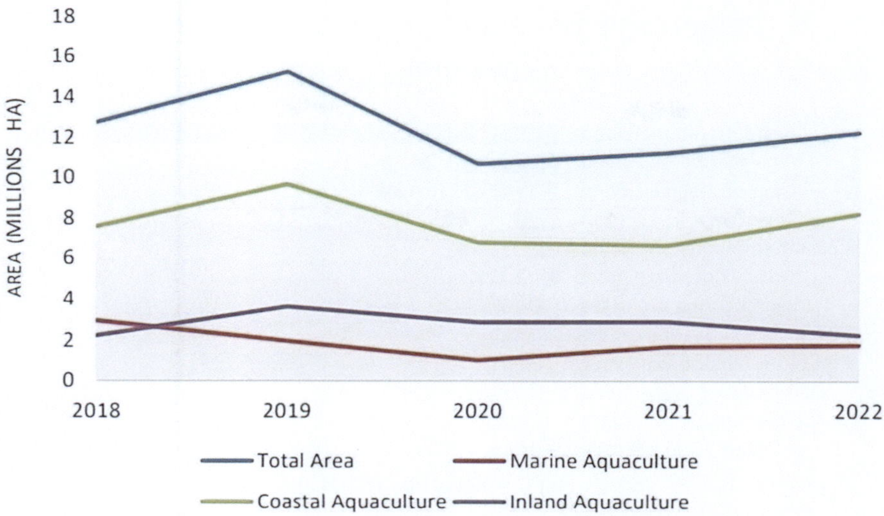

Fig. 28.1 Indonesian aquaculture area. (Source: One Data KKP 2024 (MMAF, 2024c))

decreased due to the Covid-19 pandemic outbreak (Fig. 28.1). Of the total area, coastal aquaculture, with 679,448 ha (67.1%), represents the highest, followed by inland aquaculture at 287,520 ha (18.4%) and marine aquaculture, with 102,254 ha (14.5%), the least (MMAF, 2022b). These utilized aquaculture areas represent only 6.84% of the total potential area, which is reported at 17.92 Mha, consisting of 2.8 Mha (15.8%) of inland aquaculture, 2.96 Mha of coastal aquaculture, and 12.12 Mha (67.7%) of marine aquaculture (MMAF, 2024b). The utilization levels of aquaculture areas are highest in coastal aquaculture (27.74%), followed by inland aquaculture (7.98%), and the least in marine aquaculture (1.46%) (MMAF, 2024b). Based on the figures of potency and usage, there is ample opportunity for further aquaculture development in Indonesia.

Within the respective ecosystem, a variety of fish farming systems are implemented. As illustrated in Fig. 28.2, while a dozen farming systems are listed, there is a high variation in the area utilized by respective farming systems. Aquaculture in coastal regions is dominated by traditional brackish water pond culture (61.02%), followed by semi-intensive and intensive systems, which combined accounted for an additional area of 6%. Marine aquaculture, by area, is dominated by seaweed, which accounts for 14.36% of the total, while floating-net cage and other marine farming systems account for only 0.12%. Inland aquaculture has been the most diverse farming system. They include conventional stagnant ponds, running-water ponds, rice-fish farming, fixed cages, pen systems, and floating net cages. Among these, most areas in inland aquaculture are dominated by stagnant ponds and rice-fish farming, which, in combination, account for 18.12% of the total utilized area.

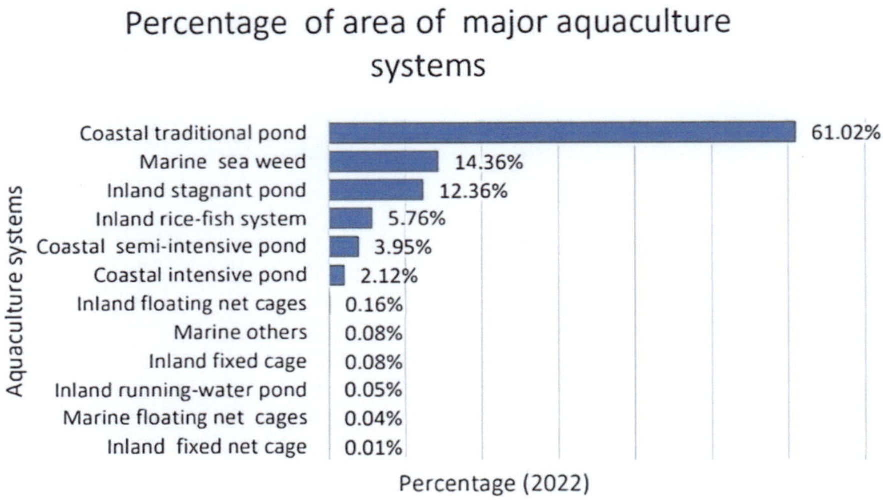

Fig. 28.2 Distribution of aquaculture production (tons) based on ecosystems and farming system. (Source: extracted from One data KKP (MMAF, 2024c))

28.2.2 Production of Top Aquaculture Commodities

The production distribution of Indonesian aquaculture commodities, irrespective of farming systems and ecosystems, is presented in Fig. 28.3. Seaweed, with 9.6 MMT, dominated production figures, accounting for 63.7%, followed by finfish with 4.2 MMT, accounting for 28%, crustacea with 0.9 MMT, accounting for 6%, and others with 0.3 MMT, accounting for 2.3%. More detailed production figures, considering the ecosystem and farming system, and more commodities are presented in Table 28.1. The table shows that out of 15 MMT total Indonesian aquaculture production in 2022, marine aquaculture contributed the most (53.5%), followed by inland and coastal aquaculture, which contributed 25.1% and 21.5%, respectively.

Based on production systems, the pond-culture system dominated coastal and inland aquaculture. In the former, it accounts for almost all fisheries production; in the latter, it accounts for 74.3%. In addition to the pond culture system, production in inland aquaculture is also contributed by floating net cage (11.6%), rice-fish culture (7.9%) and cage culture systems (5.7%). In marine aquaculture, the floating net cage has been the most common production system for finfish, while the floating longline method, off-bottom culture, and floating raft system have been widely applied for seaweed production in marine environments (Kasim et al., 2022; van der Heijden et al., 2022).

Table 28.1 lists 23 significant commodities inland, coastal and marine ecosystems and farming systems. The increase in the number of commodities listed in Table 28.1 relative to those listed in Fig. 28.3 is due to the use of more specific commodity names. For example, the collective commodity name of crustaceans in 10 primary commodities is broken down more specifically into tiger prawns, white-leg

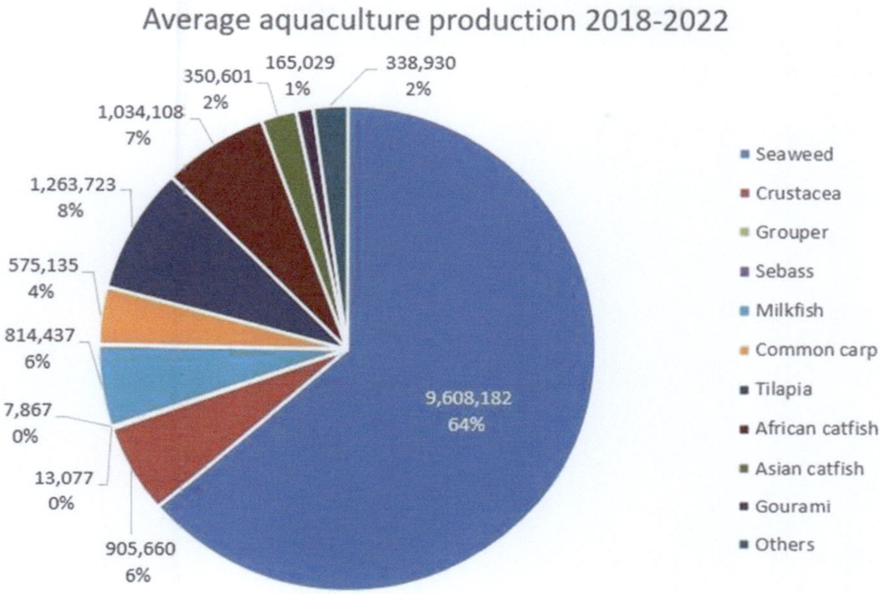

Fig. 28.3 Production (MMT) and percentage of Indonesian major aquaculture commodities during 2018–2022. (Source: One Data KKP 2024 (MMAF, 2024d))

prawns, and giant freshwater prawns. In marine aquaculture, production volume is dominated by seaweed, followed by shellfish, grouper, and giant seaperch. In addition to the brackish water system, milkfish and shrimps have been dominant in production, while seaweed ranks first. In contrast, in the freshwater system, species dominating production are African catfish, tilapia, Asian catfish, common carp, and giant gourami. However, it is essential to note that this commodity number is less than the actual number of species being cultured because some commodities consisting of probably several different species are grouped into one common or commercial name instead of the species name, as with shellfish, crabs and seaweed.

28.2.2.1 Comparative Production Volume of Native and Introduced Species

Both native and introduced species have contributed to Indonesian aquaculture production. In addition, it is also characterized by the high proportion of seaweed, the native plant aquatic taxa, which accounted for 64% of the total production. The relative contribution of native species is highly variable among different ecosystems. Considering seaweed, the native species contributed 72% of the total output. However, the distribution across different ecosystems is highly variable, with 100% in marine, 76% in coastal, and only 2% in inland environments.

Excluding seaweed and focusing on aquatic animal production, the contribution pattern of native species changes significantly. It contributes 23% of total output, with a distribution of 100% in marine, 55% in coastal, and 8% in inland

Table 28.1 Distribution of aquaculture production based on ecosystems and farming system

	Inland aquaculture					Coastal aquaculture	Marine aquaculture
Commodity	Pond	Pen culture	Rice-fish	Floating net	Cage	Pond culture	
Common carp (*Cyprinus carpio*)	294,022	1075	48,267	129,297	36,810	–	–
Java barb (*Barbonymus gonionotus*)	14,353	–	5342	–	48	–	–
Bony lip barb (*Osteochilus vittatus*)	38,451	–	5451	–	5	–	–
Nile tilapia (*Oreochromis niloticus*)	799,712	9485	150,599	236,704	69,096	–	–
Giant gourami (*Osphronemus goramy*)	162,063	97	1416	1523	602	–	–
African catfish (*Clarias gariepinus*)	1,043,048	10,025	33,635	–	10,009	–	–
Eels (*Anguilla* spp.)	92	–	424	–	–	–	–
Asian catfish (*Pangasius* sp.)	274,394	1769	–	32,053	31,942	–	–
Jelawat (*Leptobarbus hoevenii*)	615	–	–	–	1354	–	–
Pomfret (*Trachinotus* sp.)	35,951	–	–	22,215	–	–	–
Giant freshwater prawn (*Macrobrachium rosenbergii*)	5770	–	–	–	–	–	–
Asian red-tail catfish (*Hemibragus nemurus*)	–	–	–	181	–	–	–
Mozambique tilapia (*Oreochromis mossambicus*)	–	–	–	346	–	–	–

Snakehead (*Channa micropeltes*)	–	–	–	–	39,161	–	–
Marble goby (*Oxyeleotris marmorata*)	–	–	–	–	141	–	–
Milkfish (*Chanos chanos*)	–	–	–	–	–	710,678	–
Seabass (*Lates calcarifer*)	–	–	–	–	–	4210	2629
Tiger shrimp (*Penaeus monodon*)	–	–	–	–	–	128,574	–
Pacific white shrimp (*Litopenaeus vannamei*)	–	–	–	–	–	711,875	–
Crabs	–	–	–	–	–	14,142	–
Seaweed	–	–	–	–	–	1,396,324	7,837,944
Groupers	–	–	–	–	–	–	8573
Shellfish	–	–	–	–	–	–	49,076
Others	82,521	558	46,654	5535	20,538	205,752	2946
Shared percentage (FS)[1]	18.62	0.16	1.97	2.90	1.42	21.46	53.47
Shared percentage (ES)[2]	25.1					21.50	53.50
Shared percentage (FS within ES)[3]	74.3	0.6	7.9	11.6	5.7		

Source: Extracted from MMAF, 2024; 1: share of each farming system against total production; 2: share of each ecosystem against total production; and 3: share of each farming system relative to total production within the corresponding ecosystem

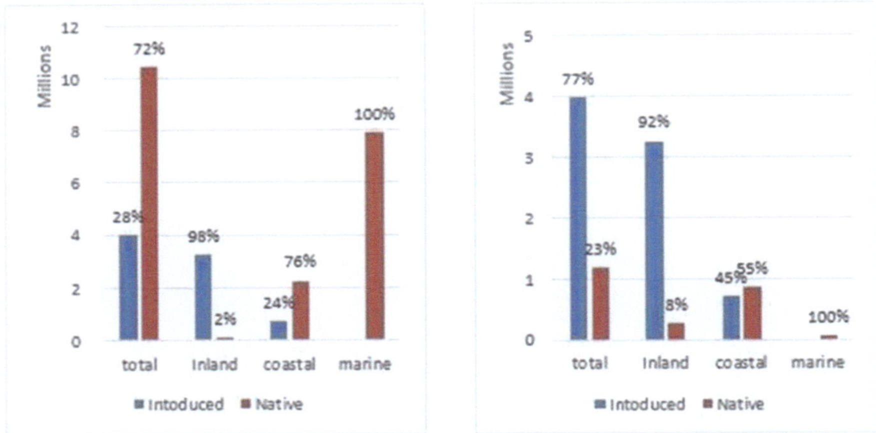

Fig. 28.4 Comparative share of native and introduced species, in volume and percentage, into the production of Indonesian aquaculture. The left figure includes seaweed, while the right excludes it. (Source: extracted and analyzed from One Data KKP. (MMAF, 2024d))

environments (Fig. 28.4). In marine aquaculture, native aquatic animal species have been exclusively used. Introducing species has significantly contributed to brackish water and freshwater systems production. The introduced Pacific white shrimp contributed almost 40% of aquatic animal production.

In contrast, in the freshwater system, the introduced species consisting of Common carp, Tilapia, African catfish, and Asian catfish shared 75% in 2012 (Nugroho et al., 2012) more than 80% of 2020 production, and more than 90% in 2022 (MMAF, 2024b). The limited number of native species recorded in fish production statistics does not mean that no native species has contributed to the production. Instead, the limited output shared by these native species has led them into collective terms, "others," making them unnoticed. The critical role of introduced species in a country's aquaculture production also occurred in major aquaculture-producing countries (De Silva et al., 2009) such as China (Liu & Li, 2010), Brazil, and the Philippines (Shelton & Rothbard, 2006). While the introduced species are expected to maintain their significant contribution toward a profitable production system, concerns about the negative impacts that these exotic species may have on the environment and native biodiversity have raised awareness on diversifying aquaculture species based on locally available germplasm to ensure long-term sustainable aquaculture (De Silva et al., 2009).

28.2.2.2 Promoting Native Germplasm for the Diversification of Aquaculture Species

In Indonesia, efforts to diversify aquaculture species through native fish germplasm have been initiated by mapping the native species of interest. An early study on mapping the potential native aquatic animal germplasms for aquaculture identified 31 species, as presented in Table 28.2 (Sukadi, 2002). A subsequent study (Nugroho

Table 28.2 List of Indonesian native species of aquaculture potential, geographic-base preference, levels of technological readiness, and present utilization in aquaculture

No	Scientific name	Local name	Geo pref[a]	TRL[b]	Ref[c]	ProdStat[d]
	Inland water species			.		
1	*Pangasius nasutus*	Patin Lokal	K	2	1, 2	NR
2	*Pangasius jambal*	Patin Jambal	–	2	1, 2	NR
3	*Leptobarbus hoevenii*	Jelawat	K	2	1, 2	R
4	*Oxyeleotris marmorata*	Betutu	K/S	3	1, 2	R
5	*Notopterus borneensis*	Belida	K/S	3	1, 2	NR
6	*Mystus nemurus*	Bating/ Baung	K/S	2	1, 2	R
7	*Helostoma temminckii*	Kapar/ tambakan	K/S	1	1, 2	R
8	*Anabas testudineus*	Pepuyu/ Betok	K	1	1, 2	R
9	*Osphronemus goramy*	Kalui	K/S	1	1, 2	R
10	*Tor douronensis*	Semah	–	1	1	R
11	*Puntius javanicus*	Tawes	S	1	1,2	R
12	*Puntius schwanefeldi*	Lampam	–		1	NR
13	*Scleropages formosus*	Arowana	–	2	1	NR
14	*Osteochilus kelabau*	Kelabau	K	2	1,2	NR
15	*Ostochilus hasselti*	Nilem	S	1	1, 2	R
16	*Clarias* spp.	Lele-lelean			1	NR
17	*Mystacoleucus padangensis*	Bilih Maninjau	S	2	1, 2	NR
18	*Thinnichthys thinnoides*	Benangin			1	NR
19	*Channa* spp.	Gabus-gabusan	K	2	1, 2	R
20	*Chanos chanos*	Bandeng		1	1	R
21	*Mugil cephalus*	Belanak		2	1	NR
22	*Macrobrachium rosenbergii*	Udang galah	K/S	1	1, 2	R
23	*Tor tambroides*	Lomi	K	2	1, 2	NR
24	*Cyclochrilichthys apogon*	Tembuwuk	K		1, 2	NR
25	*Puntius bulu*	Sanggang	K		2	NR
26	*Wallago* sp.	Tapah	K		2	NR
27	*Tor soro*	Semah	S		2	NR

(continued)

Table 28.2 (continued)

No	Scientific name	Local name	Geo pref[a]	TRL[b]	Ref[c]	ProdStat[d]
28	*Rasbora*	Wader	S		2	NR
	Marine species					
29	*Lates calcarifer*	Kakap Putih	–	2	1	Grouped under the collective names of grouper for Kerapu and Seaperch for Kakap
30	*Lutjanus argentimaculatus*	Kakap Merah	–	2	1	
31	*Lutjanus johni*	Kakap Merah	–	2	1	
32	*Cromileptis altivelis*	Kerapu Bebek	–	2	1	
33	*Siganus guttatus*	Baronang	–	2	1	
34	*Ephinephelus fuscoguttatus*	Kerapu Macan	–	2	1	
35	*Ephinephelus lanciolatus*	Kerapu Kertang	–	2	1	
36	*Ephinephelus tauvina*	Kerapu Lumpur	–	2	1	
37	*Ephinephelus itjara*	Kerapu batik		2	1	
38	*Plectropomus maculatus*	Kerapu Sunu		2	1	

[a]Geopref stands for geographic preferences, with K and S indicating Kalimantan and Sumatra, respectively
[b]*TRL* technology readiness levels, following Nugroho et al. (2012): (1) Technologies are well established and have been implemented; (2) Technologies are established and ready to be implemented; and (3) Technology needs some refinements
[c]Reference: 1. Sukadi (2002), 2. Nugroho et al. (2012)
[d]Benchmarked against 2022 national statistics of aquaculture production, denoted with R if the species production is recorded and NR if not explicitly recorded in the statistic

et al., 2012), which focused on two main Indonesian islands rich in fish biodiversity, namely, Sumatra and Kalimantan, confirmed previous results (Sukadi, 2002). The latter study also highlighted that some species were of interest in both islands, while other native species were of interest in specific islands. These people's particular interest in new aquaculture candidates seemed to be associated, in part, with local preference and endemicity of the species.

Government efforts to encourage diversification and enrichment of cultivated commodities have been carried out, among others, by establishing a breeding network. The network involves technical implementing units (TIU) of central and regional governments and private sectors. The central TIU includes those under the Directorate General of Aquaculture or the Research Agency, while regional TIUs are those under regional governments. The network's main aim was to coordinate the establishment of breeding programs to produce high-quality fish broodstock and disseminate them to stakeholders. In the last two decades, the network has released 58 farmed types (Table 28.3). Those 58 farmed types include 25 species, comprising six introduced and 19 native species. The profile of fish breeding programs

Table 28.3 List of aquaculture-farmed types produced by Indonesian fish breeding programs

No	Common name	Species	Release name	Year	GT[a]	SO[b]	BI[c]
1	Wader cakul	*Puntius binontaus*	Puntius binotatus jatimbulan	2024	1	1	2
2	Striped catfish	*Pangasianodon hypophthalmus*	Pustina	2023	2	2	1
3	Tilapia	*Oreochromis niloticus*	Sakti	2023	2	2	1
4	Tilapia	*Oreochromis niloticus*	Nirwana IV	2023	2	2	2
5	Pacific white shrimp	*Litopenaeus vannamei*	Nusadewa	2023	2	2	1
6	Giant freshwater prawn	*Macrobrachium rosenbergii*	Sijawa	2022	2	1	2
7	Gourami	*Osphronemus goramy*	Bima	2021	3	1	1
8	Cobia	*Rachycentron canadum*	Cobia	2021	1	1	1
9	Striped catfish	*Pangasianodon hypophthalmus*	Perkasa	2018	2	2	1
10	Rainbow fish	*Melanotaenia parva*	Kurumoi	2018	1	1	1
11	Gourami	*Osphronemus goramy*	Sago	2018	1	1	2
12	Common carp	*Cyprinus carpio*	Punten	2018	1	2	2
13	Kissing gourami	*Helostoma temminckii*	Takhasi	2018	1	1	1
14	Java barb	*Puntius javanicus*	Jois	2018	1	1	2
15	Milkfish	*Chanos chanos*	Gondol	2018	1	1	1
16	Gourami	*Osphronemus goramy*	Galunggung super	2018	3	1	2
17	Tilapia	*Oreochromis niloticus*	Nirwana III	2016	2	2	2
18	Common carp	*Cyprinus carpio*	Marwana	2016	2	2	2
19	Kelabau fish	*Osteochilus melanopleurus*	Kelabau	2016	1	1	1
20	Common carp	*Cyprinus carpio*	Jayasakti	2016	2	2	1
21	Common carp	*Cyprinus carpio*	Mustika	2016	2	2	1
22	Common carp	*Cyprinus carpio*	Rajadanu super Rd	2016	2	2	1
23	Abalone	*Haliotis squamata*	Abalone Haliotis squamata	2016	1	1	1
24	Snakehead fish	*Channa striata*	Haruan	2015	1	1	1
25	Common carp	*Cyprinus carpio*	Mantap	2015	2	2	1
26	Giant freshwater prawn	*Macrobrachium rosenbergii*	Si Ratu	2015	2	1	1
27	Catfish	*Clarias gariepinus*	Mutiara	2015	2	2	1
28	Tilapia	*Oreochromis* sp.	Salina	2014	3	2	1
29	Giant freshwater prawn	*Macrobrachium rosenbergii*	Gimacro II	2014	2	1	1
30	Climbing perch	*Anabas testudineus*	Climbing perch	2014	1	1	1
31	Common carp	*Cyprinus carpio*	Najawa	2014	1	2	2
32	Catfish	*Clarias* sp	Mandalika	2014	3	2	2
33	Tilapia	*Oreochromis* sp.	Srikandi	2012	3	2	1
34	Tilapia	*Oreochromis* sp.	Pandu♂, Kunti♂	2012	2	2	2
35	Red tilapia	*Oreochromis niloticus*	Nilasa	2012	2	2	2

(continued)

Table 28.3 (continued)

No	Common name	Species	Release name	Year	GT[a]	SO[b]	BI[c]
36	Tilapia	*Oreochromis niloticus*	Anjani	2012	2	2	2
37	Tilapia	*Oreochromis niloticus*	Sultana	2012	2	2	1
38	Tilapia	*Oreochromis niloticus*	Nirwana II	2012	2	2	2
39	Grouper	*Epinephelus* sp.	Cantang	2012	3	1	1
40	Kancera	*Tor soro*	Kancera	2011	1	1	1
41	Pacific white shrimp	*Litopenaeus vannamei*	Global gen	2010	2	2	3
42	Tilapia	*Oreochromis niloticus*	Larasati	2009	3	2	2
43	Pacific white shrimp	*Litopenaeus vannamei*	Nusantara	2009	2	2	1
44	Tilapia	*Oreochromis niloticus*	Best	2009	2	2	1
45	Tilapia	*Oreochromis niloticus*	Jatimbulan	2008	2	2	2
46	Lobster	*Cherax quadricarinatus*	Redclaw Huna	2006	1	1	1
47	Lobster	*Cherax quadricarinatus*	Huna	2006	1	1	1
48	Tilapia	*Oreochromis niloticus*	Gesit	2006	4	2	1
49	Tilapia	*Oreochromis niloticus*	Nirwana	2006	2	2	2
50	Striped catfish	*Pangasius* sp.	Pasupati	2006	3	2	1
51	Tilapia	*Oreochromis niloticus*	Jica	2004	1	2	1
52	Catfish	*Clarias gariepinus*	Sangkuriang	2004	2	2	1
53	Blue shrimp	*Litopenaeus stylirostris*	Blue shrimp	2002	1	1	1
54	Giant freshwater prawn	*Macrobrachium rosenbergii*	Gimacro	2001	2	1	1
55	Pacific white shrimp	*Litopenaeus vannamei*	Pacific white shrimp	2001	1	2	1
56	Striped catfish	*Pangasius jambal*	Jambal	2000	1	1	1
57	Common carp	*Cyprinus carpio*	Sinyonya	1999	1	2	1
58	Common carp	*Cyprinus carpio*	Majalaya	1999	1	2	1

Source: MMAF (2024a)

[a]GT refers to genetic technology, 1 Domestication, 2. Selection, 3. hybridizations, and 4. Genetic Sexual manipulation

[b]SO refers to the source of species, with 1 and 2 representing native and introduced species, respectively

[c]BI refers to breeding institutions, with 1 and 2 indicating TIU under central and regional governments, respectively, and 3 is the private sector

coordinated under the network (Fig. 28.5) may describe the critical role of introduced species in Indonesian aquaculture. Of 58 breeding program activities, 35 activities dealt with introduced species. Tilapia is the introduced species with the most breeding programs (13), followed by common carp (9). On average, there were six breeding programs for each introduced species. This contrasts with the breeding program profile of native species. On average, there was only one breeding program for each native species.

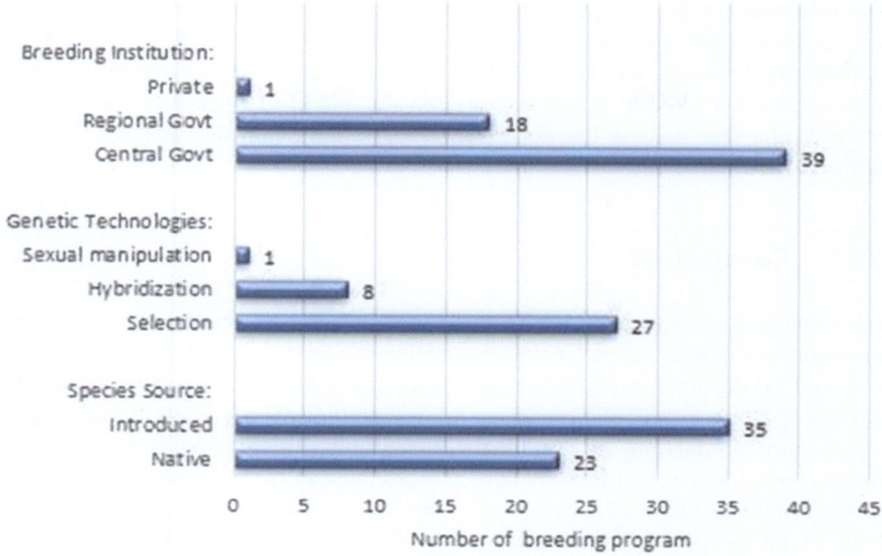

Fig. 28.5 Profile of Indonesian fish breeding programs describing species source, genetic technologies, and breeding institutions. (Source: MMAF (2024a))

Notably, while the prospective native species for diversification were identified over 20 years ago, production figures contributed by these native species have not been significant. This is evidenced by the lack of production records representing those species in national statistics. National statistics (MMAF, 2024b) show that out of 38 native species listed as prospective aquaculture candidates, only 14 species showed a significant production record. This situation may reflect several possibilities, such as a lack of technological readiness, particularly in breeding technology, limited market outreach, and economic considerations. In partial or interaction, these factors may lead to this condition. While the current profile of the production performance of new native species might look less satisfactory, it is not too surprising because the pattern at the global level looks similar. Global aquaculture production has been dominated by only a few species (Teletchea, 2019). In addition to being completely domesticated, the success of culturing a new species is also influenced by zootechnic, economic, environmental and social. For instance, when several farmed commodities have been fully domesticated, these factors would dictate farmers' choices. Farmers will evaluate comparative advantages from an economic perspective, one species over the others.

For example, a feasibility study using such economic parameters as cash flow and net present values (NPV), internal rate of return (IRR), benefit-to-cost ratio, and payback period of investment cost has confirmed the comparative advantages of snapper, grouper, black tiger shrimp, white leg shrimp, and tilapia for aquaculture. In addition, grouper and snapper had a positive outlook because of the high demand for high-quality fish in the international market (Afero et al., 2010). A fully domesticated species could be limitedly produced in a particular region because people in another area may not be familiar with that species. Likewise, a fully domesticated species may be limitedly produced because of price constraints. For instance, an esteemed food fish such as *Tor* sp., due to its high price, may not be made in massive volume due to limited supply and demand, particularly in the domestic market.

28.3 Conclusions

In summary, looking at historical records, present status, and future challenges and projections, the introduced species comprising common carp, tilapia, African catfish, and Asian catfish in freshwater aquaculture, and Pacific white shrimp in brackish water aquaculture will continue to contribute significantly to Indonesian aquaculture production. Market demand, technological readiness, domesticated nature, and economic profitability have allowed those introduced species to be globally accepted. For native species to show similar production performance, the attributes of the successfully introduced species need to be considered. While the present production share has not succeeded, using native species as an alternative and supplementary aquaculture species must be promoted. This becomes more important, particularly in response to climate change that may negatively impact aquaculture. Diversifying aquaculture commodities using native germplasm will help conserve biodiversity and build more resilient aquaculture.

References

Afero, F., Miao, S., & Perez, A. A. (2010). Economic analysis of tiger grouper *Epinephelus fuscoguttatus* and humpback grouper *Cromileptes altivelis* commercial cage culture in Indonesia. *Aquaculture International, 18*(5), 725–739.

Barros, F. (2016). Introduced species. In M. J. Kennish (Ed.), *Encyclopedia of estuaries* (pp. 361–362). Springer. https://doi.org/10.1007/978-94-017-8801-4_152

Boyd, C. E., D'Abramo, L. R., Glencross, B. D., Huyben, D. C., Juarez, L. M., Lockwood, G. S., McNevin, A. A., Tacon, A. G., Teletchea, F., & Tomasso, J. R., Jr. (2020). Achieving sustainable aquaculture: Historical and current perspectives and future needs and challenges. *Journal of the World Aquaculture Society, 51*(3), 578–633.

De Silva, S. S., Nguyen, T. T., Turchini, G. M., Amarasinghe, U. S., & Abery, N. W. (2009). Alien species in aquaculture and biodiversity: A paradox in food production. *Ambio: A Journal of the Human Environment, 38*(1), 24–28.

FAO. (2022). The state of world fisheries and aquaculture 2022. *Towards Blue Transformation.* https://doi.org/10.4060/cc0461en

Kasim, M. R., Balubi, A. M., Jalil, W., Astuti, O., Bahtiar, Muskita, W., Nurdiana, La Ola, O., Rahman, A., & Patadjai, R. S. (2022). Seaweed cultivation technologies in Indonesia: Current trends and future prospects. In *Sustainable global resources of seaweeds* (Vol. 1, pp. 75–89). Springer.

Liu, J., & Li, Z. (2010). The role of exotics in Chinese inland aquaculture. In S. S. De Silva & B. Davy (Eds.), *Success stories in Asian aquaculture* (pp. 173–185). Springer. https://doi.org/10.1007/978-90-481-3087-0

MMAF. (2022a). Kelautan dan Perikanan dalam angka tahun 2022. [Marine and Fisheries in Figures 2022]. https://kkp.go.id/setjen/satudata/page/1453-kelautan-dan-perikanan-dalam-angka/

MMAF. (2022b). Statistik KKP: Areal budidaya berdasarakan tipe budidaya. [MMAF statistics: Aquaculture area based on aquaculture types]. https://statistik.kkp.go.id/home.php?m=luaslahan&i=7#panel-footer-kpda

MMAF. (2024a). *Jaringan Dokumentasi dan Informasi Hukum; Regulasi. [Legal Documentation and Information Network: Regulations].* Legal Bureau, Ministery of Marine Affairs and Fisheries. Retrieved from https://jdih.kkp.go.id/Homedev/PeraturanAll#

MMAF. (2024b). Kelautan dan Perikanan dalam angka tahun 2024 [Marine and Fisheries in Figures 2024] (Vol. 11). MMAF. https://statistik.kkp.go.id/mobile/asset/book/Buku_KPDA_2024_050724_FINAl_FIX_FP_SP.pdf

MMAF. (2024c). *Satu Data KKP: Data Produksi Perikanan, volume produksi budidaya pembesaran per komoditas utama [One Data KKP: Fishery production data, aquaculture production volume of growing-out per main commodity].* https://statistik.kkp.go.id/home.php?m=prod_ikan_prov&i=2#panel-footer-kpda

MMAF. (2024d). Satu Data KKP: Data produksi perikanan. Areal budiaya. [OneData KKP: Fishery production data. Aquaculture area]. https://statistik.kkp.go.id/home.php?m=luaslahan&i=7

Nugroho, E. (2010). Menjadikan perikanan budidaya sebagai inkubator bisnis mandiri: Pelajaran berhargadari Taiwan. [Making Aquaculture a Self-Sufficient Business Incubator: Valuable Lessons from Taiwan]. *Media Akuakultur, 5*(1), 62–66.

Nugroho, E., & Haryadi, J. (2017). *Budiaya lele dengan sistem total akuakultur. [Catfish cultivation with total aquaculture system].* Penebar Swadaya Grup.

Nugroho, E., Sukadi, M. F., & Huwoyon, G. H. (2012). Beberapa jenis ikan lokal yang potensial untuk: Domestikasi, teknologi pembenihan, dan pengelolaan kesehatan lingkungan budidaya. [Several types of local fish for cultivation: Domestication, seed technology, and environmental health management for cultivation]. *Media Akuakultur, 7*(1), 52–57.

Phillips, M., Henriksson, P., Tran, N., Chan, C., Mohan, C. V., Rodriguez, U., Suri, S., Hall, S., & Koeshendrajana, S. (2015). Exploring Indonesian aquaculture futures.

Schwartz, M. W. (1997). Defining indigenous species: An introduction. In J. O. Luken & J. W. Thieret (Eds.), *Assessment and management of plant invasions* (pp. 7–17). Springer. https://doi.org/10.1007/978-1-4612-1926-2_2

Shelton, W. L., & Rothbard, S. (2006). Exotic species in global aquaculture – A review. *Israeli Journal of Aquaculture – Bamidgeh, 58*(1), 3–28. http://www.scopus.com/scopus/inward/record.url?eid=2-s2.0-33847660138&partnerID=40

Sukadi, F. (2002). Utilization and conservation of germplasm to increase aquaculture productivity. *Bulletin Plasma Nutfah, 8*(2), 58–65. http://repository.pertanian.go.id/bitstream/handle/123456789/12514/pelestarian_2002.pdf?sequence=1&isAllowed=y

Teletchea, F. (2019). Fish domestication: An overview. In *Animal Domestication.* https://doi.org/10.5772/intechopen.79628.

van der Heijden, P. G., Lansbergen, R., Axmann, H., Soethoudt, H., Tacken, G., van den Puttelaar, J., & Rukminasari, N. (2022). Seaweed in Indonesia: Farming, utilization and research.

Wurts, W. A. (2000). Sustainable aquaculture in the twenty-first century. *Reviews in Fisheries Science, 8*(2), 141–150.

Yue, G., Tay, Y., Wong, J., Shen, Y., & Xia, J. (2024). Aquaculture species diversification in China [review]. *Aquaculture and Fisheries, 9*, 206–2017.

MIX
Papier aus verantwortungsvollen Quellen
Paper from responsible sources
FSC® C105338

If you have any concerns about our products,
you can contact us on
ProductSafety@springernature.com

In case Publisher is established outside the EU,
the EU authorized representative is:
Springer Nature Customer Service Center GmbH
Europaplatz 3, 69115 Heidelberg, Germany

Printed by Libri Plureos GmbH
in Hamburg, Germany